IBSRAM LII

Organic-Matter Management and Tillage in Humid and Subhumid Africa

Proceedings of the Third Regional Workshop
of the *AFRICALAND* Programme
held at Antananarivo, Madagascar

9-15 January 1990

IBSRAM Proceedings no. 10

Organic-Matter Management and Tillage
in Humid and Subhumid Africa

Proceedings of the Third Regional Workshop
of the AFRICALAND Programme
held at Antananarivo, Madagascar

9-15 January 1990

IBSRAM Proceedings no. 10

The workshop was organized by

The International Board for Soil Research and Management (IBSRAM)

in collaboration with

The Ministry of Scientific Research and Development Techniques (Madagascar)
The National Centre of Applied Research for Rural Development
(CENRADERU/FOFIFA)

with the cooperation of

The Food and Agriculture Organization of the United Nations (FAO)

and cosponsored by

Bundesministerium für Wirtschaftliche Zusammenarbeit (BMZ)
Canadian International Development Agency (CIDA)
Coopération Suisse pour le Développement (CSD)
U.S. Agency for International Development (USAID)

Science Editors
E. Pushparajah
M. Latham

Publication Editor
Colin R. Elliott

Correct citation: Organic-matter management and tillage in humid and subhumid Africa. Bangkok, Thailand: International Board For Soil Research and Management, 1990. IBSRAM Proceedings no. 10.

ISBN 974-7087-00-6

Printed in Thailand

CONTENTS

Summary

IBSRAM has recently developed a number of networks for soil management research. Two of these are the IBSRAM networks under the *AFRICALAND* programme on the Management of Acid Soils in Humid Africa and Land Development for Sustainable Agriculture in Africa. The aim of these networks is to find improved soil management technologies that can be substituted for present methods where these fail to meet current production requirements or lead to soil degradation. In quite a number of FAO research projects, trials on tillage techniques and on soil organic-matter management have been included. Both the IBSRAM network projects and FAO projects are seeking to develop methods of land management which will be productive, sustainable, and acceptable to farmers.

In view of the current food shortages in tropical Africa, the question most often asked is how soil productivity in tropical environments can be sustained. An optimum resource-utilization plan should be based on scientific data obtained through well-designed and adequately equipped long-term experiments. To develop practical methods of resource use, additional research information is needed to evaluate appropriate tillage techniques and soil organic-matter management practices. This workshop brought together scientists to share experience and devise methods of research appropriate for these needs.

The meeting was held in Antananarivo, Madagascar, from 9-15 January 1990, with the aim of reviewing the work done on tillage and organic-matter management in humid and subhumid Africa, and of integrating new advances on these subjects into the existing IBSRAM/*AFRICALAND* and FAO programmes. More specifically the objectives were: to review the existing IBSRAM/ *AFRICALAND* networks, to review the FAO approach on tillage and organic-matter management, to develop complementary programmes on tillage and organic-matter management, and to enhance cooperation between these two programmes.

Eighty-three participants from sixteen countries and eight international organizations took part in the workshop (Appendix III). The country distribution of participants, in accordance with their work base, was: Botswana (1), Burundi (1), Cameroon (4), Côte d'Ivoire (2), Ethiopia (1), Federal Republic of Germany (1), Ghana (1), Kenya (1), Madagascar (37), Malawi (1), Nigeria (4), République Centrafricaine (1), Senegal (1), Tanzania (3), Togo (1), Uganda (1), USA (2), and Zambia (1). The following organizations also sent representatives to the workshop: CIRAD (2), CSIRO (1), FAO (3), IBSRAM (6), ICRAF (1), IRRI (1), ORSTOM (1), TSBF (1), and USDA/USAID (3). Ten of the representatives from international organizations were stationed in the region and had first-hand experience of the local situation.

The workshop was organized by the International Board for Soil Research and Management (IBSRAM) in collaboration with the Ministry of Scientific Research and Development Techniques (Madagascar), the National Centre of Applied Research for Rural Development (CENRADERU/FOFIFA), and the Food and Agriculture Organization of the United Nations (FAO). Financial support for the workshop was provided by the Bundesministerium für Wirtschaftliche Zusammenarbeit (BMZ), the Canadian International Development Agency (CIDA), Coopération Suisse pour le Développement (CSD), and the U.S. Agency for International Development (USAID).

In a number of sessions (see Appendix II), state of the art reports on tillage and residue management were given. The Malagasies presented their experience on soil organic-matter management and tillage for acid soils, and the role of organic matter in tropical soils was discussed. More specific subjects, notably the maintenance and management of organic inputs, cover crops and mulch, soil biological activities, agroforestry, tillage and residue management, the influence of no tillage and minimum tillage, soil conservation, reclamation of degraded lands, soil erosion, the effects of degraded lands, soil erosion, the effects of drought spells (so important in Africa), and the study of the cultivation profile, were treated in the technical papers. The AFRICALAND network participants presented progress reports and the initial results from their projects on land development and the management of acid soils.

Three working groups discussed tillage and organic-matter management (Appendix I), land development, and the management of acid soils. These often lively discussions contributed much to the understanding amongst the participants of the problems we are facing and of the way we might go about solving them. They also gave participants the opportunity to determine the best way of implementing their projects, and to decide which treatments to apply, and how to monitor the experiments.

A one-day field trip was made to the Applied Research Centre's field station at Beferona, where the experiments on the management of acid soils will be installed. The participants were shown demonstration trials based on agroforestry which have now been running for four years - and which apparently have been sustainable over this period, since yields of the upland rice have been maintained on more or less the same level (despite predictions from some experts that this would be impossible). Participants also visited the site of the acid soils, and got a good look at the soils involved, which were on a 40-60% slope and were probably clayey Kandihumults, though the classification is tentative as data are not yet available.

Although the time schedule was very tight due to the flight arrangements, participants were satisfied with the results of the meeting and were extremely happy to have been offered the chance to visit Madagascar, with its unique mixture of Asia and Africa. Also, the varied backgrounds of the participants in terms of their local and international affiliations, as well as their joint anglophone and francophone backgrounds, undoubtedly provided a stimulus for a more comprehensive development of the AFRICALAND and FAO programmes on tillage and organic-matter management in the region.

Discours d'inauguration

Opening addresses

François Rasolo

Directeur Général, FOFIFA
(FOFIFA Director General)

Monsieur le Ministre, Monsieur le représentant de la FAO, Monsieur le Directeur de l'IBSRAM, Monsieur le Secrétaire Général, Messieurs les représentants à Madagascar des organismes internationaux, honorables délégués, Mesdames et Messieurs :

C'est un grand honneur et un réel plaisir pour nous d'être parmi vous pour cette cérémonie marquant l'ouverture officielle du 3ème séminaire régional de l'IBSRAM sur le thème Travail du Sol et Gestion de la Matière Organique en Afrique Humide et Sub-humide.

En effet, la tenue de ce séminaire à Antananarivo, à l'aube de cette année nouvelle, est pour nous un sujet de fierté et un signe prémonitoire manifeste de la contribution réelle et efficace de la science au développement agricole. Le thème central du forum confirme bien ceci, et autorise même sa considération comme un gage de résolution des problèmes d'ordre technique et méthodologique, quand il s'agit des questions de gestion de fertilité.

Pour le FOFIFA, le Centre National de la Recherche Appliquée au Développement Rural, ou l'institution nationale qui s'occupe des recherches agricoles à Madagascar - pour le FOFIFA disions-nous, ces préoccupations relatives à la gestion de la fertilité figurent parmi les premières priorités de l'heure.

En effet, dans la plupart des programmes de recherche des six départements scientifiques de ce centre, et plus particulièrement au niveau du programme 'conservation des sols' du Département de Recherches Forestières et Piscicoles qui constitue le noyau fédérateur en la matière, le sujet se retrouve souvent en toile de fond et forme même un souci constant.

Ce choix et cet engagement rejoignent d'ailleurs les idées directrices émises par le Ministère de la Recherche Scientifique et Technologique pour le Développement dans ses "Réflexions sur la politique de la recherche à Madagascar", en particulier dans ce qu'il considère comme la "mise en oeuvre des technologies protectrices de l'environnement naturel".

Le ministère situe à cet effet le rôle de la recherche au niveau (nous citons) "de la mise au point des technologies permettant d'arrêter l'amplification des phénomènes d'érosion, d'appauvrissement progressif du sol et, par conséquent, de baisse du rendement agricole. Ces technologies doivent être, non seulement efficaces, mais également pratiques et accessibles à la population. On devrait donc éviter celles qui ne peuvent donner des résultats positifs que par la mise en oeuvre des moyens qu'on doit importer ou qui ne sont pas à la portée de la population".

Eu égard à tout ceci, nous osons espérer que les fructueux débats et échanges scientifiques qui auront lieu durant ce séminaire sauront dépasser le cadre du "fondamentalisme" pour signifier la participation de la science et des scientifiques à l'enjeu du développement agricole. La valorisation des résultats de recherche n'aura, en effet, atteint son plein épanouissement sans cette dimension "opérationnelle et applicable". La participation de FOFIFA dans les travaux de ce réseau IBSRAM va dans ce sens. Nous vous souhaitons donc bon courage et bon travail durant cette semaine qui promet de réflexions intensives.

Avant de terminer, au nom du FOFIFA et de tout le comité d'organisation locale, nous tenons à remercier toutes les institutions et tous les organismes donateurs, tant nationaux qu'étrangers, pour leur contribution efficace à la réalisation de ce séminaire.

Enfin, nous ne saurions terminer cette allocution sans vous faire part du traditionnel 'Tonga soa', le premier mot à apprendre dans la philosophie de l'hospitalité malagasy.

Mesdames et Messieurs,

Nous vous souhaitons donc à toutes et à tous, qu'ils foulent pour la première fois le sol malagasy ou qu'ils en sont à leur énième voyage à Madagascar, nous vous souhaitons la bienvenue et un agréable séjour à Antananarivo. Puisse ce séminaire constituer pour vous une occasion pour connaître davantage, autrement que par les rapports écrits, les réalités vécues par un de vos correspondants du réseau IBSRAM sur la gestion des sols.

Nous vous remercions de votre aimable attention.

* * * * *

Your Excellency, Minister of Scientific and Technological Research for Development, Representative of the FAO, Director of IBSRAM, Secretary-General, representatives of international organizations in Madagascar, delegates, ladies and gentlemen:

It is truly both an honour and a pleasure to be with you on the occasion of this opening ceremony of the Third Regional Workshop organized by IBSRAM, and entitled Tillage and Organic-Matter Management in Humid and Subhumid Africa.

By holding this workshop at Antananarivo, at the beginning of the New Year, you give us a feeling of pride, and to hold it at this time augers well for achieving a genuine and effective scientific contribution to agricultural development. The central theme of the meeting confirms this impression, and gives grounds for thinking of it as a pledge to resolve technical and methodological problems relating to soil management and fertility.

For FOFIFA, the National Centre for applied Research for Rural Development, which is the national institution responsible for agricultural research in Madagascar - for FOFIFA, as I was saying, issues relating to soil management and fertility are amongst the most important questions being considered at the moment.

Most of the research programmes of the six scientific departments of the centre, and notably the soil conservation programme of the Department of Forestry and Fisheries Research, which forms the central core of the department, this topic often underlies the research, and can even be considered as a subject of constant concern.

The choice of this topic and the commitment to it are in line with the guidelines laid down by the Ministry of Scientific and Technological Research for Development in 'Reflections on Research Policy in Madagascar, especially with regard to what is referred to as the "initiation of technologies which will protect the natural environment".

To achieve this objective, the ministry stipulates the role of research should be, and I quote, "to develop technologies that will help to prevent the increasing deterioration being caused by erosion, and by the gradual impoverishment of the soil, which results in a decrease in agricultural production. These technologies not only need to be effective, but also practical and amenable to the people. Consequently technologies which can only produce positive results by utilizing imported goods or items which are not accessible to the people should be avoided."

With this approach in mind, it is our hope that the discussions and exchange of scientific information engendered at this workshop will be fruitful, and will go beyond "basic" considerations, with the aim of stimulating some interaction of scientific knowledge and of scientists themselves in the agricultural development process as a whole. The validity of research efforts will not in fact be assured in a full and meaningful sense without this "practical and amenable" dimension.

FOFIFA's participation in the operations of the IBSRAM network is conceived along these lines. We extend our encouragement to you for the work ahead of you this week, and trust that it will be a time for a thorough examination of the matter in hand.

On behalf of FOFIFA and the local organizing committee, I would like to thank all the institutions who have contributed their time and effort to the workshop, and also to the national and external organizations who have provided financial backing for the workshop. These various bodies have brought to bear an effective contribution for holding this workshop.

It would not be right to end this welcoming address without the traditional 'Tonga soa', which is the first gesture of hospitality which Malagache philosophy prescribes.

Ladies and Gentlemen,

We wish you all, each one of you, all who are on Malagasy soil for the first time or making their umpteenth journey to Madagascar, we bid you welcome and a happy stay at Antananarivo. May this workshop be an occasion to learn more, beyond written

accounts, of the realities experienced by your IBSRAM network collaborators in the management of soils.
Thank you for your kind attention.

▲ ▲

Jacques Lépissier

Représentant de l'FAO
(FAO Representative)

Monsieur le Ministre de la Recherche Scientifique et Technologique pour le Développement (MRSTD), Monsieur le Directeur Général de FOFIFA, Monsieur le Directeur de l'IBSRAM, Mesdames et Messieurs les invités, Mesdames et Messieurs les participants :

Bien que très intimidé par le niveau des personnes ici présentes, c'est toujours un honneur de prendre la parole au début d'un atelier de travail, et un plaisir d'écouter parler les autres.

Vous êtes tous de redoutables spécialistes, et je ne m'aventurerais pas à aborder des aspects techniques que vous aurez largement le temps d'approfondir dans les prochains jours. Je voudrais seulement évoquer les soucis, de la FAO siège en général et de la représentation à Madagascar en particulier, à oeuvrer dans un contexte techniquement recevable et absorbable par les paysans.

Avec des moyens financiers importants, tout est possible; mais pour un paysan au bord de son champ, l'eau, les engrais, les semences sélectionnés, les pesticides sont loin - au sens propre et au sens figuré - et il est essentiellement tributaire de "Dame Nature".

Par ailleurs, la sécurité alimentaire à Madagascar, objectif majeur prévu pour 1990, est pratiquement obtenue, mais très fragile, et à long terme il faut plus que cette simple sécurité, il faut l'abondance pour satisfaire les besoins d'une population toujours plus nombreuse, mieux nourrie ... et viser en plus une exportation en quantité et de qualité.

Les sols, déjà très sollicités, vont l'être encore davantage. Pour sécuriser cette production agricole, les engrais sont à la limite la solution de facilité à laquelle tous les agriculteurs n'ont pas forcément accès.

Suite à tous ces éléments difficilement maîtrisables pour un homme seul, il convient donc, le plus vite possible, de mettre à la disposition des paysans des moyens techniques autorisant des récoltes abondantes regulières avec un moindre apport extérieur.

Ces moyens sont connus (fumure organique-amendements-compost-engrais vert, etc.). Ils existent à Madagascar, ils n'exigent pas de devises pour la plupart, ils conservent les sols pour les générations futures, ils sont à la portée des paysans. Il faut seulement les en informer et travailler plus.

Nous sommes tous là pour cela, le temps de transfert de technologie est passé, il faut tendre désormais à l'utilisation intensive de l'expérience acquise par les uns et les autres.

Vos reflexions des jours prochains vont graviter autour de cette question. Vous venez d'horizons divers avec des expériences personnelles précieuses. La FAO n'a pas le monopole du savoir, mais elle est très bien placée pour provoquer la diffusion du savoir en collaboration - cette fois-ci avec l'IBSRAM.

Nous sommes dans les meilleures conditions calendaires pour émettre des voeux. Permettez-moi d'en profiter pour vous souhaîter de fructueux travaux et que 1990 soit le début d'une longue série de décennies favorables.

* * * * *

Your Excellency, Minister of Scientific and Technological Research for Development, Director General of FOFIFA, Director of IBSRAM, guests and participants:

Although I am very conscious of addressing high-ranking dignataries and scientists, it is always an honour to be able to say a few words at the beginning of a workshop, and a pleasure to hear what others have to say.

You are outstanding specialists in the topic of this workshop, and I will not venture therefore to touch on technical matters, which at any rate you will be dealing with in some detail in the next few days. I would simply like to voice the concern of FAO headquarters in general and of the Madagascar office in particular that we should work within a context which is technically amenable and acceptable to the ordinary farmers.

When there are sufficient funds, anything can be done; but for a poor farmer viewing his land, water, fertilizers, selected varieties of grain, and pesticides are a long way off - both in a literal and a figurative sense - and he is essentially at the mercy of "Mother Nature".

Moreover, self-sufficiency in food production in Madagascar, one of the major objectives envisaged for 1990, has already been achieved; but it is a very fragile state of affairs, and in the long run more is needed than self-sufficiency of this sort. What is needed is an abundant food supply, enough food to cater for the needs of an ever-increasing population, which will provide a more nutritious diet, and enough over to export food-stuffs in quantity and quality.

Land with suitable soils is already very much in demand, and will become even more so in the future. To ensure ample food production, fertilizers may be the easiest solution. But it is a last resort solution, and not all farmers recessarily have access to fertilizers.

As a result of all these issues, which are difficult for anyone to resolve on his own, it would seem advisable to make the necessary technologies available to the farmers as soon as possible, enabling them to grow a plentiful supply of crops on a regular basis with a minimum of inputs from outside the country.

The technologies are well known (organic manure, mulches, compost, green manure, etc.). They are available locally in Madagascar, and most of them do not cost anything, they conserve the soils for future generations, and they can be procured by the farmers. All that is required is to make sure the farmers have the necessary information, and to use the technologies more frequently.

That is why we are here. The time for the transfer of technology is behind us, and from now on we need to concentrate on the intensive utilization of the means at our disposal, and on disseminating the benefits of our accumulative and collective experience.

Your deliberations in the next few days will centre around this issue. You are from countries with various perspectives and with valuable personal experiences. FAO does not have a monopoly on knowledge, but we are in a very good position to promote the dissemination of knowledge by collaborating with other organizations - in this case with IBSRAM.

At this time of the year, it is a particularly good occasion to express our good wishes. Let me, then, take this opportunity of wishing you every success for a fruitful outcome of your work together during this workshop, and also of expressing my hope that 1990 will prove to be the beginning of a series of prosperous decades.

✦ ✦

Marc Latham

Directeur, Conseil International pour la Recherche des Sols et leur Gestion
(Director, International Board for Soil Research and Management)

Monsieur le Ministre de la Recherche Scientifique et Technologique pour le Développement, Monsieur le Représentant de la FAO, Messieurs les représentants des missions étrangères, Monsieur le Directeur Général du Centre National de la Recherche Appliquée au Développement Rural, Mesdames, Messieurs :

Permettez-moi de vous exprimer mon plaisir de nous retrouver tous ici pour cette troisième réunion des réseaux IBSRAM sous le programme *AFRICALAND*, orientée plus particulièrement cette année sur le travail du sol et sur la gestion de la matière organique. Je profite de cette occasion pour remercier nos collègues malgaches qui ont bien voulu assumer la résponsabilité de l'organisation de cette réunion, et en particulier M. Jean-Louis Rakotomanana. Je tiens aussi à remercier la FAO pour avoir accepté de coorganiser cette réunion ainsi que l'Agence Canadienne pour le Développement International (ACDI), le Bundesministerium für Wirtschaflishte Zusammenarbert (BMZ) et la Coopération Suisse pour le Développement (CSD) pour leur aide financière à cette réunion.

Après Douala en 1986, Lusaka en 1987, Antananarivo en 1989 marque la troisième année d'existence des réseaux IBSRAM sur la mise en valeur des terres forestières et sur la gestion des sols acides en Afrique. Trois ans c'est court lorsque l'on regarde ce qui aurait pu être fait et qui n'a pas été réalisé. C'est suffisamment long pour indiquer la continuité d'un effort entrepris. La réalité des réseaux IBSRAM en Afrique existe, et il faut compter avec elle.

Des réseaux sur la gestion des sols en Afrique tropicale sont à la fois une nécessité, des difficultés et un espoir. Une nécessité, je pense que nous en ressentons tous le besoin. L'Afrique tropicale humide est soumise à une pression démographique inégalée dans l'histoire. La multiplication de la population, à laquelle se sont ajoutés des migrations de population en provenance des zones arides et semi-arides fortement éprouvées ces dernières années, ont amené une pression accrue sur les zones forestières. Non seulement la forêt tropicale et toute sa diversité végétale et animale sont en danger, mais les sols, support de cette richesse naturelle, sont menacés par l'érosion et les dégradations diverses. Un renforcement des recherches sur la mise en valeur durable ("sustainable") des sols forestiers est donc indispensable. Les nouveaux efforts nécessitent toutefois d'être coordonnés et d'être orientés vers des applications pratiques qui bénéficieront au paysan africain.

Les réseaux IBSRAM peuvent aider à atteindre ce but. L'échange d'expérience, comme nous allons le faire cette semaine, les discussions sur le terrain et dans les comités de réseaux, les séances de formation comme nous avons pu en avoir à l'IITA et à Yurimaguas en 1988, ainsi que la coordination de nos efforts sur le terrain vont très certainement dans la bonne voie. Elle nous permettrait d'être plus efficace et de mieux utiliser les ressources limitées mises à notre disposition.

L'expérience de ces trois dernières années nous a cependant montré qu'une nécessité reconnue n'est pas le gage d'un financement de projets nationaux. Nous étions parti sur l'hypothèse que des financements bilatéraux pourraient être aisément mobilisés pour développer le réseau. En réalité, ils ont été plus difficiles à obtenir que prévu, même lorsque le support des agences centrales étaient assurés. Ceci tient à plusieurs raisons, la petitesse des projets - et pourtant je reste persuadé que c'est la seule voie possible - les difficultés administratives au sein de certains de vos gouvernements, et aussi un manque d'effort coordonné et soutenu dans nos recherches de financement.

La recherche de financement ne peut qu'être une action commune entre vous et nous. Il faut bien voir que l'IBSRAM n'est pas le Père Noël. Les financements multilatéraux, ceux qui sont directement fournis à l'IBSRAM, sont limités. Il faut donc travailler ensemble pour rechercher des financements bilatéraux ou régionaux. Le projet malgache de gestion des sols acides est à ce sujet un modèle. Le projet a été discuté et finalisé après la réunion de Lusaka. Par la suite le FOFIFA localement et l'IBSRAM en Suisse ont approché un donateur et ont permis la conclusion du fianancement. Des actions régionales sont aussi possibles, et sont tentées auprès de l'UNDP et d'autres donateurs - mais de toute façon les recherches de financement doivent être conjointes.

Parmi les difficultés rencontrées en 1989 il faut signaler les incertitudes liées à l'implantation d'un bureau régional qui doit permettre aux coordinateurs IBSRAM d'être plus près de vous. Ce problème est maintenant en grande partie dépassé, et je suis heureux de vous annoncer que notre bureau d'Abidjan est maintenant opérationnel. Il a un téléphone, un numéro de telex, un fax, et très bientôt Bernhard Hintze va rejoindre Otto Spaargaren. Ceci ne résoudra pas tout - évidemment, mais les conditions de travail devraient s'améliorer. Hier M. Worou s'inquiétait de la lenteur d'un transfert d'Abidjan à Lomé. Les modalités administratives sont aussi à roder en vue d'éviter ce genre de retard. D'autres difficultés plus techniques peuvent se présenter, et nous espérons aussi vous aider à les résoudre.

Mais ne nous laissons pas aveugler par les problèmes. Des réseaux sur la gestion des sols en Afrique c'est aussi et avant tout un espoir. L'espoir de voir enfin le sol pris en compte dans l'expérimentation agronomique, l'espoir d'une meilleure gestion des sols qui puisse en limiter les dégradations, l'espoir enfin de préparer pour nos enfants une terre plus accueillante, qui puisse les nourir d'abondance et leur permettre de vivre en harmonie avec une nature retrouvée.

Nous avons une semaine pour discuter de ces problèmes, visiter le projet malgache, rediscuter de nos projets.

Bon courage!

* * * * *

Minister of Scientific Research and Development Techniques, FAO Representative, delegates for international organizations, Director General of the Centre for Applied Research in Rural Development, ladies and gentlemen:

I would like to say how pleased I am to be with you all today for the third conference of the IBSRAM networks under the AFRICALAND programme, which is specifically focused this year on soil tillage and the management of organic matter. May I use this occasion to thank our Malagache colleagues who were kind enough to accept responsibility for organizing this conference, and in particular to thank M. Jean-Louis Rakotomanana. My thanks are also due to the FAO for having agreed to be coorganizers of the conference, and to the Canadian International Development Agency (CIDA), the Bundesministerium für Wirtschaftliche Zusammenarbeit (BMZ), and Swiss Development Cooperation (CSD) for the financial support they provided for this conference.

After the Douala conference in 1986 and the Lusaka conference in 1987, Antananarivo in 1989 was the third year of operations for the IBSRAM networks on land development and the management of acid soils in Africa. Three years is not very long when we think of how much might have been accomplished; but it is a sufficiently long period to see the continuity of the initial enterprise. The IBSRAM networks have become an established reality, and an entity which must be taken into account.

Soil management networks in tropical Africa represent a necessity, a difficulty, and a hope. A necessity, that goes without saying. As we all realize humid Africa of the tropics is coming under increasing population pressure, of a kind which is unequalled in history. This population increase, which has been aggrevated during the past few years by migrations from arid and semiarid regions, has resulted in even more pressure on forested areas. It is not simply a case of the risk involved for tropical forests and all the diverse animal life and vegetation which they harbour, but it is also the soils themselves which are at stake, the underlying support of this natural wealth - all these things are threatened by erosion and various forms of degradation. More intensive research on sustainable land development is an indispensable requisite. At the same time, new endeavours must be coordinated and directed towards practical applications which will be of benefit to the ordinary African.

The IBSRAM networks can contribute to this objective. The exchange of experience, as we will see during the course of the week, the discussions which will take place in the field and in the network committees, the training sessions we will be having, as we did at IITA and at Yurimaquas in 1988, and also the coordination of work in the field are definitely along the right lines. This coordinated endeavour will allow us to use our limited resources more efficiently.

The experience of the past three years has, however, shown that a recognized need is not necessarily a guarantee of financial support for national projects. We started off by assuming that bilateral funds could quite easily be brought to bear on network development. In fact, the application of funds for this purpose proved to be more difficult than was at first imagined, even when the support of the central donor agencies had been assured. There are several reasons for this: the small scale of the projects - and yet I am convinced that small-scale projects are the only viable proposition - the administrative difficulties which arise in many of the governments in the region, and also a lack of coordinated and sustained effort in procuring funds.

The search for funding must be something we do together, involving both the countries in the network and IBSRAM. It should go without saying that IBSRAM is not a kind of Father Christmas. Multilateral funding - funds, that is, which are provided directly to IBSRAM - is very limited. This means we need to work together to procure bilateral or regional funds. The Malagache project on the management of acid soils is a model project in this respect. The project was discussed and finalized after the Lusaka conference. The FOFIFA from its local standpoint tried to find donor support and IBSRAM approached the Swiss government, the two actions together resulting in a satisfactory financing arrangement. Regional efforts are also possible, and can be directed towards the UNDP and other donors - but at all events, the attempt to obtain donor support must be a joint venture.

Amongst the difficulties we encountered in 1989, one of the problems was to set up a regional office which would allow the IBSRAM coordinators to be nearer to the networking countries. This problem has now been resolved to a great extent, and I am pleased to be able to announce that our Abidjan office is now functioning. It has a telephone, a telex number, and a fax number, and soon Bernhard Hintze will be there with Otto Spaargaren. This will not resolve all our difficulties immediately, of course, but it does represent a better working arrangement. Yesterday M. Worou was complaining about the time it took to get through to Lomé from Abidjan. Administrative operations *need to work more smoothly if we are going to avoid this sort of delay. Other more technical difficulties may crop up, and we hope we can help you to resolve these as well.*

But let us not only think about the problems we may encounter. Soil management networks in Africa also present us with hope. This is the hope of seeing soil taken into account in agronomical experiments, the hope of better soil management which will succeed in controlling soil degradation, the hope of leaving our children more friendly land to work, land which will feed them abundantly and allow them to live harmoniously with improved natural resources.

We have a week to discuss these problems, to visit the Malagache project, and to talk over our projects again.

May I wish you well in these endeavours!

◆ ◆

Antoine Rabesa Zafera

Ministre de la Recherche Scientifique et Technologique pour le Développement
(Minister of Scientific Research and Development Techniques)

Monsieur le Directeur de l'IBSRAM, Monsieur le représentant de la FAO, honorables invités, chers collègues chercheurs, Mesdames et Messieurs :

Lorsqu'à son retour du séminaire régional africain sur la gestion des sols acides organisé en 1988 à Lusaka (Zambie) par l'IBSRAM, le delégué de Madagascar avait annoncé la décision de l'assemblée générale du séminaire de tenir à Antananarivo une de ses prochaines réunions. Depuis, le gouvernement malgache, et en particulier le Ministère de la Recherche Scientifique et Technologique pour le Développement s'est préparé pour deux événements.

Tout d'abord et naturellement, abriter une rencontre scientifique internationale de haut niveau avec indiscutablement tous les bénéfices scientifiques attendus pour le pays; mais surtout, vivre cette forme efficace d'échanges directs entre les instituts internationaux de recherches agricoles (du groupe CGIAR) et les instituts africains de recherches agricoles, à propos d'un thème capital pour les pays tropicaux: c'est le travail du sol et la gestion de la matière organique en zones humides et sub-humides.

A vous tous, éminents spécialistes de la gestion du sol, venus de différents pays du nord comme du sud, animateurs ou participants délégués, croyez que vos homologues malgaches, le monde des chercheurs malgaches, moi-même et à travers moi le gouvernement, vous souhaitent à vous toutes et à vous tous la bienvenue à Madagascar.

Mesdames et Messieurs,

La préparation des sols et la gestion de la matière organique dans les zones humides et sub-humides tropicales, l'une des préoccupations dont l'importance en termes de développement agricole se traduit par le danger potentiel et permanent d'apparition des formes de dégradation, d'érosion chimique et mécanique du capital sol.

A Madagascar, les deux tiers de la superficie du pays sont globalement des sols acides, avec tout le concert de fragilité et de précarité que cela entraîne au niveau de la production et de la productivité agricole. En outre, ces terres correspondent aux zones les plus peuplés. Par ailleurs, à Madagascar, nous pratiquons et continuons toujours à pratiquer (malheureusement), notamment dans les zones humides et sub-humides, une agriculture itinérante, au défrichement de forêt et brûlis.

Aussi, plusieurs travaux de recherches ont-ils été conçus et entrepris depuis ces dernières années au sein de FOFIFA, en collaboration avec l'IBSRAM et avec l'aide d'organismes de financement comme la Coopération Suisse, l'USAID, l'Unesco, la Banque Mondiale, afin d'assurer à terme, une meilleure gestion des sols, c'est à dire une nouvelle forme de mise en valeur des sols, acceptable, adoptable et à la portée financière du paysan; autrement dit, rentable pour lui économiquement. La déscente sur terrain prévue ce dimanche vous permettra de constater les résultats obtenus ou en cours d'obtention.

Mesdames et Messieurs,

Le Ministère de la Recherche Scientifique et Technologique pour le Développement va accorder une attention soutenue à vos échanges d'expériences, à vos discussions, à vos différents projets de recherche pour le futur. Pendant ces quelques jours, Madagascar aura de nouvelles données de base quant à sa grande bataille pour garantir un développement national durable dans un environnement sain et equilibré.

Au nom du gouvernement, je remercie vivement par leur contribution à la tenue de cette réunion : le FOFIFA, le Conseil International sur la Recherche des Sols et leur Gestion (IBSRAM) avec la collaboration de la FAO et le co-financement de la Banque Mondiale, la Canadian International Developement Agency, la Coopération Suisse.

A vous tous venus de si loin, je souhaite un excellent séjour de travail. Je remercie vivement également nos honorables invités qui ont bien voulu rehausser de leur présence ce séminaire.

Je déclare ouverte cette réunion scientifique régionale de l'IBSRAM sur le Travail du Sol et la Gestion de la Matière Organique en Afrique Humide et Sub-humide.

Je vous remercie de votre attention.

* * * * *

Director of·IBSRAM, FAO Representative, honoured guests, fellow research workers, ladies and gentlemen:

When the Malagache delegate came back from the Lusaka (Zambia) seminar on the management of acid soils organized by IBSRAM in 1988, he told us of the plenary session decision to hold one of the forthcoming conferences at Antananarivo. Since then, the government of Madagascar, and more specifically the Ministry of Scientific Research and Development Techniques, has looked forward to two aspects of this meeting.

In the first place, of course, there was the opportunity of holding a high-level international scientific conference, with all the undoubted scientific benefits such a conference would have for the country. But the opportunity of taking part in an effective exchange scheme which puts international agricultural insitutes (from the CGIAR) and African agricultural research institutes in direct contact with one another is especially significant. The topic of this conference - tillage and organic-matter management in humid and subhumid areas - is of immense importance for tropical countries.

Distinguished specialists in soil management, coming from North and South, interested participants and delegates, I would like you all to know that your Malagache counterparts, the research community in Madagascar, and I myself on behalf of the Madagascar government, extend a very warm welcome to one and all during your visit to Madagascar.

Ladies and Gentlemen,

One of the major concerns with regard to soil preparation and the management of organic matter for agricultrual development in the humid and subhumid areas of the tropics is the potential risk of permanent soil degradation from both chemical and mechanical erosion.

In Madagascar, two-thirds of the land is occupied by acid soils, and is consequently fragile and unpredictable from the point of view of agricultural production. This part of the land is also the most densely populated area. Moreover, in Madagascar, unfortunately we still practice shifting agriculture with slash-and-burn in humid and subhumid areas.

For these reasons, a member of experiments have been undertaken in the last few years under the supervision of FOFIFA and with the collaboration of IBSRAM and the support of donor agencies like Swiss Development Cooperation, USAID, Unesco, and the World Bank, in order to ensure better soil management, with techniques that are acceptable by the farmers, and adoptable and financially viable for them, which is to say prosfitable for them. The field visit scheduled for Sunday will allow you to see for yourselves the results which have been obtained and experiments in progress.

Ladies and Gentlemen:

The Ministry of Scientific Research and Development Techniques will take a keen interest in your experiences and discussions, and in the different research projects which you have in mind for the future. During these next few days, Madagascar will be receiving some new basic information to help in its struggle to achieve sustainable national development in a sane and balanced environment.

In the name of the government of Madagascar, I extend my warm thanks to those who have helped to organize this conference: FOFIFA, the International Board for Soil Research and Management (IBSRAM), supported by FAO and financial backing from the World Bank, the Canadian International Development Agency, and Swiss Development Cooperation.

Many of you have travelled long distances to be here today, and I wish you every success in the work you have come here to do. I would also like to give special thanks to the distinguished guests who have honoured us with their presence at this seminar.

I now declare open the IBSRAM regional scientific conference on Tillage and Organic-Matter Management in Humid and Subhumid Africa.

Thank you for your kind attention.

Section 1: FAO, IBSRAM, and TSBF programmes in Africa

FAO's assistance in the development of a tillage network in Africa

C.S. OFORI[*]

Abstract

Improvements in tillage practices in the Sub-Saharan Africa region are essential for effective soil and water conservation, rationalization of farm labour and reduction in the drudgery of farm operations. Tillage research in the region must aim more at a tillage-soil-water-plant systems approach instead of individual components.

With the limited financial resources available for research and development, networking offers an effective mechanism for collaboration between countries in research, dissemination of results, and application of technological packages at farm level. Some of the approaches proposed by the Food and Agriculture Organisation of the UN in catalyzing such networks are highlighted.

Introduction

Tillage practices are important aspects of seedbed preparation and weed control for increased crop production. Various tillage methods have evolved throughout the ages, and have been greatly influenced by the different stages of improvement in traction and by the introduction of nonrenewable sources of energy in agricultural production. The development of tractors and other machines in modern agriculture has facilitated easy tillage operations, albeit

---

[*] Senior Officer (soil management), Food and Agriculture Organization of the United Nations (FAO), Via delle Terme de Caraccala, Rome, Italy.

at the cost of high energy inputs. The present awareness and concern for soil and water conservation has promoted soil management practices such as conservation tillage, thus reducing energy input, and at the same time making it possible to reduce soil degradation.

Most farmers in the tropics produce on small farms and use basic implements for soil tillage. Areas endowed with livestock benefit by animal traction, which lightens the drudgery of seedbed preparation and weeding. In the late '60s, many developing countries, in an attempt to increase their food production, resorted to large-scale mechanized farming, with the introduction of various sizes of tractors and implements which were not always the right types for their respective soils. The desired effects were often not achieved, and degradation of soils on an unprecedented scale occurred in many countries. Soil tillage is therefore regarded as one of the main areas that deserve special attention in developing mechanized crop production systems in developing countries. In this paper some possible approaches of tackling this problem, especially those recommended by the Food and Agriculture Organization of the United Nations (FAO), are discussed.

FAO's assistance

FAO's assistance to member countries in the development of tillage practices and mechanization includes technical and advisory services covering expert services in projects, consultancy missions, and support for collaborative research projects.

Panel of experts on agricultural mechanization

In 1978, FAO established a panel of experts to deal with a number of issues on mechanization in developing countries. Eighty percent of the food in most developing countries is produced by small-scale farmers. The technology adopted is still very simple but laborious. One of the major constraints in the production system still remains the inadequacy of power for tilling the land. During the past decade the panel discussed some of the issues which affect small-farmer development and production systems. These include:

- mechanization of small farms,
- soil preparation and soil conservation, and
- testing and evaluation of farm machinery and equipment in tropical regions.

The regular meetings of the panel provide an effective forum for bringing together various kinds of expertise; the outcome serves as a guideline for field project development and support.

Field projects

Field projects on animal draught power and tractorization have been developed and implemented in a number of countries. Although tillage forms a major component in all these projects, there is a need to focus more attention on specific tillage practices in relation to soils. Proven tillage practices are inadequately demonstrated under various soil conditions in most of the developing countries. There is therefore the tendency for machinery to be imported into many countries on the basis of what is wanted instead of the needs in relation to the soils to be cultivated. This has often resulted in the adoption of wrong tillage practices and the attendant soil degradation. A major role played by FAO is the provision of technical advice to member countries on these issues when requested.

Network approach

Tillage problems differ considerably from one country to another within the region. Apart from differences in soils, there are wide variations in the climate and in the cropping systems, and these all need to be considered when developing a tillage programme. Mouldboard ploughing is extensively practiced within the region. The recent approach to conservation tillage and residue management has made it necessary to consider changes in tillage practices used in the past. Moreover, the impact of fuel cost and poor economic growth in most of these countries has limited large-scale mechanization in most cases.

One of the recommendations of the FAO Panel of Experts on Agricultural Mechanization stated that: "At the present state of development, zero tillage based on mulching with or without the use of herbicides is a feasible system for the 'manual' farmer."

Minimum and zero tillage practices, although very promising, cannot be adopted without due consideration to differences in soils. Other practices such as tie-ridging and permanent-ridge systems currently used in a number of countries have proved to be efficient in moisture conservation, as well as in providing an adequate seedbed for crop growth and savings in energy and labour costs. The effect of these practices on crop yield, as well as the economics of production, still need further investigation.

A strategy for developing a tillage network will involve an approach based on agroecological zonation as well as on the presently available technology in the participating countries. In the savanna and semiarid areas, soil and moisture conservation are the major factors to be considered in tillage programmes. Residue management is an essential component of tillage practices. But in some areas of the region, because of other economic uses of crop and other residues, these factors have only played a secondary role so far in tillage programme development.

Residues in the humid tropics, on the other hand, decompose very fast when ploughed under, or are destroyed by termites when left on the surface. The development of separate tillage networks for the semiarid and humid tropics, taking into consideration climate, soils, and cropping practices, •will therefore be appropriate.

Networks

Several types of networks are possible depending on the envisaged source and the level of funding, as well as on the coordinating mechanism. It is important to select the appropriate type of network and organizational structure for effective implementation at the time of its establishment.

Regional networks comprise a number of institutions or countries within a region, with the core funding regionally controlled. Each participating country or institution is responsible for funding the activities at national level. Such activities are necessarily of national relevance and importance and are integrated into national development programmes. This will ensure continuity and appropriate budgetary provision - a condition which often is not satisfactorily fulfilled.

Another type of network is one based on national activities funded at country level, but with a central coordinating mechanism such as an international organization like FAO. The coordination agency provides opportunities for bringing together participants to discuss research results, and to develop strategies and research programmes on relevant problems. Budgetary provisions are also made for seminars and meetings.

Thirdly, a network may be formed in which the participating countries initiate and fund their own national programmes, but are supported with some modest external assistance. In addition to the external assistance, which in effect acts as a catalyst, a coordinating machinery is established to organize and finance activities such as seminars, meetings, and information exchange.

The establishment of a periodic newsletter provides an effective means of disseminating relevant information.

The availability of resources, manpower, and ongoing financial support for selected programmes is an important consideration for the successful implementation of networks. A careful selection of participating institutions is therefore essential.

There are various networks operating in Africa with different degrees of success, and it is important to identify some of the main constraints common to these networks. Often the first major constraint is a lack of budgetary provision.

Networks related to tillage, or with tillage components, have been initiated or planned at various times in Africa. The majority of these networks, however, have not addressed tillage as a specific issue. A tillage network with technical support from FAO should have the following features:

- Participating countries should either have ongoing programmes on tillage or propose to initiate work on aspects of tillage as an integral part of national programmes. The topics or aspects of tillage to be dealt with need not be identical in all countries, but should be of broad interest to ensure wider dissemination of the results.
- One national institute should assume the responsibility for the implementation of the national programme of the network in a participating country.
- A minimum of five interested countries should form the nucleus of the network. Provisions should be made for future expansion to a possible total of eight countries.
- Network participants would be encouraged and assisted by FAO to discuss their programmes together and exchange ideas, and would be given the opportunity to present their results in network consultations. Such consultations may be organized on an annual or biennial basis, and hosted on a rotational basis by participating countries.
- FAO would provide technical assistance to the network through consultancies in experimental design and the evaluation of data. Such assistance could also include dissemination of information on tillage and residue management by means of a newsletter.
- Funding possibilities would be explored through national and bilateral assistance, either directly or through FAO.
- Close collaboration should be established between the proposed tillage network and other related activities in the region to ensure complementarity.
- The network should have a minimum duration of four years. This is essential to generate adequate data for meaningful interpretation.

Preparatory work undertaken by FAO to establish a tillage network involved a mission to five countries in the region, namely Cameroon, Kenya, Zimbabwe, Malawi, and The Gambia. A common feature identified in all five countries is that although tillage research is conducted in each country visited, the research is often of limited scope and not directed to the effect of tillage in the overall tillage-soil-water-plant system for conserving soil and water resources and enhancing food production (Unger, 1986). These findings provide a good basis for the development of future tillage research programmes and essential elements for networking. The development of such a tillage programme is a big challenge to African soil scientists - not only to increase food production and rationalize labour inputs, but also to conserve the soil.

Field projects

Three countries (Cameroon, Malawi, and Nigeria) have so far been assisted by FAO to initiate field projects on tillage with a view to developing a coordinated tillage network embracing a number of countries in the region. The project in each country focuses on specific aspects of its national programme. An account of the progress made in the field work of two of the participating countries, Malawi and Cameroon, is presented below.

Malawi
A project on ridging was developed in Malawi. Ridging is labour- and energy-intensive, and with the simple technology employed by most farmers, these operations are often not completed in time for the onset of the rains for planting. Consequently planting is often late, and consequently crop yields are reduced.

The experiment was designed with the following treatments:
- **Permanent ridges (early)** This refers to ridging undertaken after harvesting, and left for a number of cropping seasons.
- **Permanent ridges (late)** This refers to ridging undertaken at the onset of rains, and left for a number of cropping seasons.
- **Nonpermanent ridges (late)** This refers to ridges reformed each year after the onset of rains.
- **Nonpermanent ridges (early)** This refers to ridges reformed each year immediately after harvesting.

Three agricultural research stations were selected - Chitedze, Lisasadzi and Chitala - and the experiments were sited on sandy clay, sandy, and clayey soils respectively. The handhoe was used for ridging, and the test crop was maize. The ridges measured 23 cm in height and 30 cm in width. The operation for such a ridge size is labour-intensive, requiring 18 man-days per hectare. At the end of the cropping season, it was observed that the ridge height had decreased to 16 cm, but the width had increased to 24 cm as a result of slight erosion and scraping of the topsoil during weeding operations.

Basal fertilizer was applied at the rate of 200 kg ha^{-1}, consisting of a compound fertilizer 20:20:0, and the maize was later topdressed with 200 kg ha^{-1} of calcium ammonium nitrate (26% N).

The first two seasons' results reported by Kumwenda (1989) are presented in Tables 1-3.

Table 1. Effect of tillage on maize grain yield (kg ha^{-1}) at Chitedze.

Treatment		1987/88	1988/89	Mean
(PE)	Permanent ridging - early	4.2	4.4	4.3
(PL)	Permanent ridging - late	4.0	4.7	4.4
(RE)	Yearly ridging - early	4.3	4.9	4.6
(RL)	Yearly ridging - late	4.4	4.9	4.7
	Mean	4.2	4.7	

Table 2. Effect of tillage on maize grain yield (kg ha^{-1}) at Lisasadzi.

Treatment		1987/88	1988/89	Mean
(PE)	Permanent ridging - early	4.2	4.3	4.3
(PL)	Permanent ridging - late	4.5	4.6	4.6
(RE)	Yearly ridging - early	4.3	4.2	4.3
(RL)	Yearly ridging - late	4.3	4.6	4.5
	Mean	4.4	4.4	

At Chitedze, the late ridging at the onset of the rains resulted in a small increase in maize yield in both seasons. These differences, however, are not significant.

Table 3. Effect of tillage on maize grain yield (kg ha^{-1}) at Chitala.

Treatment		1987/88	1988/89	Mean
(PE)	Permanent ridging - early	4.0	3.6	3.8
(PL)	Permanent ridging - late	3.6	3.6	3.6
(RE)	Yearly ridging - early	4.0	3.5	3.8
(RL)	Yearly ridging - late	4.5	3.8	4.3
	Mean	4.0	3.6	

Similar results were obtained at Lisasadzi and Chitala. The second season results, 1988/89, showed no clear trends either, even though at Chitedze the late ridging was better than the permanent late ridging. These experiments will need a few more seasons' monitoring to enable yield differences, if any, to be established. Evaluation of the labour input will be of the utmost importance in assessing the merits of the various ridging systems and the savings in terms of labour and energy that will accrue to the farmer.

Cameroon
Three tillage systems are under investigation in the experiment, namely:
- zero tillage, using herbicides to kill weeds: narrow furrows of 5 cm depth were hand-dug, using the garden hoe for seed placement (ZT);
- traditional tillage - plots were hoed (TT); and
- minimum-tillage: the soil was tilled using a self-propelled rotary cultivator (MT).

The site was originally cleared mechanically 5 years prior to the establishment of the experiment. After the first year of cropping, the site was fallowed for the succeeding four years and the weeds were cleared by cutlassing in the traditional-tillage and minimum-tillage systems. Herbicides were used in the zero-tillage treatment. The soil is a clayey Oxisol (Ferralsol - FAO/Unesco classification).

The maize test crop was planted at a spacing of 0.24 x 0.75 m, giving a plant population of 53 000 plants ha^{-1}. A compound fertilizer of 20-10-10 formulation was applied at the rate of 600 kg ha^{-1}.

Maize grain yield is presented in Table 4.

Maize yield from the zero tillage (ZT) was higher than in the traditional (TT) and minimum tillage (MT) treatments in both seasons. These differences, however, were neither significant nor economic, considering the cost of herbicides. The resource-poor small farmer can hardly afford the outlay for

Table 4. Effect of tillage systems on maize grain yield (ton ha^{-1}).

Treatment	1st season	2nd season	Treatment mean
ZT (zero tillage)	6.5 a	6.0 a	6.3
TT (traditional tillage)	5.8 a	5.4 a	5.6
MT (minimum tillage)	6.2 a	4.8 a	5.5
Mean	6.2	5.4	
CV (%)	12	14	

the purchase of herbicides. Differences in crop yield due to the various treatments must therefore be substantial to convince the farmer beyond doubt that it will be profitable to invest in herbicides.

These preliminary results underscore the need for more systematic and intensive evaluation of tillage techniques on different soils under small farm conditions. Additional data from subsequent seasons are necessary for the interpretation of the results.

Conclusion

Improvements in tillage practices in the sub-Saharan Africa are essential for increased soil and water conservation, a rationalization of farm labour, and a reduction in the present drudgery of farm operations. These improvements, should, however, go hand-in-hand with an increase in crop production. The bulk of food production in the region is in the hands of small-scale farmers with farm sizes less than 2.5 ha. Tillage practices and technologies developed under different farming systems need to be tested to ascertain their suitability and ease of adoption.

Mechanical innovations are often appropriate only to a particular size of farm. Even though tractorization is on the increase in Africa, the increased use of tractors may not completely resolve problems of appropriate tillage practices and rational use of labour by the small farmer. Attention therefore will have to be directed to the best methods.

Research on appropriate tillage practices has not received much emphasis in the region. Mechanization has so far depended heavily on the transfer of technology developed elsewhere, without adequate local testing. There are many ongoing investigations on tillage practices on similar problems in various countries with similar agroecological zones and soil conditions. With limited financial resources, it is prudent to harness resources in the region, and for aid organizations to complement each other's efforts. With this in mind, sharing of information and results through a coordinated network would certainly provide a mechanism with maximum impact. But an important prerequisite for such an approach is the development of country programmes with adequate funding which could be supported by external input.

References

AMBASSA-KIKI, R. 1987. *Soil tillage studies. Report on trials carried out on vertic and ferralitic soils.* Yaoundé: National Soils Centre.

FAO (Food and Agriculture Organization of the United Nations). 1979. Panel of Experts on Agricultural Mechanization. *First session records.* Rome: FAO

FAO. 1983. *Tillage practices on small farms in the tropics.* Panel of Experts on Agricultural Mechanization. Rome: FAO. 13p.

KUMWENDA, W.F. 1989. *Development of a permanent ridge for crop production.* Chitedze Research Station, Ministry of Agriculture. Annual Report. Malawi: MOA.

UNGER, R.W. 1986. *Mission report: Visit to African countries in connection with an FAO-supported tillage research network.* Agricultural Division, Food and Agriculture Organization of the United Nations. Rome: FAO. 25p.

IBSRAM's network on the management of acid soils in humid and subhumid Africa

OTTO SPAARGAREN[*]

Abstract

As nearly three years have pass since the IBSRAM network on the management of acid tropical soils in Africa was established, it is now possible to indicate the achievements and difficulties encountered and the prospects for the future. Training has proved to be a successful activity, whereas the procurement of bilateral funds and the establishment of experiments on the ground have been more difficult. The network, with its coordination office established in Abidjan, has approval for the funding of five projects, and should now be able to develop more smoothly. Further questions, such as the management of acid soils according to agroclimatic zones, could also be envisaged within the framework of future network activities.

Introduction

It is almost three years ago (April 1987) that we met for the first time to discuss the modalities of the network and the common strategies for implementation of this network. Since then many things have happened: we have had training courses, we have succeeded in putting some projects on the

[*] Coordinator, Acid Soils Network, Résidence Gyam, 2ème étage, Angle Bd Clozel-Av Marchand, 04 BP 252, Abidjan 04, Côte d'Ivoire.

ground, and we have learned a lot through trial and error on how to implement our experiments. It is now time to reflect on our operations, and to critically look at what we are doing and what we are trying to achieve. This evolution involves scrutinizing what we have learned during our training courses, and carefully assembling all our data into a research package that suits IBSRAM's objectives in general and the objectives of the acid soils network in particular.

What are the objectives of the network on the management of acid soils? They are, in general terms, to achieve sustained food production by smallholder farmers while conserving the soil resource. This means testing, adapting, and promoting alternative cropping systems and management technologies that fit the environmental constraints imposed on the farmer and that are socio-economically acceptable to him. In particular, we have to devise ways and means to overcome soil constraints such as aluminium toxicity, scarce nutrient supplies, and rapidly declining organic-matter levels upon cultivation. We also have to overcome the adverse soil physical properties which are often associated with acid soils in humid Africa, such as their sealing and crusting propensities, their low water-holding capacity, and their susceptibility to erosion. At the same time, we must somehow measure the socioeconomic impact of alternative technologies.

Eight fully developed proposals for national research projects now form the portfolio of IBSRAM's collaborative research network on the management of acid soils in humid and subhumid Africa - the *AFRICALAND* acid soils network. They originate from Burundi, Cameroon, Congo, Côte d'Ivoire, Ghana, Madagascar, Nigeria and Rwanda, and most of them were presented at the meeting in Lusaka, and subsequently revised and refined to suit research needs and donor requirements. An account on these projects and their progress to date will be presented later this week by the national cooperators. For the time being, we will look at the developments of the network in general since the Lusaka meeting.

Activities and developments in the framework of the *AFRICALAND* acid soils network

The activities in the framework of the network have concentrated on four issues:
- training front-line researchers in the methodological approach IBSRAM is taking in its networks;
- the procurement of funds for the projects and the coordination of the network;

- the initiation of several projects by providing "seed money" for site selection and characterization, and to get the experiments started; and
- the establishment of a regional office in Africa so as to be in a better position to serve the cooperating national research agencies.

Training

Several training courses have been organized in cooperation with other IBSRAM networks to maximize uniformity in the research approach. The training component was put into effect almost immediately after the Lusaka conference with a training course on data collection and data processing in Kasama, Zambia, in August 1987, when we were fortunate enough to have the cooperation of the Soil Productivity Research Programme.

In February/March 1988 a training course on site selection and characterization was held at IITA, Nigeria, in cooperation with the University of Ibadan. This resulted in the publication of IBSRAM's Technical Notes no. 1, and in a first approximation of IBSRAM's methodological guidelines.

In August/September, African cooperators convened in Yurimaguas, Peru, to discuss acid tropical soil management and land development practices. An account of this workshop, coorganized by the Instituto Naçional de Investigaçion Agraria y Agroindustrial of Peru, North Carolina State University, and the USAID-sponsored Soil Management Collaborative Research Support Program (TropSoils), can be found in IBSRAM Technical Notes no. 2.

Just recently, in November/December 1989, a workshop organized in collaboration with the International Council for Research in Agroforestry was concluded in Nairobi, Kenya. There are plans to publish the proceedings of this very successful workshop, provided funds can be obtained for the work.

In general, these workshops have contributed to an understanding between the cooperators in the network, and have yielded a common framework of research necessary for the transfer of data, knowledge, and experience within the network. The methodological guidelines, which are now in their third approximation after modifications and refinements made during IBSRAM workshops at ICRISAT, India, and in Chiang Mai, Thailand, are well known by the researchers, and projects are being started in accordance with the recommendations in the guidelines.

Procurement of project funds

The procurement of funds for the projects has proved to be a difficult task. There are several reasons for this. Although donors often view the network projects favourably, the funds requested for one project for the initial three years are often considered too small to be handled by agencies with large sums at their disposal. Moreover, funding of one project often requires bilateral aid relations in the field of agricultural research in order to fit in with network projects. In the case of Madagascar, it has been possible to incorporate the Malagasy project in a much broader aid programme on environmental protection and rehabilitation, funded by the Coopération Suisse pour le Développement. Funding on a bilateral aid basis for one project also requires active participation of the national cooperators, as the projects usually have to be presented during the annual donor review meetings with the respective governments.

Alternatively, grouping of projects on a regional basis seems to be more successful. Recently the Coopération Française has indicated a willingness to support several projects in Africa. This is now being followed up by drafting a project proposal combining the projects of Cameroon, Congo, and Côte d'Ivoire, and including the projects on land development in these countries. A similar arrangement might prove to be successful for Burundi and Rwanda, where core projects for research on highland agriculture in Africa could be combined; and ultimately projects in Kenya, Uganda, and Ethiopia might also be included under the same cooperative arrangement.

Funds for the Nigerian project will be submitted to the Canadian International Research Development Centre (IDRC), thanks to the efforts of one of IBSRAM's board members, Dr. Julian Dumanski. An unfortunate setback, however, is that our Nigerian cooperator, Professor Timothy Ashaye, has recently died, which is a loss for the project as well as being a sad event for all his friends and colleagues.

The proposal from Ghana has been submitted to the Dutch government for funding, but confirmation of funding has not yet been received.

"Seed money" and site characterization

So far, three countries - Cameroon, Congo, and Côte d'Ivoire - have been provided with "seed money" for site characterization and/or some initial work on their projects. In Cameroon, site characterization is presently under way, and later this week initial results will be presented. In Congo, where the project started in 1987 - almost immediately after the Lusaka conference - the "seed money" provided has not yet been used, mainly because it disappeared for a long time during the transfer period. Now that it has surfaced again, it will be used for the characterization of the new site, which was selected after the old one was abandoned for logistical reasons. The "seed money" for Côte d'Ivoire helped the cooperators to characterize the site and to start the experiments, which have been under way since March 1989. The *IBSRAM's Newsletter* no. 13 gave a report on the site characterization and the initial results of the experiments in Côte d'Ivoire.

In order to receive "seed money", cooperators have to submit a detailed budget and a short project proposal to outline the operations envisaged. When approved, a memorandum of understanding is signed and money is sent. This process of requesting and providing "seed money" has proved to be troublesome in some cases. Budgets were too general. Sometimes, too, budget items were entered that were not directly related to the site selection and characterization - as in one case when budgetary provisions were made for lawyer's fees to obtain the experimental site.

Moreover, it is often quite difficult to identify the body to administer and to account for the funds, as in a number of countries foreign exchange regulations prevent the cooperators from using the funds - to buy, for example, the necessary chemicals for the soil analysis to be carried out abroad. On top of this, we have had the experience that funds may take some time to arrive, or may arrive at a time when it would be impractical to carry out the operations they were earmarked for. We need to consider these various difficulties during the course of this meeting in order to see if there are any ways of circumventing some of the problems we have experienced in the past.

The regional office for Africa

For the last couple of years, IBSRAM has been in contact with the government of Côte d'Ivoire to get permission to establish a regional office for its *AFRICALAND* programme in Abidjan. At the end of 1988, one of the coordinators moved to Abidjan on the invitation of the Ministry for Scientific Research to get installed and to negotiate an "accord d'établissement". After a number of fruitless efforts to come to an arrangement which did not conflict with Ivorian laws, official permission was granted to open a coordination office for IBSRAM/*AFRICALAND* by a decision of the Ivorian Cabinet the beginning of November 1989. An "accord d'établissement" is presently being drafted, but it may still take some time before this is accepted, as both the Ivorian Cabinet and parliament, and the IBSRAM Board of Trustees, have to agree to the conditions of the understanding.

The future of the network

The most important task ahead of us is to get the projects on the ground. Although there is still nothing certain with regard to financing, if we are able to secure funds from Coopération Française for the projects in Cameroon, Congo, and Côte d'Ivoire, as well as securing funds from IDRC for Nigeria, five projects can be on the ground by July/August this year.

In the longer term, however, I foresee that, if the number of participating countries increases, the network will have to be restructured - preferably in relation to agroecological zones as they have been defined by FAO or have been delineated by ICRAF. One zone might comprise the humid part of Africa, the coastal zone of West Africa and Central Africa (the part with a growing season of more than 270 days), and a second zone might comprise eastern and southern Africa, with a pronounced bimodal or unimodal rainfall regime and a growing season of between 180 and 270 days, and a third zone could be a high altitude zone. The transfer of data and experience within these zones might well be more useful than a transfer of data between, say, Côte d'Ivoire and Zambia.

The future of the network also depends very much on the results that we obtain and the way they are extended to the farmer. This is one thing we have to work on right from the start of the projects. Probably the best way of extending our results is to work with farmers on the projects, notably by asking them to carry out the part of the experiment that involves the traditional system. By having farmers on the project, we will be able to discern what innovations they would be willing to accept. We, as researchers, can also learn from the farmer by studying what he is doing and why, which will give us an excellent opportunity to analyze the cropping system he is using. In the long run, this cooperation will make extension of the results much easier, as we will have more ready access to the farming community. The ultimate goal of our network is, after all, the extension and the final adaption of alternative soil management systems by the small-scale farming communities with which we are working.

The TSBF approach to soil organic-matter management research

JOHN INGRAM*

Abstract

One of the five major TSBF principles states that soil organic matter (SOM) can be separated into functional pools, each of which plays a role in nutrient release, cation exchange and aggregation. Three research objectives arise from this principle: (i) to determine the best methods for quantifying the different soil organic matter pools; (ii) to determine the relative susceptibility to management of the soil organic matter pools; and (iii) to manage soil organic matter with respect to nutrient release, cation exchange, and soil aggregation.

To help rationalize research in SOM, TSBF has adopted a general model, defined as the "Century" model, which divides SOM into three functional pools: active, slow and passive. TSBF recommends specific laboratory methodologies (which are both reproducible and cheap) for estimating "total" SOM, microbial biomass and "light fraction". Three equations relate these laboratory estimates with functional pools as defined for the general model, although only partially for the Century model.

TSBF has designed a standardized experiment (KILLSOM) to estimate the magnitude of the acitive, slow and passive pools by repeated measurement, over a period of several years, of the decline in SOM in small plots from which all organic inputs have been excluded. It is predicted that the decline will stabilize at a level corresponding to the slow + passive pools, the active pool being defined as the microbial biomass. The rate of decline relative to a control plot under natural vegetation will provide an SOM turnover index for that particular soil-climate-vegetation situation.

*   Programme Coordinator, TSBF Programme, c/o Unesco-ROSTA, PO Box 30592, Nairobi, Kenya.

It is unlikely that a given land/soil management regime will affect each pool in the same way; the management of those pools with a fast turnover time/short residence time will be in effect a management of organic inputs. Management of the slow or passive pools will more be a management of existing pools in the soil, with the emphasis being on preservation by protection.

The Tropical Soil Biology and Fertility Programme (TSBF)

TSBF has five major principles:
- **Synchrony** relates to the processes of decomposition and nutrient cycling. The aim is to gain a better understanding of the processes which underlie the release of plant nutrients from decomposing material and the subsequent uptake by plants. It is considered that such a transfer can be better synchronised with the time when plant demand is greatest, mainly through the manipulation of the input quality (e.g. nitrogen:lignin/polphenolic ratio), and the timing of application.
- **Soil organic matter** states that SOM can be separated into functional pools, each of which plays a particular role in nutrient release, cation exchange and soil aggregation. This will be further discussed in this paper.
- **Soil water** considers that the availability of soil water to plants can be improved by the better management of surface litters and SOM. Soil water dynamics are also instrumental in the synchrony theme, and are particularly pertinent in semiarid regions.
- **Soil fauna** can be manipulated to improve the physical properties of soil and regulate decomposition processes.

These four principles are all fundamental to the TSBF Programme, but are best integrated in the final principle, i.e. the integration of biological processes. In accordance with this final principle, the biological control of soil fertility is the integration of plant nutrient demand, root distribution, decomposition processes, soil fauna activities, and their interaction with soil chemical and physical properties.

The soil organic-matter principle

The soil organic matter (SOM) principle leads to three main research objectives; the first and second lead in turn to the management objective as expressed in the third:

- to determine the best methods for quantifying the different soil organic-matter pools;
- to determine the relative susceptibility to management of the soil organic-matter pools;
- to manage soil organic matter with respect to
  * nutrient release,
  * cation exchange, and
  * soil aggregation.

It is realized that this type of research is unlikely to lead to implementable management prescriptions in the short term, as expected for the synchrony research. It is nevertheless an important area of study, made the more difficult by the limited standardization of methodologies. TSBF has attempted to address this issue by advocating particular techniques for separating the so-called pools, thus defining the pools on a methodological basis. It is assumed that these are related to the time-based definitions discussed below.

The TSBF concept of soil organic matter

The evaluation of inorganic soil fertility parameters is well understood; exchangeable K or available P, for instance, correlate well with fertilizer requirements and plant growth. This is not the case for SOM, where although organic carbon and organic nitrogen levels are commonly used parameters to express SOM levels, neither is directly correlated with plant growth.

It is clear, however, that SOM plays a key role in the maintenance of a good growing environment. It is also clear that SOM levels fall when a grassland or forest is cleared for cultivation, and yield decline is parallelled by a decline in SOM status. On the other hand, some soils relatively high in "total" SOM are not as productive as would have been expected, judged on SOM status alone. What then is soil organic matter, and how can its potential be better assessed to improve soil management and cropping potential?

Clearly SOM, let alone its various functions, cannot be succinctly defined. This is because SOM represents a large range of materials, each with a particular function in the soil system. The identification of and, where possible, the quantification of the relative importance of SOM pools would undoubtably provide a valuable tool for agronomists to better manage this fundamental resource.

In recognition of the varied roles SOM plays, many workers have chosen to model SOM dynamics along the functional pool concept; the TSBF Programme similarly adopts this approach. The number of functional pools recognized varies, however, as does the estimated residence time of each pool in the soil system. Nomenclature and definition similarly vary. The definition of the pools can be made either by describing them with respect to their mean residence time in the soil or by following a stated separation procedure. The two systems should of course relate. TSBF philosophy is that the varied systems do indeed relate, and, furthermore, that the nature of the material in a given pool relates to its function in the soil.

A general model for SOM dynamics is shown in Figure 1. By way of a refinement, the recently developed Century model (Parton *et al.*, 1987) has been adopted by TSBF as a tool to help in the study of SOM dynamics (Swift, 1987).

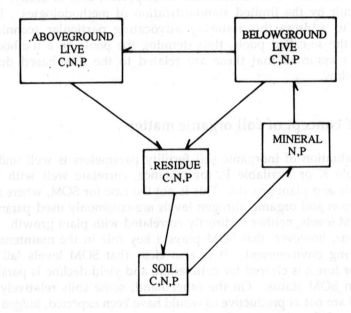

Figure 1. Flux of C, N, and P between major compartments of the plant-litter-soil subsystem: the general model.

The model proposes dividing SOM into three functional pools: active, slow, and passive (see Figure 2). Although they are expressed as discrete pools, they do in fact form a continuum - in physical nature, size class, and age. Structural and metabolic components, (differentiated by their N:lignin ratios, and hence mean residence time), are inputs to the three SOM pools. The structural and metabolic components comprise cellular material, including plant fragments; structural components can be easily recognized (usually being freshly added material) down to a size of about 0.25 mm. At the other end of the range are highly stabilized organic compounds, often found in association with clay minerals, and of a molecular size.

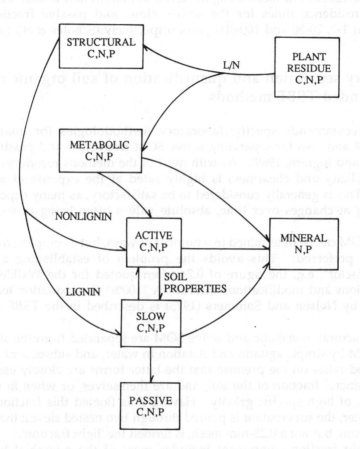

Figure 2. Major pathways of flux of C, N and P between organic compartments of litter and soil.

The decomposition rate of materials at the litter-fraction end is much faster than that towards the other end of the range. This is because a small quantity of organic carbon in each age class becomes more recalcitrant with time than the material from which it was derived (Anderson and Flanagan, 1989). It follows therefore that in a steady state situation the decomposition rate of organic residues, and hence the relative amounts in each pool, will depend as much on the nature of the original material as on edaphic, biotic, and climatic factors. Due also to the recalcitrance factor, the mean residence time of a given pool will increase as one looks along the continuum from new to old. Values for the mean residence times for the active, slow, and passive fractions are estimated at 1-2, 20-50 and 100-800 years respectively (Scholes *et al.*, 1989).

Laboratory separation and quantification of soil organic matter: recommended TSBF methods

TSBF recommends specific laboratory methodologies for quantifying "total" SOM and also for separating active SOM from slow and passive SOM (Anderson and Ingram, 1989). As with most of the methods recommended by TSBF, simplicity and cheapness is highly rated at the expense of absolute accuracy. This is generally considered to be satisfactory, as many experiments are looking at changes over time, absolute values often being of secondary interest.

Total SOM can be determined in a number of ways, but a complete oxidation method is preferred. This avoids the problem of establishing a "mean oxidation factor", e.g. the figure of 0.74 often quoted for the Walkley-Black determinations and modifications thereof. A "100%" wet oxidation technique developed by Nelson and Sommers (1975) is described in the *TSBF methods handbook*.

The structural, metabolic and active SOM are separated from the slow and passive SOM by simple agitatin and flotation in water, and subsequent sieving. This method relies on the premise that the latter forms are closely associated with the mineral fraction of the soil, and are themselves, or when in such an association, of high specific gravity. Having thus floated this fraction of the SOM in water, the supernatant is poured through two nested sieves; that which passes a 2-mm, but not a 0.25-mm mesh, is termed the 'light fraction'.

The light fraction component includes most of the microbial biomass. Biomass carbon denotes the amount of carbon in the soil biota; it does not necessarily denote its 'activity'. Biomass carbon is independently determined by a chloroform fumigation-incubation technique (Amato and Ladd, 1988), and is estimated from the ninhydrin-reactive nitrogen.

The 'heavy fraction' is defined as the total SOM less the light fraction. It is thought to comprise the slow and passive pools.

A schematic diagram for separating the three pools is shown in Figure 3.

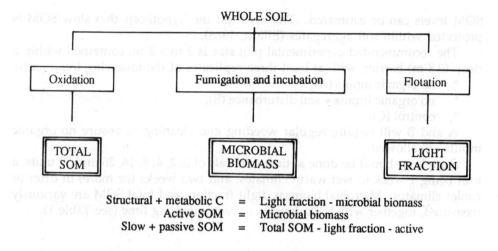

Figure 3. Fractionation of soil organic matter.

Field studies: a standardized SOM decay experiment ("KILLSOM")

The fate of SOM following land clearing and subsequent cultivation is unclear, other than that SOM levels generally decline. To enable maximum use of the SOM model advocated by TSBF, one needs to evaluate the magnitude of the active, slow, and passive pools. An SOM decay experiment has been designed by TSBF collaborative scientists (Scholes *et al.*, 1989) with three clear objectives:

- to measure the size of the slow SOM pool in different soils;
- to determine the rate of turnover of SOM in various soils under different climates and how this rate is altered by soil tillage; and
- to establish a decomposition constant for intersite comparisons, based on the *in situ* decay rate of SOM.

The experiment aims to quantify the SOM pools by repeated measurement, over a period of several years, of the decline in SOM in small plots from which all organic inputs have been excluded. It is predicted that the decline will stabilize at a level corresponding to the slow plus passive pools, the active pool being defined as the microbial biomass. The rate of decline relative to a control plot under natural vegetation will provide an SOM turnover index for that particular soil-climate-vegetation situation. By including a disturbance (i.e. tillage) treatment, an estimation of the increase in the rate of decay of

SOM levels can be estimated. This will test the hypothesis that slow SOM is protected within soil aggregates (Elliott, 1986).

The recommended experimental plot size is 2 m x 2 m, contained within a deep (0.8 m) barrier, with at least three replicates of the following treatments:

* no organic inputs (A),
* no organic inputs + soil disturbance (B),
* control (C).

A and B will require regular weeding and cleaning to ensure no organic inputs are allowed.

Sampling should be done at time intervals of 0, 2, 4, 8, 16, 26 and 52 units, a unit being 1 week in wet warm climates, and two weeks (or more) in drier or cooler climates. Microbial biomass, light fraction, and total SOM are variously measured, together with bulk density, at each sampling time (see Table 1).

Table 1. Timetable for taking soil samples at different depths for the various analyses.†

Depth (cm)	Sample time						
	0	2	4	8	16	25	52
0-5	Mic N‡	Mic N	Mic N	Mic N	Mic N	Mic N	Mic N
	LF§			LF			LF
	Tot#			Tot			Tot
5-10	Mic N	Mic N	Mic N	Mic N	Mic N	Mic N	Mic N
	LF			LF			LF
	Tot			Tot			Tot
10-20	Mic N	Mic N	Mic N	Mic N	Mic N	Mic N	Mic N
	LF			LF			LF
	Tot			Tot			Tot
20-40	LF			LF			LF
	Tot			Tot			Tot
40-80	LF			LF			LF
	Tot			Tot			Tot

† The dry mass of soil taken for each depth interval at each sample time must be recorded, along with dimensions of the sample hole, to allow the bulk density of the soil to be calculated. Each unit is one week in wet warm climates, and two weeks in drier or cooler climates.

‡ Mic N - microbial nitrogen (microbial carbon and phosphorus are optional).

§ LF - light fraction.

Tot - total organic carbon, total nitrogen, and organic and inorganic phosphorus.

Experimental details and a full rationale are given as referenced for Scholes *et al.* (1989) above.

Soil organic-matter management options

It should be noted that the third TSBF research objective listed above aims directly at the management issue. It has been proposed that different functional pools contribute variously to nutrient release (active), cation exchange (slow and/or passive) and aggregate stability (passive). If this is indeed found to be the case, and that a particular pool can be managed - i.e. "promoted" (if desirable) at best, or its decline slowed at worst - certain soil management options are implicit. From the recalcitrance factor considered in an earlier section (the TSBF concept), and considering that soil does not have very high levels of SOM, it could well follow that once the slow and passive pools are oxidized, it will take a very long time to reestablish them. This, however, is not clear.

It is unlikely that a given land/soil management regime will affect each pool in the same way; the management of a given SOM pool will depend on the pool in question. The management of those pools with a fast turnover time/ short residence time will be in effect a management of organic inputs. Management of the slow or passive pools will more be a management of existing pools in the soil, with the emphasis being on preservation by protection. The results from the KILLSOM experiment described above will help soil "users" decide on an overall management strategy, the vulnerability of a given pool (or pools) being better understood.

The proposed study at the Institut National Agronomique at Yoaundé, Cameroon, will be the first IBSRAM/TSBF collaborative experiment to address the SOM dynamic and aggregate stability question. In that TSBF is in essence a methodology, it should be possible to introduce the philosophy discussed in this paper to other sites, thereby further testing the various hypotheses put forward.

References

AMATO, M. and LADD, J.N. 1988. Assay for microbial biomass based on ninhydrin-reactive nitrogen in extracts of fumigated soils. *Soil Biology and Biochemistry* 20: 107-114.

ANDERSON, J.M. and FLANAGAN, P.W. 1989. Biological processes regulating organic matter dynamics in the Tropics. In: *Dynamics of soil organic matter in tropical ecosystems.* Hawaii: University of Hawaii Press.

ANDERSON, J.M. and INGRAM, J.S.I., eds. 1989. *Tropical soil biology and fertility: handbook of methods.* Wallingford, UK: CAB International. 180p.

ELLIOTT, E.T. 1986. Aggregate structure and carbon, nitrogen and phosphorus in native and cultivated soils. *Soil Science Society of America Journal* 50: 627-633.

NELSON, D.W. and SOMMERS, L.E. 1975. A rapid and accurate method for estimating organic carbon on soils. *Proceedings of the Indiana Academy of Science* 84: 456-462.

PARTON, W.J., SCHIMEL, D.S., COLE, C.V. and OJIMA, D.S. 1987. Analysis of factors controlling soil organic matter levels in Great Plains grasslands. *Soil Science Society of America Journal* 51: 1173-1179.

SCHOLES, R.J., PALM, C.A., PARTON, W.J., ELLIOTT, E.T. and SANCHEZ, P.A. 1989. Soil organic matter decay experiment. In: Report of the IVth Interregional TSBF Workshop, ed. J.S.I. Ingram and M.J. Swift, *Biology International*, (Special Issue) 20: 35-37.

SWIFT, M.J., ed. 1987. TSBF Interregional Planning Workshop. *Biology International* (Special Issue) 13: 10-13.

Section 2: Tillage and residue management in Africa

Tillage practices and residue management in Tanzania

P.L. ANTAPA and T.V. ANGEN*

Abstract

Tillage systems and residue management in Tanzania have developed over the centuries. Due to the agroecological diversity within the country, different systems have evolved to produce husbandry practices which conserve the soil and effectively use available water. Over the past century, mechanized systems have been introduced, to a limited extent, in both the private and public sectors.

This paper describes some of the main systems which have developed in Tanzania, as well as some introduced mechanized systems. The evolution of tillage practices and residue management to control soil erosion and conserve soil moisture on a large state farm in northern Tanzania is highlighted.

Introduction

Tanzania, with a population of approximately 24 million people, lies between latitude 1°0" to 11°50"S and longitude 29°30" to 40°30"E. Over 80% of the economy depends upon agriculture, of which 80% is subsistence farming, where the main tool for cultivation is a handhoe.

* Respectively: Senior Agricultural Research Officer, Agricultural Engineering and Soil Conservation Department, Selian Agricultural Research Institute, PO Box 6024, Arusha, Tanzania, and: Land Resource Scientist, CIDA, Selian Agricultural Research Institute, PO Box 6160, Arusha, Tanzania.

The rainfall within the country is diverse, and in some areas is extremely variable from year to year. The temperature regimes have a wide span, since they are highly correlated to altitude - which ranges from sea level to 5895 m asl. Sixty four agroecological zones have been identified at a national planning scale (1:2 000 000), and are based on temperature regimes and moisture characteristics during the growing period (de Pauw, 1984). The dependable growing period varies from less than 2 months to 8-12 months. The most successful tillage and residue-management practices tend to conserve soil moisture and maintain a good residue cover to control excessive soil erosion. A brief overview of some traditional and introduced tillage and residue management practices in Tanzania is given below.

Tillage practices - the Tanzanian experience

Tillage can be defined as the operation or practice taken to prepare the soil surface for the purpose of crop production. For any given location, the choice of a tillage practice will depend on one or several of the following factors (FAO, 1980, 1984):

A) **Soil factors**
Relief (slope)
Erodibility
Erosivity
Rooting depth
Texture and structure
Organic-matter content
Mineralogy

B) **Climatic factors**
Rainfall amount and
distribution
Water balance
Length of growing season
Temperature (ambient and soil)
Length of rainless period

C) **Crop factors**
Growing duration
Rooting characteristics
Water requirements
Seed

D) **Socioeconomic factors**
Farm size
Access to power source
Family structure and
composition
Labour situation
Access to cash and credit
facilities
Objectives and priorities

E) **Others**
Government policies

Dominant tillage practices

In Tanzania, the tillage practices used are somewhat specific to a particular agroecological zone, as well as to the crop being grown. The systems can be broadly separated into clean and conservation tillage systems (Antapa, 1989).

Conservation tillage systems

The main feature of this system is that residue is retained on the soil surface and/or efforts are made to control the surface water movement. This allows more time for water to infiltrate, and subsequently soil erosion is decreased. Some tillage systems in this category are:

Mulch (stubble-mulch) tillage

These are practices for maintaining crop residues on the soil surface with minimum or reduced tillage operations, also referred to as mulch farming. The mulch reduces evapotranspiration, increases infiltration rates, and suppresses weed growth. This practice is carried out by small- to large-scale farmers for both perennial crops like coffee and banana, as well as annual crops such as wheat and barley.

When carried out on mechanized farms, V-shaped tillage implements like sweeps, chisel ploughs, and rod-weeders are among the best implements to conserve surface residues (Table 1). The small-scale farmer generally uses handhoes for weeding and therefore maintains crop residues on the surface.

Table 1. Effects of tillage machines on surface residue remaining after each tillage operation.

Tillage machine	Residue maintenance (%)
Subsurface cultivator	90
Mixed-type cultivator	75
Mixing and inverting disc (one way)	50
Inverting type, i.e. mouldboard	10

Zero tillage (no-till)

This is the extreme side of reduced tillage, where mechanical tillage is completely eliminated. Weed control is achieved through the application of herbicides. Examples of this practice can be found in the mechanized wheat fields in northern Tanzania and on some perennial crops, e.g. coffee plantations.

Ridging or listing

In this practice, alternating furrows and ridges are formed on the land. The farmers are advised by extension staff to follow the contour, but this is not always done. The practice is commonly applied on lowland areas with relatively high moisture regimes. In such areas, the crop is grown on the ridge. Crops commonly grown in these areas include maize, cassava, sweet potatoes and cotton. These structures are generally made by using handhoes and animal- or tractor-drawn ridgers.

Tie ridging or basin listing

This resembles the practices described above, but has additional cross ridges which form small basins in the furrow (FAO, 1987). It is among the most effective ways of decreasing surface runoff, thus maximizing moisture availability for the crop, as well as reducing soil erosion by water. The practice is very common in areas of low rainfall and/or moderate slopes, and the system is dominantly used for growing sorghum, groundnuts, and cotton. Handhoes and animal- or tractor-drawn ridges are used as tillage implements.

Matengo pits ('ngoro')

The practice is commonly found in the southwestern part of Tanzania in the hilly areas of Mbinga District. Land is prepared in the same manner as in the tie-ridge system, except that the grass is cut and piled in squares of 2 x 2 m. Then, using handhoes, the grass is covered with soil from all sides, thus creating round basins. The crops grown under this system include wheat, maize, and beans.

Clean tillage systems

In clean tillage systems, all plant residues are covered or removed, and the growth of all vegetation is prevented, except for the desired crop. Unwanted vegetation is usually controlled by ploughing, and followed by some form of secondary tillage operation (FAO, 1980).

The dominant implements used are discs and mouldboard ploughs (Table 1), and the system is mainly used in the cultivation of annuals, especially maize, wheat, barley, and some kinds of oil seed. The practice is widely used throughout Tanzania.

The advantages of the system are that it achieves much better weed control, and is less troublesome than other cultural practices. Its major disadvantages are that it requires more energy, and it tends to induce greater erosion hazards (Table 2).

Table 2. Power/fuel requirements for different tillage implements.

Operation	Fuel and soil draught requirements		
	Low ($L\ ha^{-1}$)	Medium ($L\ ha^{-1}$)	High ($L\ ha^{-1}$)
Subsoiling (35 cm depth)	12.2	19.6	27.6
Mouldboard ploughing (20 cm)	10.8	17.3	24.3
Chiselling (20 cm)	7.0	11.7	16.4
Offset discing	5.6	8.9	12.6
Field cultivation (ploughed ground)	5.1	5.6	6.1
Cultivation (sweeps)	2.8	3.3	3.7
Cultivation (rolling tines)	2.8	3.3	3.7
Rotary hoeing	2.3	2.3	2.3

Source: Griffith and Parson (1981).

Crop-residue management

A wide variety of crops are being grown in the country, and the secondary products - their residue - are handled in a variety of ways, depending on location, climate, and environmental conditions.

Management options

The various ways of dealing with crop residues in accordance with different situations are detailed below.

Mulching

Surface mulch tends to minimize wind and water erosion, increase soil moisture, suppress weed growth, and moderate soil temperatures (Acland, 1971). In areas where tillage operations are done with V-shaped blades, a high percentage of trash is left on the surface for conservation purposes. Large mechanized wheat farms and some coffee estates are good examples of these practices.

Incorporation during tillage

Residue incorporation ensures a complete return of nutrients to the soil for plant use (Table 3). Disc implements are the most widely used tillage tools in Tanzania, and are known to incorporate over 50% of the residue after each tillage operation. The operation is often carried out in maize fields where disc harrows are used.

Table 3. Mean nutrient concentrations of various crop residues.

Source of crop residue	Nutrient concentration (%)					
	N	P	K	Ca	Mg	S
Millet (stalk)	.65	.09	1.82	.35	.23	.15
Sorghum (stalk)	.58	.10	1.51	.21	.13	.10
Maize (stalk)	.70	.14	1.43	.36	.11	.12
Wheat straw	.62	.12	1.72	.27	.15	.12
Rice straw	.58	.13	1.33	.20	.11	-
Groundnut leaves	2.56	.17	2.11	1.98	.68	-
Cowpea leaves	1.99	.19	2.20	3.16	.46	-
Cotton stalks and leaves	1.33	.27	2.35	1.27	.25	-

Source: FAO (1980).

Burning

Burning of trash is sometimes done to facilitate seedbed preparation, especially in areas of marginal rainfall such as the central regions of the country. In addition, some burning operations are aimed at controlling pests and diseases, e.g. cotton stems and maize stalks in the eastern and western zones.

Domestic use (fuel and construction)

In crops like pigeon peas, bullrush millet, and maize, the residue is utilized as fuel (firewood) for cooking, etc. Millet stalks are used for making house doors in some rural areas. Roofing and fencing materials are among other domestic uses of residues.

Animal feed

Residues from crops like beans, maize, wheat and barley are used for animal feed. This is carried out by grazing the fields, or the residue is harvested for feed in situations where animals are in confinement.

Composting and farm manure

Although composting has not gained momentum in the country, a proportion of the crop residue is utilized in this way. For instance, residues are heaped around banana plants for mulching, and subsequently are used as fertilizer after decomposition.

Specific example - wheat production under large-scale mechanization

In the early seventies, a large state-managed rainfed wheat production scheme was initiated with assistance from the Canadian International Development Agency. The soils of the area are fine- and medium-textured Vertisols, Mollisols, and Inceptisols developed on calcareous volcanic tuff. The scheme presently occupies approximately 28 000 ha and is at an elevation ranging from 1600 to 1850 m asl. The rainfall in the area is monomodal with an average precipitation of about 600 mm a season. Rainfall intensity is typically high, and rates in excess of 50 mm h^{-1} have been recorded (Fenger *et al.*, 1986).

Tillage practices and residue management

During the initial stages of the development, land was completely cleared of vegetation, and farm operations were done on macroblocks of not less than 60 ha (Antapa, 1989). The tillage implements utilized at the onset of production were primarily discs, which incorporate most of the residues into the soil. In addition, field operations were often carried out along the slope, which encouraged the downward flow of surface water.

Conservation tillage practices

Several techniques have been introduced to improve field management on the farms. The first step was the introduction of conservation implements. This included the introduction of tilling with V-shaped blades (sweeps), which are known to retain trash on the soil surface. Disc implements are used to a limited extent on lower slope positions which are dominated by Vertisols. In addition to these tillage practices, air seeders have been introduced in an attempt to reduce the number of tillage operations between wheat crops.

Conservation structures

In an attempt to manage the surface water flow, various types of conservation structures were installed. The distance between them was based on rational formulae in conjunction with multiples of selected field implements (seeders). The design of the system was based on soil and land conditions. Structures which were installed included grass strips, as well as graded and absorption channels. To date, the system has had both positive and negative effects. On the positive side, the structures force all cultivation to run along the contour, which in itself decreases excessive erosion. On the other hand, structural failures have occurred, producing highly erosive forces which cause large gullies down the length of the field (Angen, 1989).

Current situation

A pilot project (80 ha) was initiated in 1988 to evaluate broad-based terrace structures in conjunction with a minimum tillage system. The aim of the project was to develop an improved system of surface water management to decrease soil loss and increase the amount of water available to the crop. Initial results after one season are encouraging, and farm managers are eager to adopt the technology.

In addition to developing conservation tillage practices, land-clearing operations during the last decade have differed from the initial programme. All "new land" has been cleared, leaving strips of natural vegetation along the contours at predetermined intervals down the slope. Natural vegetation has also been maintained along existing waterways. This system of clearing, in conjunction with conservation tillage systems, has proved to be an effective method of rainfed wheat production in the northern highlands of Tanzania.

Summary and conclusions

Tillage and residue management systems which have developed in Tanzania have been governed by both environmental and socioeconomic conditions. Some traditional practices tend to conserve soil and moisture by maintaining crop residue on the soil surface. The initial introduction of mechanized tillage systems promoted erosion by incorporating surface residue into the soil. Farmers are presently adopting mechanized systems which are conservative in nature.

Tillage and residue management systems, whether labour-intensive or mechanized, should maintain the soil structure and retain residues on the surface to prevent excessive soil loss and to ensure the most effective use of available water. This will ensure sustainable productivity for future generations.

References

ANGEN, T.V. 1989. *Evaluation of the conservation program at the Hanang Wheat Complex*. Selian Agricultural Research Institute. Report no. 5. Arusha, Tanzania: SARI.

ANTAPA, P.L. 1989. Development of soil and water conservation measures at the Hanang Wheat Complex, Arusha, Tanzania. Report presented to the 16th Annual Scientific Conference of the Tanzania Society of Agricultural Engineers, Mbeya, September 1989.

ANTAPA, P.L. 1989. Tillage. Lecture notes given to the National Agriculture, Livestock and Extension Rehabilitation Program, Arusha, Tanzania, October 1989.

ACLAND, J.D. 1971. *East African crops*, 72-73. Rome: FAO/London: Longmans.

DE PAUW, E. 1984. *Consultants' final report on the soil physiography and agro-ecological zones of Tanzania*, 1-41. Tanzania: Ministry of Agriculture/Rome: FAO.

FAO (Food and Agriculture Organization of the United Nations). 1980. *Tillage systems for soil and water conservation*, 33-240. FAO Soils Bulletin no. 50. Rome: FAO.

FAO. 1984. *Organic recycling in Africa*, 74-79. FAO Soils Bulletin no. 43. Rome: FAO.

FAO. 1986. *Soil conservation for developing countries*, 50-80. FAO Soils Bulletin no. 30. Rome: FAO.

FAO. 1987. *Soil and water conservation in semi-arid areas*, 49-73. FAO Soils Bulletin no. 57. Rome: FAO.

FENGER, M., HIGNETT, V. and GREEN, A. 1986. *Soils of Basotu land and Balangida land Lelu areas of Northern Tanzania*, 17-18. Canada: CIDA/Tanzania: SARI Wheat Program.

GRIFFITH, D.R. and PARSON, S.D. 1981. Energy requirements for tillage-planting systems. In: *Proceedings of the Conference on Crop Production with Conservation in the '80s*, 272ff. American Society of Agricultural Engineering. Chicago, Illinois: ASAE.

Tillage and organic-matter management for dryland farming in Botswana

K. MONAGENG, N. PERSAUD, D.C. CARTER, L. GAKALE,
K. MOLAPONG, G. HEINRICH, J. SIEBERT, and N. MOKETE[*]

Abstract

Tillage and organic-matter management in dryland farming in Botswana are discussed in the context of prevailing agroclimatic, soil, and socioeconomic interrelationships. Results from past and present research on tillage/soil fertility are presented, and these show that it is possible to attain much higher levels of production than are presently obtained by dryland farmers in Botswana. The organization and role of the recently initiated National Tillage Research Program to further improve and promote dryland tillage/soil fertility technology in Botswana are discussed. Crop residues and animal manure are the main available organic-matter resources in Botswana, but are not generally used in arable agriculture. Compared to tillage, much less attention has been given to research on organic-matter use and management in Botswana.

[*] Respectively: Senior Technical Officer (tillage), Soil Management Specialist, Agronomist (sorghum and millet), Director of Agricultural Research, Agricultural Research Officer (chemistry/soil fertility), Agronomist (farming systems), Agronomist (farming systems), and Assistant Agricultural Research Officer (agronomy), Department of Agricultural Research, Pvt. Bag 0033, Gaborone, Botswana.

Introduction

Agricultural production practices reflect the farmers' adaptation to prevailing interrelationships between agroclimatic, soil, and socioeconomic factors, as shown in the following Venn-type diagram.

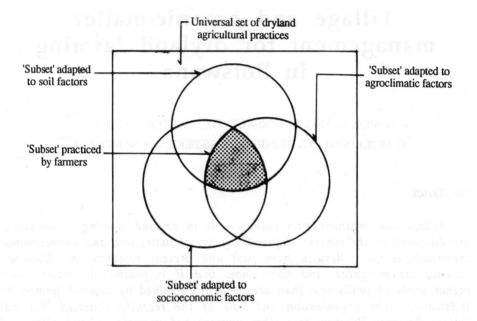

Figure 1.   Schematic representation of agricultural production practices.

In Botswana agricultural practices are still, to a large extent, based upon its farmers' collective, traditional, semiempirical understanding of these interrelationships. By itself, traditional knowledge evolves too slowly to keep pace with internal and external political, economic and demographic changes. Experience has demonstrated that effective research, geared to scientific understanding and application of these interrelationships, can accelerate the pace of change and improvement in agriculture in Botswana. This paper discusses dryland tillage and organic-matter management practices in Botswana in this context.

Dryland farmers and their farms in Botswana

Dryland farmers are concentrated on lands along a narrow north-south strip of Botswana's eastern border and around the Okavango Delta. Between 80-90% of dryland farms are subsistence or marginal enterprises contributing on the average less than 10% of the farmer's total income in cash and kind. Cattle and off-farm activities provide most of the farmer's income. Surveys have shown that these farmers' main concerns in rough order of importance are drought, bird damage, shortage of draught power, labour shortage, livestock damage, lack of seed, poor soil fertility, pests and diseases. Table 1 provides some general socioeconomic data for Botswana.

Table 1. Socioeconomic data for Botswana.

---

1. **Population data from 1981 census:**
     Population: Total: 967 000, Urban: 17.7%, Rural: 82.3%
     Household size: Urban: 4.5, Rural: 5.8
     Female heads of household: Urban: 33%, Rural: 50%
     Population density/km^2: Western: 0.14, Eastern: 6.40, Gaborone: 17.2
     Projected population: 1991: 1 356 000, 2001: 1 894 000
     Population growth rate: 3.4%

2. **Agricultural sector data**
     Total land area: 581 730 km
     Agriculture sector % of GDP: 1966: 39.3%, 1982: 7.4%
     Potential arable area: 1 360 000 ha.

Source: Ministry of Finance and Development Planning (1985).

These farms receive an irregularly distributed, mean annual rainfall of 400-600 mm, most of it from November until February. There is a high risk of seasonal soil moisture deficits and crop failure. Botswana farmers have responded to this production environment by employing low-cost, risk-averse, opportunistic soil and crop management techniques which give a fair return only in the infrequent years with high and well-distributed seasonal rainfall. Crops produced include sorghum, maize, millet, jugo beans, groundnuts, cowpeas, sunflower, and various melons.

Land tenure is communal, i.e. the land is allocated to the farmer and he practically owns it, but he cannot sell it. In Botswana land is traditionally allocated for housing and social activities, for crop production and for grazing livestock. These three holdings are usually kilometres apart and farm families divide their time and resources at these locations as appropriate. The average size of the cropland holding is 4-6 ha, and farmers rarely live on or close to their croplands.

Traditional dryland farms are either fully rainfed or 'molapo' farms. Molapo farming is a form of recession agriculture practiced in the Ngamiland area of the Okavango deltaic floodplain in northern Botswana. Channels and floodplains, termed 'molapo' in Setswana, are cultivated after the floods recede. The location of fields and the area cultivated from year to year depends on the extent and time of flooding and flood recession, and on the rainfall. Little effort is made to control the flooding and flood recession by building bunds or other control structures.

Improved dryland farms are more commercially oriented. Farmers live on or near their farms and use improved soil and crop management practices, including winter ploughing, early planting, fencing, row planting, improved seeds, weed control, fertilizers, etc. Improved molapo farms use bunds with inlet structures to control flooding. Average yields on improved traditional farms are 2 to 3 times that of traditional subsistence farms. A high concentration of improved fully rainfed farm is found in the zone inhabited by the Barolong, one of Botswana's many ethnic groups.

Mechanized commercial dryland farms exist in Botswana, and are usually leasehold or freehold, with good access to markets, supplies, machinery, etc. They are mostly concentrated in the Mpandamatenga area in northern Botswana, the Tuli Block area of eastern Botswana along the Limpopo River, and in the Barolong farms in southeastern Botswana.

Table 2, which is based on a farm survey in Gomare in northern Botswana, is

Table 2. Results of a farm survey of Gomare, Botswana.

	%
Households headed by males	54
Cattle owners	67
Plant every year	92
Plough with oxen	92
Own a plough	59
Own a harrow	1
Someone with employment	42
Sell cattle	53
Sell grain	17

Source: Department of Agricultural Research (1983)..

typical of dryland molapo farms and farmers in Botswana. In eastern and southern Botswana the percentages are different, e.g. a much lower percentage of farmers use oxen for ploughing.

Tillage

Importance of tillage in Botswana

Primary and secondary tillage are essential for crop production in Botswana in order to ensure proper physical seedbed and rootbed conditions for crop establishment and growth and to control weeds.

Tillage operations are the most costly inputs for dryland crop production in Botswana. A national area of 200 000 ha ploughed and planted to sorghum at 50 pula ha^{-1} would annually cost Botswana's farmers 10 million pula. This would amount to close to 50% of the value of the national sorghum harvest at a mean yield of 300 kg ha^{-1} and a market price of 25 pula per sack of 70 kg.

Tillage practices

The single-furrow mouldboard plough is the implement used by most traditional dryland farmers. Ploughing in of broadcast seed is the most prevalent planting method used in about 80-90% of the cropped area. Very few farmers row-plant their crops. An average of about 4-5 ha are ploughed per family, mostly by cattle or donkey teams, although the use of tractors is increasing. Farmers often share the available trained draught-animal teams and equipment. It takes about 2 to 3 days working about 4 to 5 hours per day to plough/plant a hectare with animal draught. Farmers therefore plough/plant their fields progressively in portions during the growing season whenever there is adequate rainfall and whenever draught animals and time are available, until the entire field is planted. Fields are not fenced, and the cattle are usually not herded until the majority of farmers in a given area begin ploughing. Thus farmers are often unwilling to take advantage of early rains because cattle may destroy unprotected crops. Consequently, the annual cropped area fluctuates widely and is mostly dependent on rainfall. It can vary from 160 000 to 290 000 ha. Crop establishment and crop stands in a given field are variable due to the differences in planting dates and soil moisture conditions. Portions of most fields are planted to sole sorghum or maize or in mixtures with melons or cowpeas. Crops are weeded once or not at all. Kraal manure, fertilizers, pest control, crop rotations, and crop residues are rarely used.

Tillage in relation to climate and soil

In semiarid dryland farming, tillage practices can be used to conserve and maximize available soil moisture utilization by increasing the ratios of infiltration/precipitation, uptake/infiltration, and transpiration/evapo-transpiration in the following catenary pathway:

```
        (infiltration)        (uptake)          (transpiration)
Rainfall ---------------> soil ----------> plant ----------------> atmosphere
                     (runoff)      ↓------> evaporation --------↑
```

Botswana's soils have low natural productivity. Eighty-five percent of Botswana is covered by Kalahari sands, which are currently not used for arable farming. Most of the arable soils fall into the FAO units Arenosols, Luvisols, and Vertisols, located along the eastern side of Botswana. All assessments of these soils in the arable part agree that they are marginal for arable farming. The Arenosols and Luvisols have a relatively high sand content and are texturally classified as either sands, loamy sands, or sandy loams. These soils are "hard-setting" i.e. they become structureless, dense, and hard when dry. In the north of Botswana, silty clays are found around the Okavango Delta, and in the Mpandamatenga area there is a major area of Vertisols. Many of these arable soils are shallow, have a low moisture-retention capacity, and are weakly developed. Soil moisture deficits are frequent, and in many places the soils form surface crusts after heavy rains, inhibiting infiltration and promoting runoff. Most soils are very deficient in phosphorus, have a generally low but variable levels of mineral nitrogen, are low in organic matter, and have a low cation-exchange capacity. Table 3 shows fertility levels of major soil groups of Botswana.

Tillage research, past and present

Tillage methods

Previous tillage research in Botswana was concerned with two broad themes:

Table 3. Average chemical analyses of topsoil for major groups.

	Topsoil depth (cm)	No. pits	pH	% OC	ppm P	CEC	Ca	Mg	Na	K	% clay
						------ cmol kg⁻¹ soil ------					
1. Arenosols											
Cambic Arenosols	23	12	5.7	0.3	10	5.5	4.2	1.1	0.1	0.4	8
Ferrallic Arenosols	17	118	5.2	0.3	3	3.3	2.0	0.4	0.2	0.2	5
Calcaric Arenosol	18	3	6.6	0.6	3	8.4	6.9	1.1	0.1	0.5	10
Eutric Arenosols	19	128	5.0	0.2	3	3.3	2.4	0.6	0.1	0.2	4
Luvic Arenosols	19	34	5.2	0.2	3	3.4	1.8	0.6	0.1	0.2	4
2. Luvisols											
Chromic Luvisols	18	73	4.8	0.3	4	7.1	4	1.3	0.1	0.5	14
Ferrallic Luvisols	18	141	5.1	0.3	3	5.6	3.0	0.9	0.1	0.4	13
Calcic Luvisols	18	133	6.1	0.4	4	8.5	9.7	1.8	1.1	0.6	13
3. Vertisols											
Pellic Vertisols	10	123	6.5	0.7	5	47.0	33.4	11.9	0.4	1.1	60
Chromic Vertisols	15	7	7.0	0.8	5	42.1	44.4	9.3	0.6	1.1	51

Notes: pH in 0.01M CaCl₂, P by Bray 2, CEC by neutral ammonium acetate, OC by Walkley/Black.

Source: Molapong and Schalk (1989).

- Evaluation of alternative tractor-adapted tillage methods to the traditional system. The energy requirements of these alternative systems and their effects on soil moisture conservation, soil physical properties, and crop yields formed the main focus of these studies.
- Evaluation of alternative tillage implements to the mouldboard plough. Efforts were concentrated on animal-drawn tillage implements that can reduce tillage time over the traditional single-furrow mouldboard plough, permit row planting over the traditional broadcasting, and permit weed control and fertilizer application.

The research was done over a number of years on a single site. Site characterization, soil moisture, and other measurements were not undertaken to the degree that would have permitted fully explaining how and why the treatments influenced crop yields over the years. Observed yield responses generally indicated that increasing the depth of tillage increased yields. However, this response would vary with the seasonal rainfall pattern and any other management factor - such as type of implement, weed control, etc. that affect soil moisture. Yield response to tillage seemed to depend on both soil-moisture availability and soil-moisture accessibility.

Mouldboard ploughing was shown to require more than twice the energy compared to chiselling, discing, or precision strip tillage. Double ploughing was effective in increasing yields in most, but not all, situations. The first ploughing can be done either between seasons during autumn or winter, or with the beginning of the rains in early spring. Quantitative data are needed to understand the mechanism of the observed yield responses.

Current studies on tillage are aimed at obtaining more comprehensive data on crop responses to tillage, and specifically on:

- the influence of tillage practices on soil physical properties with time during the cropping season;
- the effect of the soil type and seasonal rainfall pattern on crop growth and yield response to tillage; and
- the interrelationships between tillage system/soil-moisture availability/soil fertility that can affect crop establishment and crop yields.

These ongoing studies are conducted under the National Tillage Research Program initiated in 1988. The overall objective of this programme is to improve tillage technology for Botswana's farmers. Figure 2 shows schematically how the programme is currently organized.

During 1988/89, the NTRP initiated a multilocational, multiyear tillage trial to evaluate five core tillage systems. Of the nine sites, eight were on farmers' fields, and the five core tillage treatments were as follows:

T1 = **Conventional:** Single mouldboard ploughing on day of planting. Crop row-planted to obtain 50 000 plants per hectare (±10%).

T2 = **Double ploughing:** Early spring mouldboard ploughing with first rains followed by second ploughing on day of planting. Crop row-planted to obtain 50 000 plants per hectare (±10%).

T3 = **Deep ripping:** Deep ripping to 50 cm on 150-cm centres as soon as possible after harvest, followed by shaping and other secondary tillage as appropriate with this system. Crop row-planted along riplines to obtain 20 000 plants per hectare (±10%).

T4 = **Ploughing and cultivation:** Early ploughing as in T2, followed by tined cultivation on day of planting. Crop row-planted to obtain 50 000 plants per hectare (±10%).

T5 = **Conventional with wide row spacing:** As for T1, but with 150 cm row spacing as in T3. Crop row-planted to obtain 20 000 plants per hectare (±10%).

At individual sites other treatments were added as appropriate to suit specific needs and objectives.

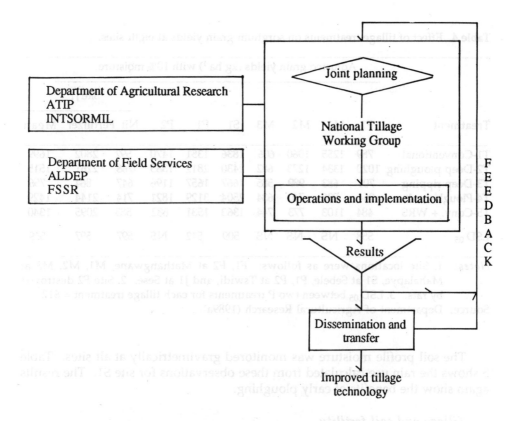

Notes: ATIP - Agricultural Technology Improvement Project.
        INTSORMIL - International Sorghum and Millet Research Program.
        ALDEP - Arable Lands Development Program.
        FSSR - Farming Systems Southern Region Project.

Figure 2. Organization of the National Tillage Research Program.

Table 4 shows the adjusted sorghum grain yields for the treatments at the various sites. Adjusted yields are the yields from all mature heads harvested, corrected for heads already harvested or grazed heads. Such adjustments were required at a few sites.

The results in Table 4 do indicate a beneficial effect of early ploughing and fertilizer on yields. The fertilizer used at site J1 was TSP at 15 kg P ha^{-1}. Stability analysis comparing each treatment with T1 showed, not suprisingly, that the conventional method appeared to be the most stable over the locations.

Table 4. Effect of tillage treatments on sorghum grain yields at eight sites.

Treatment	Adjusted grain yields (kg ha⁻¹) with 10% moisture									
								Site J1		
	F1	M1	M2	M3	S1	P1	P2	Nil	Fertilizer	Mean
T1-Conventional	789	1255	1060	606	1836	1351	1470	592	2802	1699
T2-Deep ploughing	1020	1334	1273	682	2430	2812	1365	768	2261	1515
T3-Deep ripping	701	683	989	585	1667	1657	1196	647	864	756
T4-Plough + cult.	964	938	740	634	2504	3129	1821	714	2144	1429
T5-Conv. + WRS	484	1103	773	744	1363	1531	882	585	2095	1340
LSD.05	558	NS	NS	NS	509	512	NS	597	597	569

Notes: 1. Site locations were as follows: F1, F2 at Mathangwane, M1, M2, M3 at Mahalapye, S1 at Sebele, P1, P2 at Tswidi, and J1 at Sese. 2. Site F2 destroyed by rats. 3. LSD.05 between two P treatments for each tillage treatment = 612.
Source: Department of Agricultural Research (1989a).

The soil profile moisture was monitored gravimetrically at all sites. Table 5 shows the rain use calculated from these observations for site S1. The results again show the benefits of early ploughing.

Tillage and soil fertility
Lack of adequate P is second to soil moisture as the most limiting plant growth factor in Botswana's arable soils. Table 6 shows the overall mean effect of applied P in farmers fields on traditionally broadcast/plough-planted sorghum and maize.

Studies on soil fertility and its management are being incorporated in NTRP to complement those on soil moisture conservation tillage. Observations made at the end of the 1988/89 growing season of the rooting system of fertilized and unfertilized sorghum plants at sites J1 and P1 show that the roots of fertilized plants grew approximately twice as deep and twice as wide as those of unfertilized plants. The NTRP is conducting studies to:
- evaluate fertilizer/tillage interactions;
- calibrate soil tests to determine their ability to predict P uptake patterns from the soils of Botswana; and
- develop fertilizer response curves from field trials to determine the critical levels of P and other major nutrients in the soils of Botswana, and to make fertilizer recommendations.

Table 5. Rain use for various tillage treatments used in the national tillage trial at site S1, Sebele.

| | Water storage and use (in mm) estimated for the following intervals: | | | | | |
| | 05/12/88 to 08/02/89 | | 08/02/89 to 09/03/89 | | Σ | kg grain ha^{-1} |
Treatment	ΔS	ΔP-ΔS	ΔS	ΔP-ΔS	ΔP-ΔS	Σ ΔP-ΔS
T1-Conv.	24.3	116.7	3.8	150.9	267.6	6.861
T2-DP	15.7	125.3	-11.4	166.6	291.9	8.325
T3-Deep rip	-2.3	143.3	8.5	146.2	289.5	5.758
T4-Plough + cult.	-14.8	155.8	11.4	143.3	299.1	8.372
T5-Conv. + WRS	45.9	95.1	-2.6	157.3	252.4	5.400
Interval in days	65		29			
ΔP values	141.0		154.7			
0.5 x Penman PET	156.4		58.6			

Notes: Planting date = 1 December 1988.
  ΔS values = difference in soil water stored to 1 m at later/earlier date in each interval.
  ΔP values = difference in cumulative rainfall at later/earlier date for each interval.
  Σ denotes "the sum of".
Source: Department of Agricultural Research (1989b).

Table 6. Mean sorghum and maize yield as a response to phosphate fertilizer by district and by year in eastern Botswana.

District	No. of sites	No. of years	% yield increase over control
Barolong	39	5	59
Ngwaketse	46	5	60
Bamalete	14	4	48
Kweneng	17	4	45
Kgatleng	16	2	30
Mahalapye	19	5	28
N.E. districts	19	3	30

continued

Table 6. (cont'd)

Year		Type of year	
1969/70	16	Dry	34
1970/71	20	Avg. rainfall	51
1971/72	60	Avg. rainfall	48
1972/73	6	Dry	47
1973/74	68	Wet	49

Source: Jones (1984).

Tillage, weeds, and soil-moisture conservation

Left uncontrolled, weeds can nullify the effects of moisture-conservation tillage. The NTRP has initiated studies to evaluate the effects of tillage systems on weed burden and weed types and to evaluate various mechanical and/or chemical techniques for their effectiveness in controlling weeds.

Tractor vs. animal draught

The government of Botswana subsidizes both animal and tractor draught for tillage operations. The NTRP has initiated studies on the relative merits of animal vs. tractor draught for tillage operations to help orientate government assistance to the smallholder farmers in Botswana. Specifically, studies are being conducted to assess:

- whether there is a significant yield gap between animal vs. tractor draught tillage operations.
- whether precision strip tillage, which is currently considered a possible soil moisture conservation practice, can be adapted for animal draught.
- whether, from a soil-conservation perspective, tractor draught produces any deleterious long-term effect on soil physical properties and productivity.

Maize and moisture-conservation tillage

Moisture-conservation tillage systems require more labour and capital inputs than traditional systems. As shown in Table 7, growing maize with these systems may provide better cost/benefit ratios than sorghum and save foreign exchange. The NTRP is evaluating several moisture-conservation tillage systems using maize as the indicator crop. One of these systems, the precision strip-tillage or deep-rip system, has been examined for several years for its potential to grow maize. The cultivated strips are permanently laid out

1.5 mm apart, and are 30-40 cm wide. The strips are ripped to 50-60 cm deep. The area between the strips is never cultivated except for shallow scraping with a blade harrow to control weeds. The area between the strips serves two purposes: runon and traffic area. The area is compacted by traffic when preparing the land. The compacted soil reduces infiltration and encourages rainwater to run onto the strip, and increases the amount of water in the strip. Kraal manure is applied in the strips at a rate of 153 kg ha^{-1}, along with single superphosphate (SSP) and lime at a rate of 50 and 70 kg ha^{-1} respectively. As shown in Table 8, yields of 1 to 3 t ha^{-1} are possible with this system.

Table 7. Information on cereals in Botswana for 1987/88.

	Maize	Sorghum	Millet
Total requirements (x 1000 t)	103	57	3
Import (x 1000 t)	91	7	-
Domestic production 1987/88	15	75	6
Area in ha, 1987/88 (x 1000)	45	211	30
Yield (kg ha^{-1})	330	350	190
Price (pula/12.5 kg)	7.5	3.5	-

Source: SADCC (1988).

Table 8. Maize yields at two planting dates with different forms of precision strip tillage.

Treatment	Grain yield kg ha^{-1}			
	DOP1†	DOP2‡	Difference	Overall mean
Single autumn rip (SAR)	1433	3195	1762	2314
Autumn + spring rip	1306	3020	1714	2163
SAR + discing	1548	3141	1593	2345
SAR + landshaping	1078	2757	1679	1917
Mean	1341	3028		2185

† Without manure, SSP, and lime.
‡ With manure, SSP, and lime.
Source: Department of Agricultural Research (1989c).

Organic-matter management

Crop residues

The common practice on traditional dryland farms is to graze crop residues after harvest. Crop residues can possibly conserve moisture, improve soil fertility and structure, prevent erosion and surface crusting, etc. The shortage of fodder during the dry winter season overrides these potential benefits, and crop residues are used primarily for fodder in Botswana. Levels of available crop residues are presented in Table 9.

Table 9. Dry-matter yield of crop residue for various crops.

Crop	Fertilizer status	DM (ton ha^{-1})
Sorghum (var. Segaolane)	No	1.45
	Yes	5.0
Maize (var. KEP)	Yes	4.0
Millet	No	2.1
	Yes	3.5

Source: Mosienyane (1983).

Little research has been done on the possible benefits of crop residues. A recent multiyear, multisite study on mulching with crop residues recommended that sorghum or millet should not be mulched under dryland conditions in eastern Botswana. Yield responses were highly variable across years and sites. It was argued that with such variability mulching would find little application among risk-averting subsistence farmers in Botswana.

Animal manure

Animal manures incorporated into the soil can affect soil structure and augment soil nutrients, especially in such countries as Botswana where soils are poor. In Botswana, animal manures accumulate in kraals, but are rarely used. Farmers perceive kraal manure as increasing weeds in their fields and as being heavy and bulky, and uneconomical to transport and apply. Kraal manure is largely mixed with soil, as shown in an analysis of samples from three kraals (Table 10).

Table 10. Characteristics of kraal manure.

Kraal	% moisture	% soil†	%N†	%P†
1	26.4	66.2	0.37	0.18
2	18.2	70.0	0.48	0.16
3	22.5	63.3	0.63	0.16

† % soil, N, and P are on moist weight basis.

Results from a multiyear study on the use of kraal manure on sandy Arenosols are shown in Table 11. Grain and stover yields on these soils were significantly improved by the addition of manure. It is possible that this may have been due to the effect of N and P added with the manure.

Table 11. Effect of kraal manure on sorghum grain and stover yields across years on a Ferrallic Arenosol at Goodhope, Botswana.

Year	Grain (t ha⁻¹)		Stover (t ha⁻¹)	
	no KM†	KM†	no KM†	KM†
1985/6	0.14	0.21	0.50	0.80
1986/7	0.35	1.00	0.85	1.44
1987/8	0.80	2.10	1.61	2.73

† KM = Kraal manure.

Source: Carter (1987).

References

CARTER, D.C. 1987. Sorghum and millet production using stover and manure amendments on different soil types in Botswana. *Bulletin of Agricultural Research in Botswana* 6:1-11.

DEPARTMENT OF AGRICULTURAL RESEARCH. 1983. *Report of the Agricultural Development of Ngamiland Project 1982/83.* Gaborone, Botswana: Government Printing Officer.

DEPARTMENT OF AGRICULTURAL RESEARCH. 1989a. *Summary report on national tillage trials 1988/89*. Gaborone, Botswana: Government Printing Office.

DEPARTMENT OF AGRICULTURAL RESEARCH. 1989b. *National tillage trials 1988/ 89 - Report for site SI at Sebele*. Gaborone, Botswana: Government Printing Office.

DEPARTMENT OF AGRICULTURAL RESEARCH. 1989c. *Report on deep ripping presented to the National Working Group on Tillage Research*. Gaborone, Botswana: Government Printing Office.

JONES, M.J. 1984. The experimental basis of current sorghum and maize fertilizer recommendations for eastern Botswana. *Bulletin of Agricultural Research in Botswana* 2: 3-17.

MINISTRY OF FINANCE AND DEVELOPMENT PLANNING. 1985. *Botswana National Development Plan VI*. Gaborone, Botswana: Government Printing Office.

MOLAPONG, K.F. and SCHALK, B. 1989. Preliminary fertilizer recommendations for selected agronomic crops in Botswana. Paper presented at the SARCCUS 15th regular meeting of the Standing Committee for Soil Science, Gaborone, Botswana, 24-26 October 1989.

MOSIENYANE, P. 1983. Crop residues for animal feeding. *Bullentin of Agricultural Research in Botswana* 1: 3-9.

SADCC (Southern African Development Coordination Conference Secretariat). 1988. *SADCC Food Security Bullentin* (October-November 1988). Gaborone, Botswana: SADCC.

Tillage and residue management on small farms in Malawi

WELLS F. KUMWENDA*

Abstract

The paper briefly discusses current tillage systems in Malawi, which include the 'chososo', 'magodi' (or 'marongo'), 'mphima', 'matutu' systems. On another point, there is a description of the use of crop residue, which is mainly used as animal feed and bedding, with the end products being used as manure.

Tillage operations, which are energy-intensive and time-consuming, and the shortage of labour at the critical time are a constraints in ensuring the timeliness of planting. An investigation at three sites over two growing seasons shows that semipermanent ridges could be a possible solution.

Background

Malawi is situated between 9 to 17° latitude and 32 to 34° longitude, and has a population of eight million people. The total area of Malawi is 11.8 million ha, of which 2.5 million ha are under water, leaving 9.4 million ha of land. The area of land which could be cultivated is 5.3 million ha, which is 56% of the total land area. The rest of the land is permanently waterlogged, shallow, and has over 12% of steep slopes. This land is reserved for natural vegetation, afforestation, wildlife, and provides water catchment areas.

Almost 90% of the population is based in the rural areas, and earn a living from small land holdings. The economy of the country is highly dependent on

---

* Farm Machinery Unit, Ministry of Agriculture, Chitedze Research Station, P.O. Box 158, Lilongwe, Malawi.

agriculture, which provides 40% of the GDP, at least 90% of the foreign exchange earnings, and employs about 85% of the labour force. Agricultural production is derived from smallholder subsector farming on customary land and from estate subsector farming on leasehold land. The smallholder subsector is dominated by subsistence farming, although there is a shift towards cash cropping. The soils and climatic conditions are favourable for agriculture in many parts of the country. The main food crops grown are maize, sorghum, cassava, pulses, rice, groundnuts, millet and potatoes. Cash crops include tobacco, cotton, tea, coffee, and guar beans. The average holding size for smallholder farmers is less than 1.0 ha, and over 70% of this is devoted to maize. The main priority for the estates is the production of high-value export crops for foreign exchange earnings. The major crops that are grown by the estates are cashew nuts, macademia nuts, tea, coffee, sugarcane, and tobacco.

Present tillage systems

The 'chisoso' method

This method involves intense burning of the soil using dry trees, and in some countries the same method is known as the chitemene system. The place chosen for cultivation is known as a 'chaliro'. Trees are cut liberally on areas adjacent to the chaliro and laid in portable heaps from July onwards.This operation is referred to by farmers as 'kugumula mahlahla' or 'kutema nthebere', and some farmers refer to the operation as 'kusosa'. When the trees have wilted and dried, but before the leaves have commenced to fall, they are dragged on to the chaliro, which is now called 'chisoso' or 'libibi'. Trees are piled high over the area and burned before the rains, which usually arrive in November. Subsequent hoeing after the rains is very shallow, merely sufficient to make a seedbed for finger millet (*Elensive coracana*).

The 'magadi' or 'marongo' system

This is a system of opening fallow land, and cultivation commences in February during the rains when the ground is soft. Hoeing is deep, and the main idea is to turn the sod and bury the grass. To further decompose the grasses, the sods are often heaped together grass downwards in small mounds called 'matutu'. Clearing the plot of trees takes place later during August, and these are heaped together and burned in November before the rains. As no extra firewood is carried on to the magadi, the heaps of trees cut on the magadi are not contiguous, and the burnt patches thus created are called 'michinga'. Also, as bush growth is seldom extensive on the magadio soils, the fires are never intense. To form a seedbed the sods are broken up by a handhoe after the rains. The undecomposed grass and rubbish are gathered into piles to be used as manure for the crop in the following year. Finger millet (*Eleusine coracana*), locally known as 'lupuko', is sown between the rubbish heaps after the first heavy rains.

The 'mphuma' system

In this system, hoeing commences in November-December after the first light rains, but as the ground is hard it is a shallow hoeing in contrast to the deep early hoeing used for magadi tillage. Trees are cut down and burnt as for magadi tillage, and after a week or two the area is planted with finger millet. This method is used to enlarge an existing garden, especially in the second year. Occasionally a chisoso garden can be enlarged by the mphuma method at the time of planting.

A garden during the first year is either chisoso, magadi or mphuma, and is usually planted with finger millet. It is only in the second year when maize, groundnuts or beans have been planted that it is called 'munda'. In the second year a chisoso garden is called 'chasalala', and is usually planted with finger millet again. Fields are usually discarded after the fourth season because of heavy weed infestation. Discarded gardens are called 'sala' or 'lifusi'.

The chisoso, magadi or mphuma tillage methods are commonly used on the upland for opening up new farmlands, especially in areas with plenty of space and trees. Wilson (1941) reports that these tillage systems are employed by farmers among the Angoni-Tumbuka tribes of Mzimba. These systems are still used in very limited areas because of increasing land pressure and current agricultural recommendations, which discourage any form of tillage that will lead to soil erosion and deforestation.

The 'matutu' system

This is a system of making moulds using grass trash and soil. It is commonly used in cassava-growing areas along the lakeshore. In some cases the same system is used for growing crops in waterlogged areas. Several planting stations can be made on one mound, and in the case of root crops several sticks are planted in one mound.

Relay cropping

This system is mainly practiced in the southern region of Malawi - in Blantyre, Zomba, Chiradzulu, Thyolo, and Mulanje districts. In these areas, farmers have small fields, but they generally receive prolonged rainfall. Farmers plant their main crop on ridges in November, and this is usually maize, sometimes mixed with pigeon peas. Before the main crop is fully matured and removed from the field, a second crop is planted on small ridges made in between the main ridges. Planting of a second crop is done in February and March when there is still a lot of moisture in the soil. Beans, cassava, wheat, and potatoes are used as the second or relay crops.

Modified ridging

It happens in some cases that a farmer is late in making new ridges. After the first rains, the farmer digs some soil from the old ridge into the furrow and makes a new planting station. After two weeks, he comes again to make ridges along the furrow. This ridging operation combines the first weeding operation. This type of tillage can only be practiced if a farmer is using a handhoe, and cannot be carried out with animal-drawn implements.

Crop-residue management

There are a number of ways in which crop residues are utilized on small farms in Malawi.

Livestock feed

Crop residues in the form of maize stalk, groundnut stover, and rice/wheat straw are collected at the time of harvest and given to the animals as feed and bedding. The animals eat these as a supplement to their daily grazing, or during the dry season. Sometimes the same feeds are given during fattening when the animals are zero grazed. There is one rainy season in Malawi, and this feature affects the availability, quality, and volume of natural pastures. Crop residues are very abundant and form an important source of feed, but unfortunately not all the farmers know how to preserve them. Many farmers leave the residues in the field too long where they overdry, and lose palatability and nutrients. There is considerable scope for more effective utilization of crop residue in order to increase digestibility, crude protein, and metabolizable energy, which could well benefit animal production.

Making manure

The cost of chemical fertilizers has gone up so much that most smallholder farmers are finding it difficult to buy them without government subsidies. Some farmers use crop residues for making farmyard manure. Farmers who have livestock use crop residues as bedding for the animals. The residues rot and are mixed with dung and urine from the animal. The manure greatly improves soil conditions and nutrient contents, hence increasing crop productivity. Farmers who do not have animals simply plough in crop residues at the end of the rainy season so that by the time the rains come these will have decomposed. Manure helps farmers to cut down on their chemical fertilizer requirements, and this increases their farm incomes. The process of incorporating crop residues into the soil also greatly accelerates water infiltration and promotes soil micro-organisms.

Making domestic soda

Women in the rural areas use crop residues for making domestic soda for seasoning special relish. Maize stover or groundnut haulms are sun-dried and burned. The ashes are put in a container with perforations on the bottom. Water is then passed through the ashes into a container at the bottom. This resulting water which is yellowish in colour can be used directly, or evaporated by boiling in order to get solid soda.

Sterilizing the soil

Some farmers use crop residues for sterilizing the soil instead of using chemicals like ethel dibromide. Many tobacco farmers use maize stover for sterilizing their nursery beds before sowing. A heap of dry maize stover about 22.50 cm thick is laid along the bed, and is covered with 7.5 cm of soil. When the stover is burnt, the soil around the stover is sterilized. This process is carried out during October, when both the soil and the stover are dry.

Recommended tillage system

Smallholder farmers in Malawi are advised to start preparing their gardens just after the rains. The farmer is supposed to cut his partly dry crop and stock it in heaps, either inside his garden or along the edges of his field. This then gives him a chance to work in the field while the ground is still moist. The crop finishes drying while in heaps.

Although most farmers cut and stook their crops, they rarely prepare their gardens at the same time because of other engagements on the farm. It so happens that during this time farmers have to cut grass for thatching their houses, to prepare crops for the market, and to make storage arrangements.

Farmers who have draught animals are advised that they should plough their fields just after the rains, and that they should ridge their fields during the months of September to October so that planting can be done with the first rains.

Current tillage research work

Small-scale farmers mostly use handhoes for field cultivation, but some use draught animals. There are high yield losses of major food crops because of late planting, weeds, diseases, and pests. Agricultural production by small-scale farmers is carried out overwhelmingly on farms where labour is scarce during a certain period of production.

Ridging of land for crop production has been in practice for many years in this part of the world (Unger, 1986). In Malawi, ridging has been recognized as an effective soil and water conservation measure, especially in places where slopes are critical (more than 5%). There is an outstanding recommendation to ridge across the slope on the contour, or to combine it with tied ridges. Brown (1963) reported that for reasons of soil conservation, all maize in Malawi was grown on ridges about 0.91 m apart from crest to crest laid out on the contour. Cultivation was done mostly by handhoe, although a few of the more progressive farmers were using ox-drawn implements.

Ridging is done by alternating the ridge and furrow positions, resulting in a major soil loosening, and this method affords the opportunity for incorporating plant residues and manure. The disadvantage of making ridges every year is that they require a lot of labour and energy, which smallholder farmers cannot provide in a limited time. The time required for various operations is given in Table 1. This causes delays in planting, resulting in low yields, and in some cases farmers are forced to plant a smaller field than desired (Lungu, 1971).

Table 1. Minimum time required for various operations.†

Operation	Treatment‡			
	PE	PL	RE	RL
Land clearing	24	63	39	56
	(4)	(8)	(5)	(7)
Ridging	18.6	93	84	82
(splitting ridges, wet season)	(11)	(11.7)	(10.5)	(10.2)
Planting	19	19	19	19
	(2.4)	(2.4)	(2.4)	(2.4)
First weeding	53	52.6	46	49
	(6.6)	(6.6)	(5.8)	(6)
Second weeding	36	53	52	50
	(45)	(6.6)	(6.4)	(6.3)
Basal dressing	16.6	16.6	16.6	16.6
	(2)	(2)	(2)	(2)
Top dressing	39	39	39	39
	(5)	(5)	(5)	(5)
Banking	52	.51	51	52
	(6.5)	(6.6)	(6.4)	(6.5)

† The figures show minimum hours per hectare required, but those in brackets are
the number of 8-h days required for one man using a handhoe.
‡ PE = permanent early; PL = permanent late; RE = ridge early; RL = ridge late.

An alternative to complete renewal of the ridges each year is to use a
permanent-ridge system for crop production, which in addition to reducing
labour and energy requirements has the potential for improving soil conditions
for plant root growth, and consequently makes it possible to have more timely
planting (Unger, 1986).

Zero tillage or no-till is not practiced by smallholder farmers in Malawi,
partly because of weather which does not permit crop residues to decompose
before the next rains, and secondly because it was thought herbicides would be

too expensive for the farmers. This idea needs to be reviewed from time to time along with other studies on the cost of labour.

In Guyana, Trinidad, studies done by Simpson and Gumbs (1985) in the Department of Soil Science of the University of the West Indies comparing the effect of three tillage methods - reduced tillage, conventional tillage, and deep tillage - showed no significant differences in plant height and yield of maize. At Chitedze Research Station, yields of maize grown on the same ridges for more than four seasons with an undersown pasture legume did not decline (Dzowela, 1986).

Most farmers find it difficult to weed their gardens because this operation has to be done within a specific time period. Farmers can prepare their land, plant, and harvest during the dry season when they have more time, although it is very hard to prepare land during this period; but they must finish their first weeding before the end of twenty-one days after planting. Each year many hectares of weed-smothered maize can be seen in many parts of the country. In addition to this, several hectares of maize are planted late in the season because farmers spend their limited time in making new ridges. Use of permanent ridges would greatly assist small farmers to alleviate the above problems.

Crop rotation is encouraged in Malawi for fertility and disease-control purposes. A farmer who is growing maize, tobacco, and groundnuts (which are common crops for smallholder farmers) on one hectare or more could be advised to divide his field into four plots and follow the suggested rotation system (Table 2). Farmers with less than a hectare, but not too small a field, could also follow the rotation system. This would enable the farmer to maintain his ridges for three seasons before making new ones, but he would need to change crops in these fields. Rotation is not possible on fields that are too small. Farmers with such fields normally practice mixed cropping (the growing of more than one crop in one field at any given time), or simply split their fields into two - the largest portion being allocated to a food crop.

Table 2. Suggested four-year crop rotation.

Field	Year 1	Year 2	Year 3	Year 4
1	T	M	M	G
2	M	M	G	T
3	M	G	T	M
4	G	T	M	M

Key: T = tobacco; M = maize; G = groundnuts.

FAO/MOA experimental work

In view of the above observations, it was decided to initiate a three-year experiment with the assistance of FAO on the use of permanent ridges, which after being made could be used for a minimum of three seasons before fresh ones were made. This tillage system was consistent with current recommendations of planting on the ridges, and at the same time allowed farmers to plant early. Labour and energy requirements would also be reduced.

Objectives

The purpose of the study was to investigate the use of permanent ridges for maize production under rainfed conditions, and more specifically the objectives were as follows:

- to determine the effect of a permanent ridge system on the labour used for tillage, especially land preparation;
- to determine the effect of the time and method of ridging on the timeliness of planting and crop establishment; and
- to determine the effect of the time and method of ridging on crop growth and grain yield.

Materials and methods

The treatments were evaluated in a field study with three sites: (i) the Chitedze Research Station, representing medium-textured (sandy clay) soils; (ii) the Chitala Research Station (Salima), representing fine-textured (hard clay) soils; and (iii) the Lisasadzi Residential Training Centre (Kasungu), representing coarse-textured (sandy) soils. Each treatment was replicated three times. Plots were 15 x 15 m with ridges 0.91 m apart and planting stations at 0.91 m. Ridges were made across the slope using a handhoe. Planting stations were made by a handhoe and four seeds were planted in each station. These were later thinned to three. Each station had three plants giving a plant population of 37 000 plants ha^{-1}.

Grain yield data were collected from a 5 m x 5 m net plot of the centre ridges. Maize seed (var. NSCM 41) was planted immediately after the rains. Fertilizer was applied by the dollop method, i.e. punching two holes, each 10 cm deep and 10 cm away from the plants according to the present recommendation of 200 kg ha^{-1} 20:20:0 as basal dressing and 200 kg ha^{-1} of calcium ammonium nitrate as a top dressing. Data were collected for plant root depth and volume, number of leaves, grain yield, time taken to ridge, weed, and apply fertilizer; ridge height and width, soil profile characteristics, rainfall, and plant roots.

The following treatments were evaluated:

- permanent ridge (early) - height reduction and residue burying done immediately after harvest;
- permanent ridge (late) - height reduction and residue burying done after the onset of the rainy season;
- nonpermanent ridge (early) - ridges reformed each year immediately after harvest; and
- nonpermanent ridge (late) - ridges reformed each year after the onset of the rainy season.

The type of cultivation equipment used in this experiment was a handhoe, because most smallholder farmers in the country use handhoes to prepare their fields. Animal-drawn equipment was not used because of the limited financial resources for the experiment.

Results and discussion

There was adequate rainfall on all the three sites in the first season, but excessive rainfall was experienced in the second season.

Plant roots could, over time, contribute to soil organic-matter improvement, and help to ensure soil conservation. At a hundred days, a number of plants were dug out to find out how much of the plant remained in the soil in the form of roots. Although not all the roots could be removed, it was found that a mature plant contributes between 0.11 to 0.16 kg, representing approximately 4.07 to 5.92 t ha^{-1}.

Ridge height and width changed significantly in the fresh ridges of the first season. At the beginning of the season, the ridges were 22 cm high and 30 cm wide. At the end of the season, they were 16 cm high and 34 cm wide on average. This change was attributed to soil erosion from the top of the ridge to the bottom, and to soil compaction - especially early in the season when the ground was not covered by the plants. Ridge height was measured from the furrow by putting a straight edge across two ridges and a rule vertically from the centre of the furrow. It was found that this method was not very accurate in the middle of the season when the furrows were covered with trash and some soil from the top of the ridges.

Weeds deprive crops of plant nutrients, moisture and light, causing a reduction in yield, and this makes it important that frequent weeding is done to reduce competition. In Malawi, two weedings are necessary, the first being more important than the second because there is no ground cover and weeds grow more vigorously.

The time required to do various farm operations varies from person to person, and also depends on the sex, age, and health of the farmers, and on the weather, the soil moisture, and the condition of the soil. The figures shown in Table 1 are averages from healthy men working under supervision. Some women, children, and weak men are likely to do the same work more slowly.

The usable crop yield was very comparable between treatments over the two seasons of this work (Table 3). This shows that perhaps small farmers could save their time and energy by using semipermanent ridges over some seasons.

Table 3. Averages for usable grain (kg ha^{-1}) at 12.5% MC.

Treatment - ridges	Chitedze		Lisasadzi		Chitala		Overall mean
	87/88	88/89	87/88	88/89	87/88	88/89	
Permanent early (PE)	4 155	4 440	4 229	4 308	3 982	3 600	4 119
Permanent late (PL)	4 002	4 668	4 517	4 599	3 649	3 600	4 172
Ridge yearly early (RE)	4 275	4 852	4 316	4 236	4 012	3 492	4 197
Ridge yearly late (RL)	4 449	4 928	4 335	4 628	4 514	3 796	4 442
Mean	4 220	4 722	4 349	4 443	4 045	3 622	4 233

Acknowledgement

The author expresses his gratitude to the FAO, the Department of Agricultural Research of the Ministry of Agriculture, and various scientists for their financial, logistics, and technical assistance in carrying out this study.

References

BROWN, P. 1963. Maize growing in Nyasaland: Distribution and cultivation. *Empire Journal of Experimental Agriculture* 31(131).

DZOWELA, B.H. 1986. *Understanding and utilization of climbing forage legumes in maize crops.* Chitedze Research Station, Ministry of Agriculture. Pasture commodity research programme review. Lilongwe, Malawi: MOA.

LUNGU, N.F. 1971. Effect of time of planting on maize yield. Ph.D. thesis, University of Malawi, Bunda College, Lilongwe, Malawi.

SIMPSON, L.A. and GUMBS, F. 1985. Comparison of three tillage methods for maize (*Zea mays*) and cowpeas (*Vigna unguiculata* L. Walp.). Production on a coastal clay soil in Guyana. *Tropical Agriculture* (Trinidad) 62(1): 25-29.

UNGER, P.W. 1986. Report on a visit to African countries in connection with the FAO supported tillage network. Rome: FAO, Mimeo.

WILSON, S.G. 1941. Agricultural Practices among the Angoni-Tumbuka tribes of Mzimba. *East African Agricultural Journal* (October 1941).

Tillage practices and their effect on soil productivity in Nigeria

O.A. OPARA-NADI[*]

Abstract

Tillage practices can have profound effects on the soil environment through their influence on the physical, chemical, and biological properties of the soil. These changes affect the productivity of the soil, and thus affect the growth and yield of crops. The role of tillage as a soil management practice is particularly vital in those tropical soils having a thin surface horizon, low water-holding capacity, supraoptimal soil temperature regimes, particularly during the seedling stage of crop growth, and high susceptibility to soil erosion. Consequently the main objectives of tillage research in Nigeria and elsewhere in the tropics are to develop appropriate tillage methods that will preserve and sustain soil productivity, maintain ecosystem stability, and become acceptable to the farmers.

Most research results on tillage methods in Nigeria show that local specificity of these results lead to seemingly contradictory conclusions. Consequently the adaptability of any tillage method from one soil and agroecological environment to another should be done with consideration for soil properties and rainfall regimes. It is also important to investigate the long-term effects of a wide range of tillage methods to determine the most suitable systems needed to alleviate specific soil constraints in any agroecological environment. This report reviews research information on tillage practices in Nigeria. Through collation and evaluation of available

[*] College of Agriculture and Veterinary Medicine, Imo State University, P.M.B. 2000, Okigwe, Nigeria.

research data, knowledge gaps can be identified and appropriate research priorities defined for the future.

Introduction

In Nigeria and some other areas of the humid tropics, increasing pressure on land has made it necessary to abandon the traditional system of shifting cultivation and its related bush-fallow rotation in favour of more realistic intensive farming systems. The traditional hoe-farming method is slowly being replaced by more sophisticated and less back-breaking tillage methods and other cultural practices, especially in large-sized farms. Consequently tillage practices are becoming an integral part of soil management and conservation techniques in many parts of the humid and subhumid tropics. However, the degradation of the soil structure, the formation of compacted layers near the soil surface, accelerated soil erosion, supraoptimal soil temperatures, and decreased soil organic matter, nutrients, and water-holding capacities are some of the soil management problems often encountered under certain tillage practices (Stibbe, 1970; Lal, 1974, 1975, 1976a; Opara-Nadi and Lal, 1984, 1986). It is commonly accepted knowledge that soil productivity in the tropics declines rapidly with intensive, continuous farming (Kang and Juo, 1984; Sanchez *et al.*, 1982; Sobulo and Osiname, 1986; Lal, 1981c) even when supplementary fertilizers are in use (Stephens, 1969; Le Mare, 1972). However, the reasons for the decline in productivity vary from one soil to another, and may be due to ecological conditions or to growing and management strategies (Lal, 1985c).

Soils management practices such as tillage systems and residue management have both direct and indirect effects on soil physical, chemical, and biological properties, and thus may affect the growth and yield of crops. Appropriate tillage practices are those that curtail the degradation in soil properties and decline in crop yields (Lal, 1985a). So the main objectives of tillage research in Nigeria and elsewhere in tropical Africa are to develop appropriate tillage methods that will preserve and sustain soil productivity, maintain ecosystem stability, optimize the biophysical environment, and alleviate soil-related constraints to crop production (Lal, 1981b, c; 1982, 1984b; Greenland, 1981).

Soil of different physical and chemical properties respond differently to different tillage methods. Consequently, the choice of a tillage method should be based on several considerations, and notably such factors as the initial soil conditions, alterations in soil properties caused by different tillage methods, and the previous and intended land use (Lal, 1982). Appropriate tillage practices for specific soils and crops aimed at preventing soil degradation and sustaining high crop yields form one of the key factors for developing stable and productive soil management practices in Nigeria and other parts of the tropics. A knowledge of the magnitude of changes in the soil environment brought about by tillage methods is therefore important in planning for cultural operations and in developing appropriate soil management systems to alleviate soil-related constraints to crop production. A review of tillage research relating crop responses to different tillage methods for different soils and agroecological regions of Nigeria is important at this stage in the agricultural development of the country. The failure of numerous large-scale mechanized farms in various regions of tropical Africa is evidence of land misuse and soil mismanagement. Choosing an appropriate tillage method and other land-use practices should drastically curtail, and even prevent, soil degradation. Only through collation, analysis, and the application of available research findings can knowledge gaps be identified and appropriate research and development priorities for the future defined. Consequently the objectives of this article are (i) to review, evaluate and assess available research information on tillage methods and their effects on soil productivity in Nigeria, and (ii) to define research priorities for future tillage research.

Climate, vegetation, and soils

Nigeria, with a total land area of about 923 768 km^2, lies within the coordinates 4°N-14°N and 2°E-14°E. The vegetation of the area is marked by two distinct types - the rainforest to the south, and the savanna to the north (Figure 1). The rainforest, which occupies less than one-third of the total land area of Nigeria, comprises mainly tall trees in a dense cluster with very few patches of grass. The savannas, which make up more than two-thirds of the total land area, comprise mainly short grasses and scattered short trees. The number of trees diminishes from the southern Guinea savanna to the Sahel zone. The derived savanna or forest savanna, separating the rainforest from the fire-swept savanna, and occupying about 8% of the total land area, has an intermediary vegetation which is between that of the rainforest and that of the typical savanna.

The climate of Nigeria is tropical, and is characterized by two distinct seasons, the wet season and the dry season. The wet season, which is hot and humid, starts from March and ends in October in the rainforest zone, while in the savanna zone it starts in May and ends in September. On the other hand, the dry season begins in November and ends in February in the rainforest zone, while in the savanna zone it starts in October and ends in April. In the rainforest zone, the mean annual rainfall ranges from about 3500 mm at the extreme to about 1000 mm at the northern limit of the lowland rainforest zone. In the savanna zone, the mean annual rainfall ranges from about 1300 mm at the southern limit to as low as 350 mm at the northern limit of the savanna belt. Temperatures are generally high and fairly evenly distributed throughout the year. For instance, mean annual temperature is approximately 26°C while mean monthly temperatures vary between approximately 20°C and 31°C.

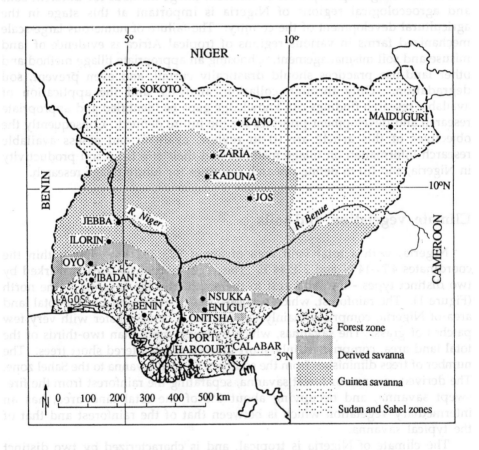

Figure 1. Map of Nigeria, showing the main vegetation zones.

The bulk of the potential arable land in Nigeria, consisting of Alfisols, Ultisols and Oxisols, have low inherent fertility, low-activity clay, and various nutrient imbalances - as well as physical constraints such as poor structural stability, low water-holding capacity, and high susceptibility to erosion and other soil-degrading factors. The arable food crop production potential of the soils is limited by those physical and chemical constraints which affect their productivity and yield-sustaining capabilities. For example, the drought stress frequently experienced by crops grown on Alfisols is due to low available-water reserves and high soil-surface temperatures. On the other hand, the acid Ultisols and Oxisols have various nutritional disorders due to low pH, deficiencies of major nutrients, and aluminium and manganese toxicity. Intensive and sustainable crop production on the Alfisols, Ultisols, and Oxisols require good soil management practices in order to prevent yield decline as a result of soil degradation.

Tillage research in the rainforest zone

Tillage studies on Alfisols

Constraints to intensive land use in Alfisols are predominantly physical, and comprise low water-holding capacity, and susceptibility to erosion and soil compaction (Lal, 1985a). Aware of these constraints and the limits to food production imposed by the traditional system of shifting cultivation and the related bush-fallow rotation in the tropics, scientists at the International Institute of Tropical Agriculture (IITA) in Nigeria started research on no-tillage or mulch-tillage systems in 1970 (Rockwood and Lal, 1974; Lal, 1973, 1974, 1976b). No tillage, minimum tillage, zero tillage, and mulch tillage are terms synonymous with conservation tillage (Willis and Amemiya, 1973; Rockwood and Lal, 1974; Lal, 1976b; Greenland, 1981). According to Greenland (1981), Lal (1973, 1974, 1975, 1976b), Rockwood and Lal (1974), Lal and Hahn (1973), the no-till system of cultivation with crop residue mulches is the basis of conservation farming because it conserves water, prevents erosion, maintains organic-matter content at a high level, and sustains economic productivity. In addition, there are savings in machinery investment and in the time required for seedbed preparation (Lal, 1974, 1985b; Rockwood and Lal, 1974).

Since the early seventies, other scientists working in national research institutes and universities in Nigeria have also conducted studies on Alfisols to compare the effects of different tillage methods on soil properties, crop growth, and yield (Agboola and Fayemi, 1972; Agboola, 1981; Aina, 1979; Aina et al., 1976; Osuji et al., 1980; Osuji and Babalola, 1982; Osuji, 1984; Wilkinson and Aina, 1976). The results of these studies agree with data obtained by scientists at IITA. Experimental data on Alfisols in Nigeria show that no-tillage soil management with crop residue may maintain the productivity of some tropical soils by eliminating soil erosion and preserving soil structure, soil organic matter, and the water- and nutrient-holding capacities of the soil, as well as maintaining favourable soil temperatures. Rockwood and Lal (1974) reported that a layer of dead crop residue 1-2 cm deep on the soil surface of no-till plots decreases the minimum soil temperature (Table 1) and improves soil moisture conditions (Table 2). The data in Table 1 show differences of 8-10°C in maximum soil temperature between no-tillage and ploughed plots under different crops. Soil moisture retention at 0-10 cm depth under different crops was higher by 1-4% in no-till plots than in ploughed plots (Table 2).

Table 1. The effect of tillage on maximum soil temperature at 5-cm depth under different crops two weeks after planting (1 May 1973).

Treatment	Maximum soil temperature (°C)			
	Maize	Pigeon pea	Soybean	Cowpea
Ploughed	41.4	40.0	41.4	41.8
No tillage	31.6	32.4	32.4	33.4
Difference	9.8	7.6	9.0	8.4

Source: Rockwood and Lal (1974).

Table 2. The effect of tillage on soil moisture retention at 0-10-cm depth under different crops two weeks after planting.

Treatment	Moisture retention in soil (%)			
	Maize	Pigeon pea	Soybean	Cowpea
Ploughed	9.7	10.8	7.3	12.3
No tillage	13.3	12.1	10.6	15.4

Source: Rockwood and Lal (1974).

In a recent study, Lal (1986a) investigated the effects of eight methods of preplanting tillage systems on maize growth and yield for eight consecutive crops grown from 1980 to 1983 on an Alfisol in southwest Nigeria. The results from this study show that the 15.00-hrs soil temperature on the ridged seedbed at 5 cm depth exceeded 40 °C, and was about 10 °C more than in no-till with mulch (Table 3). In a comparison between no-till with residue mulch and mouldboard ploughing followed by two harrowings, mulching a ploughed treatment decreased the 15.00-hrs soil temperature by 4 to 5 °C. In a comparison between no-till with residue mulch and no-till without residue mulch, the removal of crop residue mulch from no-till plots resulted in an increase of 3 to 4 °C in soil temperature (Table 3). Similar effects of mulching, no-till, and ridging have been reported by other workers (Lal, 1973; Manrya and Lal, 1980).

Table 3. Seedling emergence of maize and soil temperature as affected by different methods of seedbed preparation during the first-season crop (1983).

Treatment	Seedling emergence (count 10 m row^{-1})	Mean soil temperature (°C)	
		08.00 hrs	15.00 hrs
A	28	29.9	34.1
B	34	28.9	34.9
C	31	29.6	42.1
D	29	29.4	42.7
E	20	29.3	37.9
F	28	29.7	43.7
G	31	30.5	37.3
H	6	28.3	43.9
LSD (0.05)	8	0.8	1.4

A = no-till residue mulch; B = no-till with chiselling in the row zone to 50 cm depth; C = mouldboard ploughing followed by harrowing; D = disc ploughing followed by rotavation; E = no-till without residue mulch; F = mouldboard ploughing at the end of rains; G = mouldboard ploughing followed by two harrowings with residue mulch; H = mouldboard ploughing followed by two harrowings and ridgings.
Source: Adapted from Lal (1986a).

Rockwood and Lal (1974) and Lal (1974) observed that biological (e.g. earthworm) activity was stimulated by a decrease in maximum soil temperature and a favourable moisture regime in no-tillage plots. Whereas ploughed plots have a tendency to form a hard crust, no-tillage plots maintain better soil tilth, due primarily to enhanced earthworm activity. Earthworm activity in no-till plots approaches that found under forest fallow. As many as 2400 earthworm casts m^{-2} were counted in IITA no-tillage plots, compared with fewer than 100 casts m^{-2} for the ploughed plots (Rockwood and Lal, 1974; Lal, 1974). In a study of an Alfisol in western Nigeria, Osuji (1984) observed that water use was similar for crops under the different tillage practices studied, but water-use efficiency, which for the early-season maize production ranged from 118.65 kg ha^{-1} cm^{-1} in 1978 to 76.34 kg ha^{-1} cm^{-1} in 1980, was significantly higher in zero tillage than in other tillage treatments most of the time (Table 4). The

Table 4. Effect of tillage practices on water use, maize yield, and water-use efficiency (early season).

Year	Treatment	Water use (cm)	Grain yield (kg ha^{-1})†	Water-use efficiency (kg ha^{-1} cm^{-1})†	Plant population at harvest (per plot)†
1976	Conv. tillage	32.15	3106 a	96.61 a	180 a
	Plough	29.64	2923 a	98.62 a	179 a
	Zero tillage	30.44	2639 b	86.70 b	160 b
	Manual	29.19	2692 b	92.22 ab	188 a
1977	Conv. tillage	48.19	5240 a	108.74 a	203 a
	Plough	47.64	5067 a	106.36 a	198 ab
	Zero tillage	49.20	5123 a	104.13 a	186 b
	Manual	48.00	4612 b	96.08 b	194 ab
1978	Conv. tillage	49.60	5533 a	111.55 a	207 a
	Plough	50.01	4998 b	99.94 b	205 a
	Zero tillage	50.14	5949 c	118.65 c	203 a
	Manual	49.01	4303 d	87.69 d	199 a
1979	Conv. tillage	49.54	5259 a	106.16 a	200 a
	Plough	48.92	5174 a	105.76 a	206 a
	Zero tillage	49.69	5887 b	118.47 b	198 a
	Manual	49.01	4103 c	83.72 c	204 a
1980	Conv. tillage	49.62	5384 a	108.50 a	202 a
	Plough	49.98	5238 a	104.80 a	199 a
	Zero tillage	49.21	5678 b	115.38 b	205 b
	Manual	48.64	3713 c	76.34 c	197 a

†   Means followed by the same letter in the same column are not significantly different at the 5% level by Duncan's multiple range test.
Source: Osuji (1980).

author also reported that from the point of view of yield, the cultivated treatments had an initial yield advantage over zero tillage. However, over a longer period, yields from the zero-tillage plots equalled and even exceeded the yields from the cultivated treatments, especially during the drier late seasons.

Studies on the effects of mechanized tillage systems on the properties of a tropical Alfisol in watersheds cropped to maize (Lal, 1985c) showed that soil physical properties and chemical constituents declined substantially in the ploughed watershed after six years of continuous mechanized farming and twelve maize crops, while the decline in these properties were decidedly less in the no-tillage watershed (Table 5).

Table 5. Effects of motorized tillage methods on soil chemical properties at 10-20-cm depth, 5 and 6 years after tillage treatments were imposed.

Property	5 years				6 years			
	No tillage		Ploughed		No tillage		Ploughed	
pH (1:1 in H_2O)	5.4	±0.18	4.9	±0.4	5.5	±0.3	4.9	±0.2
Organic carbon (%)	1.05	±0.15	1.25	±0.13	1.00	±0.20	1.0	±0.14
Total nitrogen (%)	0.130	±0.04	0.168	±0.04	0.154	±0.04	0.186	±0.032
Bray P (ppm)	8	±3	18	±7	10	±4	33	±12
Exchangeable cations (cmol+ kg^{-1})								
Ca^{2+}	18	±4	22	±5	25	±9	19	±6
Mg^{2+}	4.9	±1.2	4.1	±1.2	3.7	±2.0	1.3	±0.3
K^+	1.5	±0.4	0.5	±0.1	2.3	±0.2	1.6	±0.3
Na^+	0.4	±0.1	0.5	±0.1	2.3	±0.2	2.2	±0.2
Mn^{2+}	0.02	±0.01	0.5	±0.4	0.4	±0.3	1.7	±1.0
Total acidity	0.6	±0.4	1.6	±1.3	2.5	±0.2	7.2	±4.2

Source: Lal (1985c).

In a long-term study (10 years) the effect of no tillage and conventional ploughing with and without crop residue mulch on soil properties and yield of maize on an Alfisol in southwest Nigeria, Opara-Nadi and Lal (1986) observed that total porosity, moisture retention, saturated and unsaturated hydraulic conductivity, and the maximum water-storage capacity increased under no-tillage mulch more than they did in other treatments. Similar observations were also made with regard to Alfisols in western Nigeria by Aina (1979). Experiments on an Alfisol on twin watersheds of about 5 ha each cropped to maize (Lal, 1984a) show that cumulative runoff in 1979 was 10 times higher and erosion 42.2 times higher from a ploughed watershed than from a no-till watershed (Table 6). The infiltration capacity 5 years after land development was 3.8 cm h^{-1} for the ploughed and 10.4 cm h^{-1} for the no-till watershed.

Table 6. Runoff and erosion measurements with an H-flume for 1979 on watersheds of about 5 ha.

Parameter	No-till			Conventional tillage		
	First season	Second season	Total	First season	Second season	Total
1979						
Runoff (mm)	3.60	17.93	21.53	173.53	51.42	225.06
Erosion (t ha^{-1})	0.012	0.118	0.130	4.91	0.59	5.50
Rainfall (mm)	583.5	257.8	841.3	583.5	257.8	841.3
1980						
Runoff (mm)	25.13	9.31	34.44	122.80	30.16	152.96
Erosion (t ha^{-1})	0.307	0.025	0.33	1.61	0.286	1.89
Rainfall (mm)	621	279	900	621	279	900

Source: Lal (1984a)

Opara-Nadi and Lal (1986, 1987c) reported that a mulch-based no-tillage technique was superior to conventional tillage for postclearing soil management under different land-clearing methods for Alfisols in western Nigeria. Fertilizer response under no tillage and conventional ploughing has been demonstrated for an Alfisol in western Nigeria. Lal (1979) observed that yields of maize grain at low levels of nitrogen application were significantly lower under no tillage than with conventional ploughing, though yields were the same at high levels of nitrogen application. These results contrasted with those reported earlier from the same plots in the first and second year after imposing tillage treatments (Lal, 1974). The author attributed the differences in crop response to tillage treatments on the same plot within 4-5 years to the progressive increase or improvement of organic matter, the nitrogen and phosphorous levels, the cation-exchange capacity in the no-till plots, and to a decline in these soil characteristics in conventionally ploughed plots due to erosion (Lal, 1974).

From a series of experiments reported for the rainforest ecological environment, with predominantly Alfisols of medium fertility, having a coarse-textured surface horizon and susceptible to erosion and drought, there are no beneficial effects of mechanical tillage over no-tillage or residue-tillage systems (Aina, 1979; Aina et al., 1976; Amon et al., 1981; Couper et al., 1979; Juo and Lal, 1979; Kang and Yunusa, 1977; Kang et al., 1980; Osuji and Babalola, 1982; Osuji, 1984; Agboola, 1981; Agboola and Fayemi, 1972; Lal, 1973, 1974, 1975, 1976a, b, 1979a, b, 1980, 1981a, b, c, 1983a, b; Lal and Hahn, 1973; Lal et al., 1974; Lal et al., 1978a, b; Lal and Oluwole, 1983). Rather some of these studies (Opara-Nadi and Lal, 1986; Lal, 1985c, 1986a) indicated that the no-till with residue mulch maintained more favourable soil properties and produced higher yields of maize grain under continuous maize than under mulching after ploughing (Table 7).

Table 7. Maize grain yield (t ha⁻¹) for eight different seasons, as influenced by different methods of seedbed preparation.

Treatment§	1980†		1981†		1982†		1983†	
	I‡	II‡	I‡	II‡	I‡	II‡	I‡	II‡
A	2.5 a	2.4 a	3.6 a	0.8 bc	4.6 a	1.2 a	4.4 a	0.5 c
B	2.0 a	1.8 a	3.4 a	1.0 bc	4.3 ab	0.9 a	4.2 a	1.4 b
C	2.7 a	2.0 a	3.1 ab	1.3 ab	3.8 ab	1.1 a	3.7 a	2.3 a
D	2.4 a	2.1 a	3.4 a	1.0 abc	3.8 ab	0.8 a	4.6 a	1.8 ab
E	2.1 a	1.8 a	2.8 ab	0.8 bc	4.2 ab	1.0 a	3.6 a	1.8 ab
F	2.6 a	1.8 a	3.4 a	0.7 c	3.8 ab	0.9 a	3.8 a	2.1 ab
G	2.1 a	2.0 a	2.9 ab	1.0 abc	3.9 ab	1.0 a	4.4 a	1.6 b
H	2.5 a	2.1 a	2.2 b	1.4 a	3.7 b	0.8 a	3.4 a	1.3 bc
LSD (0.05)	0.8	1.0	1.2	0.5	0.7	0.4	1.6	0.8

† Means followed by the same letters in the same column are not significantly different at the 5% level by Duncan's multiple range test.
‡ I = first season, II = second season.
§ See legend in Table 3.
Source: Lal (1986a).

Other studies (Lal, 1984a, 1986c) show that chiselling in the row zone may have an ameliorative effect if the soils are initially compacted, as often happens in mechanized farming. The data reported in some studies (Lal, 1973, 1986a) show that the ridges often used as a conventional method of seedbed preparation have no beneficial effects over a no-till or flat-planting system. Most studies on Alfisols in Nigeria indicate that in the rainforest environment the most useful tillage system for grain crops combines the provision of crop-residue mulch on the soil surface and the elimination of all mechanical tillage before planting.

Tillage studies on Ultisols

Ultisols not only have physical constraints limiting crop production, but in contrast to Alfisols also have various nutritional disorders - notably low pH, deficiencies of major nutrients, and aluminium and manganese toxicity. Studies on the effects of tillage systems on soil properties and crop yield on Ultisols in Nigeria are sparse, and the results are rather contradictory (Opara-Nadi and Lal, 1987a, b; Hulugalle et al., 1985, 1986, 1987; Maurya and Lal, 1979a; Rodriguez and Lal, 1979; Okigbo, 1979; Ogunremi et al., 1986a). In southern Nigeria, Okigbo (1979) observed that no-tilling of a Nkpologu soil (an Ultisol) had more adverse effects on cassava yield and performance than any of the other preplanting cultivations. Maurya and Lal (1979b) observed that for an acid Ultisol in southeastern Nigeria, there was no significant difference in cassava tuber yield between no-till and ploughed plots. The mean tuber yield for two consecutive years was 8 t ha^{-1} for no-till plots and 9 t ha^{-1} for ploughed plots. This study also showed that soil pH declined with continuous cropping from an initial value of 4.5 to 4.0 in ploughed plots and to 4.4 in no-till plots.

Studies on the effects of no-till and disc ploughing with and without residue mulch on the properties of a tropical Ultisol under root crops (Opara-Nadi and Lal, 1987b, Hulugalle et al., 1985, 1987) showed that tillage reduced soil bulk density (Table 8), but infiltration increased only when tillage and mulch were combined. Soil temperature was lowest under untilled, mulched plots, and soil chemical properties improved most in those plots. Mulching increased cassava yield by 32.3% in untilled plots (Table 9), but the yield of tilled plots was not similarly affected. Data in Table 9 also show that ploughed-mulch treatments produced the highest yam tuber and cocoyam cormel yields. Recent studies by Opara-Nadi et al. (in press) show that at a depth of 0-10 cm, diffusivity and unsaturated hydraulic conductivity functions were highest for mulched plots of no-till and ploughed treatments (Tables 10 and 11). Soil moisture reserves at a depth of 0-15 cm was highest for the no-till with mulch, whereas the ploughed mulched treatment had the highest moisture reserve at a depth of 15-45 cm.

Table 8. Effects of tillage and mulching on bulk density at 0-10 cm depth of a tropical Ultisol in southeastern Nigeria.

Treatment	Bulk density (Mg m^{-3})			
	1984		1984	
	Yam†	Cassava†	Cocoyam†	Cassava†
No-till, with mulch	1.38 a	1.40 a	1.33 a	1.37 a
No-till, without mulch	1.42 a	1.40 a	1.33 a	1.41 a
Conventional tillage with mulch	1.29 b	1.24 b	1.24 b	1.26 b
Conventional tillage without mulch	1.31 b	1.34 a	1.27 b	1.36 a

† Means followed by the same letters in the same column are not significantly different at the 5% level by Duncan's multiple range test.
Source: Adapted from Opara-Nadi and Lal (1987b), Hulugalle et al. (1985, 1987).

Table 9. Effects of tillage methods and mulching on yields of root tubers at Onne in southeastern Nigeria.

Treatment	Yam† 1983	Cocoyam† 1984	Cassava† 1983/84
No tillage, with mulch	11.1 b	1.10	16.8 a
No tillage, without mulch	9.8 c	0.30 b	12.7 b
Conventional tillage, with mulch	13.9 a	1.90 c	13.1 ab
Conventional tillage, without mulch	10.9 bc	0.40 b	14.5 ab

† Means followed by the same letters in the same column are not significantly different at the 5% level by Duncan's multiple range test.
Source: Adapted from Opara-Nadi and Lal (1987b), Hulugalle et al. (1985, 1987).

In a study on the effects of tillage and seeding methods on soil physical properties and the yield of upland rice for an Ultisol in southeast Nigeria, Ogunremi et al. (1986a) observed that in direct-seeded rice, the maximum grain yield was obtained in ploughed treatments (Figure 2a). The application of crop-residue mulch decreased grain yield by 43.1%, 27.1% and 12.3% on compacted, no-till, and ploughed plots. In transplanted rice, the application of crop-residue mulch increased the rice yield by 50.0%, 183.8%, and 92.5% on compacted, no-till, and ploughed plots respectively (Figure 2b). The highest grain yields of 6.3 and 6.1 t ha^{-1} respectively in ploughed plots for the first and

Table 10. Diffusivity functions at 0-10-cm depth under different tillage methods and mulching.

Treatment†	Diffusivity, $D(\theta)$ $(cm^2\ min^{-1})$							
	Volumetric moisture content, v $(cm^3\ cm^{-3})$‡							
	0.39	0.36	0.33	0.30	0.27	0.24	0.21	0.18
NT - M	0.35 a	0.23 a	0.13 a	0.038 a	0.0054 a	0.004 a	0.003 a	0.002 a
NT + M	0.70 b	0.59 b	37.00 b	0.170 ab	0.0890 ab	0.037 a	0.027 a	0.016 ab
CT - M	0.58 ab	0.38 ab	0.23 ab	0.110 a	0.0530 a	0.040 a	0.027 a	0.012 ab
CT + M	0.70 b	0.45 ab	0.38 b	0.260 b	0.1600 b	0.120 b	0.099 b	0.041 b

† NT - M = no-till, without mulch; NT + M = no-till, with mulch; CT - M = conventional tillage, without mulch; CT + M = conventional tillage, with mulch.
‡ Means followed by the same letter in the same column are not significantly different at the 5% level by Duncan's new multiple range test.
Source: Opara-Nadi et al. (in press).

Table 11. Unsaturated hydraulic conductivity-moisture content relationship at 0-10 cm depth under the four treatments.

Treatment†	Diffusivity, $D(\theta)$ $(cm^2\ min^{-1})$							
	Volumetric moisture content, v $(cm^3\ cm^{-3})$‡							
	0.39	0.36	0.33	0.30	0.27	0.24	0.21	0.18
NT - M	1.87 a	0.91 a	0.42 a	0.74 a	0.010 a	0.0076 a	0.0002 a	0.00012 a
NT + M	3.20 b	2.68 b	1.36 b	0.45 bc	0.160 ab	0.0290 b	0.0017 ac	0.00130 c
CT - M	2.27 ab	1.34 ab	0.60 a	0.27 ac	0.130 ab	0.0170 ab	0.0029 bc	0.00078 b
CT + M	0.08 ab	1.39 ab	0.88 ab	0.59 b	0.370 b	0.0310 b	0.0055 b	0.00044 ab

† NT - M = no-till, without mulch; NT + M = no-till, with mulch; CT - M = conventional tillage, without mulch; CT + M = conventional tillage, with mulch.
‡ Means followed by the same letter in the same column are not significantly different at the 5% level by Duncan's new multiple range test.
Source: Opara-Nadi et al. (in press).

and second seasons were associated with a greater uptake of P, Na, Fe, and Zn at flowering and of N, Mg, K, Mn, and Cu at both the maximum tillering stage and at the flowering stage. For this Ultisol, the best rice growth and yield were obtained with ploughing, irrespective of the seeding method.

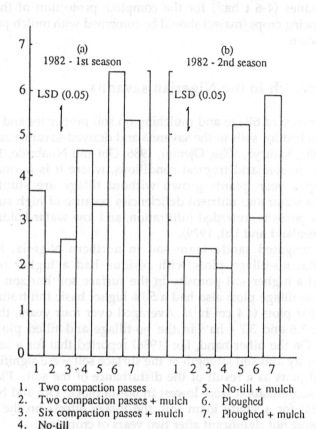

Figure 2. Effects of tillage, mulching, and soil compaction on the grain yield of rice for (a) direct seeding and (b) transplanting methods (Ogunremi et al., 1986a).

1. Two compaction passes
2. Two compaction passes + mulch
3. Six compaction passes + mulch
4. No-till
5. No-till + mulch
6. Ploughed
7. Ploughed + mulch

Studies on Ultisols in southeastern Nigeria (Opara-Nadi and Lal, 1987a; Hulugalle et al., 1985, 1986, 1987, Opara-Nadi et al., 1990) indicate that the adoption of no-tillage or conventional tillage methods on Ultisols in the high-rainfall region of southeastern Nigeria must be carried out in combination with crop-residue mulch for maximum crop yield and an improvement in soil physical properties. In the humid and subhumid tropics of Africa and South

America, no-tillage systems with crop-residue mulch of 4-6 t ha^{-1} have proved successful (Mannering and Meyer, 1963; Lal, 1982; Roth *et al.*, 1988). To produce enough residues (4-6 t ha^{-1}) for the complete protection of the soil surface, mulch-producing crops (maize) should be combined with mulch-producing cover crops in rotation.

Tillage research in the Nigerian savanna

The influence of tillage and mulching on soil properties and crop yield has also been studied for soils in the savanna and derived-savanna zones of Nigeria (Adeoye, 1982; Maurya, 1986; Ojeniyi, 1986; Obi and Nnabude, 1988; Ike, 1986, 1987, 1988). In semiarid tropical conditions, where it is normal to have one rainfed crop a year, plants grown without tillage are stunted and show symptoms of water and nutrient deficiencies because of high surface-soil bulk density, low porosity, retarded infiltration, and low water-holding capacity of the soil (Greenland and Lal, 1979).

For an irrigated sandy loam soil in northern Nigeria, Maurya (1986) observed that no-tillage plots with residue had a higher organic carbon content, and a higher soil porosity in the surface soil horizon than the tilled plots. The no-tillage plots also had a 50% higher basic infiltration rate (6.6 cm h^{-1}) than tilled plots (4.4 cm h^{-1}). Averaged over four years, the mean wheat yields were 3.6 and 3.7 t ha^{-1} in the no-tillage and tilled plots respectively (Table 12). On the other hand, Ike (1988) reported that for a sandy loam soil, the bulk density and soil strength of the surface soil were significantly reduced in ploughed plots as a result of the disturbance of the soil. Plant height and dry-matter accumulation were lowest in no-till plots. Obi and Nnabude (1988) observed that for a sandy loam soil, the tillage effects on the measured soil properties were not significant after two years of cropping.

From the results of a previous study (Ike, 1987), it was observed that on medium-textured ferruginous soils of the tropical savannas, expensive tillage practices can be avoided without decreasing yields significantly. Hence, except perhaps where hardpans exist in the soil profile, deep tillage of the sandy loam soil of the tropical savanna has no yield advantage over the reduced-tillage systems. From this study, it was concluded that no tillage would be an attractive soil management concept for increased food production in the savanna region.

Table 12. Wheat grain yield (t ha⁻¹) as affected by tillage and residue management practices during the 1982-1985 cropping seasons.

Treatment		Irrigation (days)	Yield (t ha⁻¹)			
			1982	1983	1984	1985
No tillage	Residue	8	3.9	3.4	4.5	3.5
		16	3.6	3.1	4.2	3.5
	No residue	8	3.7	3.6	4.2	3.0
		16	3.7	3.5	4.1	2.9
	Mean		3.7	3.4	4.2	3.2
Tillage	Residue	8	2.8	4.0	4.3	4.0
		16	3.1	3.7	4.4	3.6
	No residue	8	3.0	3.7	3.6	4.1
		16	3.1	3.4	3.6	3.4
	Mean		3.0	3.7	4.0	3.8
LSD (5%)						
	Tillage		0.18	NS	NS	0.28
	Residue		NS	NS	0.20	NS
	Irrigation		NS	NS	NS	NS

Source: Maurya (1986).

Tillage methods and wetland utilization

In recent times, tropical wetlands are rapidly being developed for the needed increase in rice (*Oryza sativa* L.) production. However, knowledge is still limited concerning the optimum soil and crop management practices (Rodriguez and Lal, 1985). In southern Nigeria, the wetlands comprise Inceptisols of relatively high fertility and with a favourable soil moisture regime, but hitherto underutilized or unutilized because of health hazards and a lack of appropriate technologies (Lal, 1986b). In general, research information on soil and crop management of the wetlands in Nigeria is scanty. Lal (1986b) reported that tillage requirements for irrigated rice (*Oryza sativa*) vary among soils of varying percolation rates and textural and structural properties. For the same soil, rice response to tillage methods depends on exogenous factors, notably microclimate, supplemental irrigation, crop-residue management, and the rate and method of fertilizer application (Lal, 1986b).

Some research conducted in Nigeria on the development of reduced-tillage techniques for wet paddy rice (Maurya and Lal, 1979b; Lal, 1983b, 1986b; Rodriguez and Lal, 1985; Ogunremi et al., 1986b) have indicated satisfactory grain yield provided weeds are adequately controlled. Rodriguez and Lal (1985) reported that no-till rice in which weeds were adequately controlled produced grain yields comparable to those produced after conventional puddling. Lal (1986b) reported that differences in yield between no-till rice and rice grown after puddling were insignificant at fertilizer application rates of 100-150 kg N ha^{-1}. In addition, the continuous no-till system created favourable soil physical and nutritional properties of the surface layer.

For soils with a relatively high clay content, there were no obvious advantages in rice yield in puddling over the no-till method of seedbed preparation. Puddling (wet tillage) is recommended in soils with rapid percolation to decrease water and nutrient losses (Rodriguez and Lal, 1985). Ogunermi et al. (1986b) observed that reduction in percolation can be better achieved by mechanical compaction than wet tillage. Generally, an objective of tillage operations in highly permeable coarse-textured soils in relation to rice is the reduction of percolation and leaching losses (Lal, 1986b). Experiments conducted in western Nigeria on such soil (Lal, 1983b) indicated an equivalent or lower rice yield on unpuddled soil, depending on the crop-residue management. Other studies (Maurya and Lal, 1979b; Rodriguez and Lal, 1985) showed that the adverse effects of no-till, probably due to anaerobic decomposition of rice straw, are usually not observed in the first one or two crops. Although the rate of crop-residue decomposition is generally high in the rainforest ecological environment (Jenkinson and Ayanaba, 1977; Maduakor et al., 1984), nevertheless in wet paddy rice production the crop residue should either be removed or burnt in situ to facilitate transplanting and other farm operations (Lal, 1986b).

For wetland utilization in Nigeria, the following observations based on available research information can be made:

- at present, research data are scanty regarding the long-term effect of tillage and methods of seedbed preparation on rice yield and soil properties (Lal, 1986b);
- tillage requirements for lowland rice production should be evaluated in terms of soil texture and water management practices (Ogunremi et al., 1986b);
- tillage may be required at first for land development and to level the land in order to facilitate proper water management, but it does not seem necessary to go through the dry and wet tillage operations before every crop of rice is transplanted (Rodringuez and Lal, 1985);

- the possibility of combining conventional puddling with zero-tillage techniques should be studied for the potential productivity of wetlands for paddy rice production (Lal, 1986b; Rodringuez and Lal, 1985).

Conclusion

The productivity of soils in the tropics declines under intensive and continuous cropping. This decline in productivity is a result of soil degradation, which in term can be attributed to land misuse and soil mismanagement. Studies conducted in Nigeria and elsewhere indicate that the productivity of tropical soils can be maintained provided that the soils are properly managed in such a way that the soil and environmental equilibrium status attained under forest fallow is not drastically altered during cropping. Improved soil-surface management practices that can be substituted for the present methods, which lead to soil degradation or do not meet current production requirements, are necessary to increase food production and alleviate soil-related constraints for intensive and sustainable crop production in the tropics. Appropriate tillage methods tailored to suit particular soils, crops, and agroecological zones form part of the package of soil management strategies for enhancing and maintaining soil productivity. Research information on appropriate tillage methods can be obtained through well-designed and adequately equipped long-term experiments.

The experimental data presented in this review show that for productive and sustainable food crop production in tropical Africa, tillage methods have to be tailored to suit particular soils, crops ,and agroecological environments. Since crop responses to tillage methods depend on soil properties, the rainfall regime and the crop grown, it is necessary to investigate the effects of a wide range of tillage methods in different agroecological environments in order to determine the most suitable method for any particular soil. There is a need to assess tillage requirements for different soil conditions, agroecological regions, and the crops to be grown. Recommendations to farmers regarding appropriate methods should be based on scientific data obtained through well-designed and adequately equipped long-term experiments.

Numerous research findings indicate that zero tillage or reduced tillage with residue mulch may be appropriate for Alfisols in the humid tropics. However, more research is needed to make the system more versatile. Specific problems associated with no tillage that should be investigated include the development of (i) alternate and cheap systems of weed control, (ii) suitable crop rotations that include cover crops, and (iii) appropriate equipment for planting and fertilizer application.

Ultisols present a different problem in terms of tillage requirements because of their low pH and acidity-related constraints, such as Al and Mn toxicities. The application of no-till or conventional tillage must take cognizance of the lime requirements of these soils; and the methods, rates, and frequency of liming for different tillage methods need to be studied. This review shows that organic mulch treatments, whether of undisturbed killed sod on no-tillage plots or as crop residue introduced on conventional tillage plots, are important as soil management practices. Lime-tillage and lime-organic matter interactions in different agroecological zones need to be studied.

Experiments on tillage and residue management should be carried out in the different ecological zones in Nigeria, with a view to determining the following points:

- the interactions between tillage methods, residue management, and cropping systems;
- the rooting characteristics of crops in combination and in crop rotations, and the exploitation of these characteristics to maximize nutrient utilization;
- soil and water management and conservation under different tillage methods and residue management techniques;
- nutrient dynamics from mineral and organic fertilizers under different tillage methods;
- lime-tillage interactions for acid soils;
- soil-surface management after annual burning for seedbed preparation.

The low-input technologies suitable for farmers in tropical Africa include legume intercropping, efficient crop-residue management, and alley cropping and planted fallow (agroforestry) combinations under appropriate tillage methods. These practices emphasize both the importance of biological processes of soil fertility maintenance and the intensive and continuous use of the land at a low economic cost without overexploiting this natural resource base. Soil and water management under these cultural practices, need to be studied for different soils and ecological zones.

References

ADEOYE, K.B. 1982. Effect of tillage depth on physical properties of a tropical soil and on yield of maize, sorghum and cotton. *Soil Tillage Research* 2: 115-231.

AGBOOLA, A.A. 1981. The effects of different soil tillage and management practices on the physical and chemical properties of soil and maize yield in a rainforest zone of Western Nigeria. *Agronomy Journal* 73: 247-251.

AGBOOLA, A.A. and FAYEMI, A.A. 1972. Effect of soil management on corn yield and soil nutrients in rain forest zones of Western Nigeria. *Agronomy Journal* 64: 641-644.

AINA, P.O. 1979. Soil changes resulting from long-term management practices in Western Nigeria. *Soil Science Society of America Journal* 43: 173-177.

AINA, P.O., LAL, R. and TAYLOR, G.S. 1976. Soil and crop management in relation to soil erosion in the rainforest of western Nigeria. In: *Soil erosion: prediction and control*, 75-84. Soil Conservation Society of America, Special Publication no. 21. Washington, DC: SCSA.

ARMON, M.R., LAL, R. and OBI, M. 1981. Effects of tillage systems on properties of an Alfisol in south west Nigeria. *Ife Journal of Agriculture* 3: 1981.

COUPER, D.C., LAL, R. and CLASSEN, S. 1979. Mechanized no-till maize production on an Alfisol in tropical Africa. In: *Soil tillage and crop production*, ed. R. Lal, 147-160. IITA Proceedings no. 2. Ibadan, Nigeria: IITA.

GREENLAND, D.J. 1981. Soil management and soil degradation. *Journal of Soil Science* 32: 301-322.

GREENLAND, D.J. and LAL, R. 1979. Towards optimizing soil physical characteristics for sustained production from soils in the tropics. In: *Soil physical properties and crop production in the tropics*, ed. R. Lal and D.J. Greenland, 529-530. Chichester, UK: John Wiley and Sons.

HULUGALLE, N.R., LAL, R. and OPARA-NADI, O.A. 1985. Effect of tillage system and mulch on soil properties and growth of yam (*Dioscorea rotundata*) and cocoyam (*Xanthosoma sagittifolium*) on an Ultisol. *Journal of Root Crops* 11: 9-22.

HULUGALLE, N.R., LAL, R. and OPARA-NADI, O.A. 1986. Effect of spatial orientation of mulch on soil properties and growth of yam (*Dioscorea rotundata*) and cocoyam (*Xanthosoma sagittifolium*) on an Ultisol. *Journal of Root Crops* 12: 37-45.

HULUGALLE, N.R., LAL, R. and OPARA-NADI, O.A. 1987. Management of plant residue for cassava (*Manihot esculenta*) production on an acid Ultisol in southeastern Nigeria. *Field Crops Research* 16: 1-18.

IKE, I.F. 1986. Soil and crop responses to different tillage practices in a ferruginous soil of the Nigerian savanna. *Soil Tillage Research* 6: 261-272.

IKE, I.F. 1987. Influence of tillage practice and nitrogen and phosphorus fertilizer rates on crop yields in the tropical savanna. *Soil Science* 143: 213-219.

IKE, I.F. 1988. Soil compaction under different tillage methods and its effect on growth and yield of cowpea. Paper presented at the 16th Annual Conference of the Soil Science Society of Nigeria held at Minna, 27-30 November 1988. 6p.

JENKINSON, D.S. and AYANABA, A. 1977. Decomposition of carbon 14 labelled plant material under tropical conditions. *Soil Science Society of America Journal* 41: 912-915.

JUO, A.S.R. and LAL, R. 1979. Nutrient profile in a tropical Alfisol under conventional and no-tillage system. *Soil Science* 127: 168-173.

KANG, B.T. and JUO, A.S.R. 1986. Effect of forest clearing on soil chemical properties and crop performance: VI, Soil management. In: *Land clearing and development in the tropics*, ed. R. Lal, P.A. Sanchez and R.W. Cummings Jr., 383-394. Rotterdam, Boston: Balkema.

KANG, B.T. and YUNUSA, M. 1977. Effects of tillage methods and phosphorus fertilization on maize in the humid tropics. *Agronomy Journal* 69: 291-294.

KANG, B.T., MOODY, K. and ADESINA, J.O. 1980. Effects of fertilizer weeding in no-tillage and tilled maize. *Fertilizer research* 1: 87-93.

LAL, R. 1973. Effects of seed bed preparation and time of planting of maize in Western Nigeria. *Experimental Agriculture* 9: 304-313.

LAL, R. 1974. No-tillage effects on soil properties and maize (*Zea mays* L.) production in western Nigeria. *Plant and Soil* 40: 129-143.

LAL, R. 1975. *Role of mulching techniques in tropical soil and water management.* IITA Technical Bulletin no. 1. Ibadan, Nigeria: IITA. 38p.

LAL, R. 1976a. No-tillage effects on soil properties under different crops in western Nigeria. *Soil Science Society of America Journal* 40: 762-768.

LAL, R. 1976b. *Soil erosion problems on an Alfisol in western Nigeria and their control.* IITA Monograph no. 1. Ibadan, Nigeria: IITA. 208p.

LAL, R. 1979a. Influence of six years of no-tillage and conventional plowing on fertilizer response of maize on an Alfisol in the tropics. *Soil Science Society of America Journal* 43: 399-403.

LAL, R. 1979b. Effective conservation farming systems for the humid tropics. In: *Proceedings of a Conference on Soil Conservation in the Tropics* (Colorado, 1979), 57-76. Madison, W1: ASA.

LAL, R. 1980. Soil conservation: preventive and control measures. In: *Soil conservation: problems and prospects,* ed. R.P.C. Morgan, 175-181. Chichester, UK: John Wiley and Sons.

LAL, R. 1981a. Soil erosion problems on an Alfisol in western Nigeria and their control: VI. Effects of erosion on experimental plots. *Geoderma* 25: 215-230.

LAL, R. 1981b. Soil conditions and tillage methods in the tropics. *Proceedings of the WAWSS/IWSS Symposium on No-tillage and Crop Production in the Tropics* (Liberia, 1981).

LAL, R. 1981c. Soil management in the tropics. In: *Characterization of soils of the tropics: classification and management,* ed. D.J. Greenland. UK: Oxford University Press.

LAL, R. 1982. Tillage research in the tropics. *Soil Tillage Research* 2: 305-309.

LAL, R. 1983a. Soil erosion and its relation to productivity in tropical soils. *Preserve the Land* (Conference, Hawaii, 1983).

LAL, R. 1983b. *No-till farming.* IITA Monogram no. 2, Ibadan, Nigeria: IITA. 64p.

LAL, R. 1984a. Mechanized tillage systems effects on soil erosion from an Alfisol in watersheds cropped to maize. *Soil Tillage Research* 4: 349-360.

LAL, R. 1984b. Soil erosion from tropical arable lands and its control. *Advances in Agronomy* 37: 183-248.

LAL, R. 1985a. No-till in the lowland humid tropics. In: *The Rising Hope of our Land* (Conference, Georgia, 1985), 235-241.

LAL, R. 1985b. A soil suitability guide for different tillage systems in the tropics. *Soil and Tillage Research* 5: 179-196.

LAL, R. 1985c. Mechanized tillage systems effects on properties of a tropical Alfisol in watershed cropped to maize. *Soil Tillage Research* 6: 149-162.

LAL, R. 1986a. Effects of eight tillage treatments on a tropical Alfisol: I. Maize growth and yield. *Journal of the Science of Food and Agriculture* 37: 1073-1082.

LAL, R. 1986b. Effects of 6 years of continuous cultivation by no-till or puddling systems on soil properties and rice yield of a clayey soil. *Soil Tillage Research* 8: 181-200.

LAL, R. 1986c. No-tillage and minimum tillage systems to alleviate soil related constraints in the tropics. In: *No-tillage and minimum tillage systems,* ed. M.A. Sprague and G.B. Triplett, 261-317. New York: John Wiley and Sons.

LAL, R. and HAHN, S.K. 1973. Effects of methods of seedbed preparation, mulching and time of planting on yam in western Nigeria. In: *Proceedings of the International Society of Tropical Root Crops* (Nigeria, 1973). Ibadan, Nigeria: IITA.

LAL, R. and OLUWOLE, J.O. 1983. Physical properties of earthworm casts and surface soil as influenced by management. *Soil Science* 135: 114-122.

LAL, R., KANG, B.T., MOORMANN, F.R., JUO, A.S. and MOOMAW, J.C. 1974. *Soil management problems and possible solutions in western Nigeria.* 20p.

LAL, R., MAURYA, P.R. and OSEI-YEBOAH. 1978a. Effect of no-tillage and plowing on efficiency of water use in maize and cowpea. *Experimental Agriculture* 14: 113-120.

LAL, R., WILSON, G.F. and OKIGBO, B.N. 1978b. No-till farming after various grasses and leguminous cover crops in tropical Alfisol: I. Crop performance. *Field Crops Research* 1: 71-84.

LE MARE, P.H. 1972. A long-term experiment on soil fertility and cotton yield in Tanzania. *Experimental Agriculture* 8: 299-310.

MADUAKOR, H.O., LAL, R. and OPARA-NADI, O.A. 1984. Effects of methods of seedbed preparation and mulching on the growth and yield of white yam (*Dioscorea rotundata*) on an Ultisol in southeast Nigeria. *Field Crops Research* 9: 119-130.

MANNERING, J.V. and MEYER, L.D. 1963. The effects of various rates of surface mulch on infiltration and erosion. *Soil Science Society of America Proceedings* 27: 84-86.

MAURYA, P.R. 1986. Effect of tillage and residue management on maize and wheat yield and on physical properties of an irrigated sandy loam soil in northern Nigeria. *Soil Tillage Research* 8: 161-170.

MAURYA, P.R. and LAL, R. 1979a. No-tillage system for crop production on an Ultisol in eastern Nigeria. In: *Soil tillage and crop production,* ed. R. Lal, 207-220. IITA Proceedings no. 2. Ibadan, Nigeria: IITA.

MAURYA, P.R. and LAL, R. 1980. Effects of no-tillage and plowing on roots of maize and leguminous crops. *Experimental Agriculture* 16: 185-193.

OBI, M.E. and NNABUDE, P.C. 1988. The effects of different management practices on the physical properties of sandy loam soil in southern Nigeria. *Soil Tillage Research* 12: 81-90.

OGUNREMI, L.T., LAL, R. and BABALOLA, O. 1986a. Effects of tillage methods and water regimes on soil properties and yield of lowland rice from a sandy loam soil in southwest Nigeria. *Soil Tillage Research* 6: 223-234.

OGUNREMI, L.T., LAL, R. and BABALOLA, O. 1986b. Effects of tillage and seeding methods on soil physical properties and yield of upland rice for an Ultisol in southeast Nigeria. *Soil Tillage Research* 6: 305-324.

OJENIYI, S.O. 1986. Effect of zero-tillage and disc ploughing on soil water, soil temperature and growth and yield of maize (*Zea mays* L.). *Soil Tillage Research* 7: 173-182.

OKIGBO, B.N. 1979. Effects of pre-planting cultivation and mulching on the yield and performance of cassava (*Manihot esculenta*). In: *Soil tillage and crop production*, ed. R. Lal, 75-92. IITA Proceedings no. 2. Ibadan, Nigeria: IITA.

OPARA-NADI, O.A. 1987b. Effects of no-till and disc plowing with and without residue mulch on tropical root crops in southeastern Nigeria. *Soil Tillage Research* 9: 231-240.

OPARA-NADI, O.A. and LAL, R. 1984. Diurnal fluctuations in hydro-thermal regime of a tropical Alfisol as influenced by methods of land development and tillage systems. *Zeitschrift für Pflanzinernähreung und Bodenkunde* 147: 150-158.

OPARA-NADI, O.A. and LAL, R. 1986. Effects of tillage methods on physical and hydrological properties of a tropical Alfisol. *Zeitschrift für Pflanzenernähreung und Bodenkunde* 149: 235-243.

OPARA-NADI, O.A. and LAL, R. 1987a. Influence of method of mulch application on growth and yield of tropical root crops in southeastern Nigeria. *Soil Tillage Research* 9: 217-230.

OPARA-NADI, O.A. and LAL, R. 1987c. Effects of land clearing and tillage methods on soil properties and maize root growth. *Field Crops Research* 15: 193-206.

OPARA-NADI, O.A., LAL, R. and HULUGALLE, N.R. In press. Mulching and tillage methods on soil properties of a tropical Ultisol under root crops.

OSUJI, G.E. 1984. Water storage, water use and maize yield for tillage systems on a tropical Alfisol in Nigeria. *Soil Tillage Research* 4: 339-348.

OSUJI, G.E. and BABALOLA, O. 1982. Tillage practices on a tropical soil: I. Effects on soil physical and chemical properties. *Journal of Environmental Management* 14: 343-358.

OSUJI, G.E., BABALOLA, O. and ABOABA, F.O. 1980. Rainfall erosivity and tillage practices affecting soil and water loss on a tropical soil in Nigeria. *Journal of Environmental Management* 10; 207-217.

ROCKWOOD, W.G. and LAL, R. 1974. Mulch tillage: a technique for soil and water conservation in the tropics. SPAN: *Progress in Agriculture* 17: 77-79.

RODRINGUEZ, M. and LAL, R. 1979. Comparison of zero and conventional tillage systems in an acidic soil. In: *Soil tillage and crop production*, ed. R. Lal, 197-206. IITA Proceedings no. 2. Ibadan, Nigeria: IITA.

RODRINGUEZ, M.S. and LAL, R. 1985. Growth and yield of paddy rice as affected by tillage and nitrogen levels. *Soil Tillage Research* 6: 163-178.

ROTH, C.H., MEYER, B. FREDE, H.G. and DERPSCH, R. 1988. Effect of mulch rates and tillage systems on infiltrability and other soil physical properties of an Oxisol in Parana, Brazil. *Soil Tillage Research* 11: 81-91.

SANCHEZ, P.A., BANDY, D.A. VILLACHICA, J.H. and NICHOLOIDES, J.J. 1982. Soils of the Amazon Basin and their management for continuous crop production. *Science* 216: 821-827.

SOBULO, R.A. and OSINAME, O.A. 1984. Soil properties and crop yields under continuous cultivation with different management systems. In: *Land clearing and development in the tropics*, ed. R. Lal, P.A. Sanchez and R.W. Cummings, Jr. Rotterdam, Boston: Balkema.

STEPHENS, D. 1969. Changes in yields and fertilizer responses with continuous cropping in Uganda. *Experimental Agriculture* 5: 263-269.

STIBBE, E. 1970. An approach to tillage research in the semiarid climate in Israel. In: *Tillage research methods* (Silsoe, UK, 1970), ed. N.J. Brown, D.E. Patterson and G. Spoor.

WILKINSON, G.E. and AINA, P.O. 1976. Infiltration of water into two Nigerian soils under secondary forest and subsequent arable cropping. *Geoderma* 50: 51-59.

WILLIS, W.O. and AMEMIYA, M. 1973. Tillage management principles. In: *Conservation tillage* (Iowa, 1973), 22-42. Washington, DC: SCSA.

Des expérimentations en riziculture pluviale menées dans le Moyen Ouest malgache

JACQUELINE RAKOTOARISOA

Résumé

A Madagascar, l'exploitation des sols de tanety - sols ferrallitiques - a pris ces dernières décennies de plus en plus d'importance. La production du riz de bas-fonds ne suffit plus en effet à nourrir la population sans cesse croissante. Les cultures sèches voient cependant leur productivité limitée par la pauvreté naturelle de ces sols de tanety. Le pouvoir de fixation et de libération du phosphore reste le premier facteur limitant de ces types de sol.

Il est en effet connu qu'en milieu acide, l'aluminium se combine à l'acide phosphorique pour donner des phosphates d'aluminium insolubles et inassimilables. Différents anions organiques comme l'acide humique sont susceptibles d'empêcher la combinaison aluminium-phosphate. Aussi, l'apport de fumier, source de matière organique a été considéré comme un des moyens permettant de réduire notablement l'inassimilabilité de l'acide phosphorique.

Partant de cet hypothèse, une expérimentation visant à évaluer le rôle que joue la matière organique dans le mécanisme de blocage du phosphore par l'aluminium a été conduite.

Les résultats de cette expérimentation ont mis en évidence que l'apport de fumier de parc contribue à améliorer la capacité de production de ces sols ferrallitiques.

Abstract

Rainfed rice experiments conducted in midwest Madagascar

The "tanety" soils - upland ferrallitic soils - have been used more extensively in Madagascar in recent years than they were previously. In fact, lowland rice production is no longer sufficient to feed the increasing population. But rainfed crops have limited productivity potential due to the inherently poor fertility of "Tanety" soils. Phosphorus fixation and its release is the primary limiting factor for the agricultural use of these soils.

It is well known that in an acidic environment, aluminium combined with phosphoric acid produces insoluble aluminium phosphates which can not be assimilated by plants. Different organic anions such as humic acid can prevent the aluminium-phosphate from combining. Therefore the input of manure, a source of organic matter, is considered to be one of the means of reducing the nonassimilability of phosphoric acid.

Taking this hypothesis as a base, an experiment was carried out on the role played by organic matter in blocking phosphorus by aluminium.

The results showed that manure input contributes to the improvement of the production potential of these ferrallitic soils.

Introduction

L'exploitation des sols de tanety (terrains exondés) a pris ces dernières décennies de plus en plus d'importance. La production de riz de bas fonds ne suffit plus en effet à nourir la population sans cesse croissante. Les cultures sèches voient toutefois leur productivité limitée par la pauvreté naturelle des sols de tanety. Par ailleurs, les techniques agricoles traditionnellement pratiquées pour leur exploitation (culture continue sans apport de fertilisants autre que du fumier au cas où le paysan peut en disposer) font perdre rapidement à ces sols leur faible potentiel de production. Pour permettre aux agriculteurs d'exploiter continuellement ces sols de tanety, il s'est avéré indispensable de mettre au point des techniques de mise en valeur adéquate.

Des études menées antérieurement ont mis en evidence l'effet bénéfique de l'apport d'une forte fumure minérale de redressement en première année et d'une fertilisation moyenne d'entretien les années suivantes. Par rapport aux ressources financières actuelles des paysans, et face à la hausse exhorbitante des prix des engrais, ces fumures sont disproportionnées. C'est ainsi que la recherche de pratiques culturales plus économiques s'est fait sentir. Sachant que les sols du type ferrallitique qui constituent la majeure partie de ces sols de tanety présentent en général des carences graves en calcium (Ca^{++}), en potassium (K^+), en magnésium (Mg^{++}) et en phosphore assimilable d'une part, et une teneur en matière organique assez élevée en surface mais avec un rapport C/N croissant en profondeur (au-delà de 20 cm) indiquant une faible évolution de la matière organique d'autre part, une expérimentation visant à étudier l'intéraction fumier phosphore a été mise en place. L'apport de phosphore a été conçu dans le but d'améliorer la capacité de production de ces types de sols.

Présentation générale de la zone d'étude

L'expérimentation a été mise en place à la station de Kianjasoa dans le Moyen Ouest malgache.

Cette région du Moyen Ouest présente un type de paysage caractérisé par une succession de plateaux plus ou moins étendus, de 900 à 1000 m d'altitude, dont les flancs sont souvent abrupts. Celle-ci est souvent découpée :
- soit par des vallées à fond plat, mal drainées,
- soit par des vallées en 'v' étroites, qui constituent le lit de parcours d'une rivière.

La zone d'étude s'inscrit dans la région centrale à influence occidentale, et présente un climat du type tropical sec, caractérisé par une longue saison sèche de 5 à 6 mois, un peu fraîche, et une saison pluvieuse et chaude accusant une pluviométrie moyenne annuelle de 1500 mm.

Les conditions climatiques et géologiques de la région ont donné naissance à :
- des sols brun-rouges ferrallitiques moyennement desaturés, formés sur migmatites, couvrant les sommets aplanis des collines et les pentes;
- des sols hydromorphes organiques tourbeux se rencontrant dans les bas-fonds perchés.

Le trait essentiel du relief est dominé par les émergences de plateaux plus ou moins étendus, si bien que les sols ferrallitiques représentent 75 à 80% des sols de la région.

Le Moyen Ouest, présentant encore une grande disponibilité en terre, offre une potentialité énorme, non seulement pour l'agriculture mais aussi pour l'élevage. C'est donc une région permettant aisément la mise en oeuvre d'un système de production intégrant l'agriculture et l'élevage. Sur sol de tanety, le type d'assollement le plus pratiqué de la région comporte le manioc, le riz pluvial, le maïs et l'arachide.

Démarche méthodologique

Le pouvoir de fixation et de libération du phosphore, premier facteur limitant des sols ferrallitiques, est fonction du pH, qui règle la précipitation des phosphates par l'aluminium échangeable. Il est connu qu'en milieu acide, l'aluminium se combine à l'acide phosphorique pour donner des phosphates d'aluminium insolubles et inassimilables. Différents anions organiques sont susceptibles d'empêcher la combinaison aluminium-phosphate. L'acide humique paraît être particulièrement efficace. Aussi, l'apport de fumier, source de matière organique à la portée des paysans, devrait permettre de réduire notablement l'inassimilabilité de l'acide phosphorique.

Le fumier pourrait, par ailleurs, améliorer les propriétés physiques du sol, dont dépendent la capacité de retention en eau et le développement racinaire de la plante.

Partant de cet hypothèse, une expérimentation, se fixant comme principal objectif d'appréhender le rôle que joue la matière organique dans le mécanisme de blocage du phosphore par l'aluminium, a été conduite dans le cadre d'un système de culture test avec une rotation de légumineuses et du riz suivi par du maïs. Le tableau 1 détaille les traitements comparés dans cet essai. Ils sont au nombre de 8.

Il convient particulièrement de noter la très faible dose d'engrais minéraux adopté dans le souci de placer cet essai dans le contexte socio-économique où vit les paysans.

Une durée de trois ans est prévue pour cette experimentation. Les résultats que nous présenterons ci-après correspondent à ceux de la deuxième année de la succession culturale (riz pluvial). Les paramètres étudiés sont encore limités aux composants de rendement de la culture (nombre de talles herbacées, nombre de panicules essentiellement), puis à la production. A l'issu de la troisième année qui est actuellement en cours et comporte du maïs, nous essayerons de compléter les paramètres étudiés par des données sur les principales caractéristiques physico-chimiques du sol en vue de dégager leur évolution par rapport à la première année d'installation.

Interprétation des résultats

Le tableau 1 présente le rendement en paddy sec correspondant à chaque traitement.

Tableau 1.   Les différents traitements comparés, et les rendements en paddy sec obtenus.

No.	NPK Unité ha^{-1}	Dolomie kg ha^{-1}†	Fumier t ha^{-1}	Rendement en paddy sec kg ha^{-1}
0	0-0-0	500	0	1830 ab
1	15-30-30	500	0	3040 bc
2	15-0-0	500	5‡	3040 bc
3	15-30-30	500	5	3230 c
4	15-0-30	500	20	3685 c
5	15-0-30	500	-	1690 a
6	0-0-0	0	5	3180 c
7	0-0-0	0	0	1850 ab

†   La dolomie apportée sur la culture précédente agit ici en arrière action.
‡   5 t de fumier contiennent 20 kg de N, 12 kg de P_2O_5, 22 kg de K_2O.

Il apparaît d'après ces résultats que :
-    sans apport d'engrais minéraux, notamment phosphatés, et en absence du fumier, le rendement en paddy plafonne autour de 1, 5 - 1,8 t ha^{-1}.
-    le fumier seul, à la dose de 5 t ha^{-1}, fait passer le rendement en paddy de 1,8 à 3,2 t ha^{-1}, soit un supplément de production de plus de 1 t ha^{-1}.
-    l'apport d'une faible dose de fertilisation minérale NPK en addition à 5 t ha^1 de fumier n'a que peu d'effet sur les rendements en paddy.
-    une dose faible d'engrais azotés et potassiques ajoutée à 5 t ha^{-1} de fumier procurent le même rendement en paddy qu'une faible dose de fertilisation minérale NPK.
-    en l'absence d'engrais minéraux phosphatés, 20 t ha^{-1} de fumier permettent de dégager près de 2 t ha^{-1} de supplément de production par rapport à NK seul.

Ces faits semblent confirmer que le phosphore reste le premier facteur limitant de la productivité de ces sols ferrallitiques, et que le fumier atteune l'effet de la déficience en phosphore de ces sols non seulement par son apport P_2O_5 mais aussi surtout par le rôle essentiel que joue la matière organique dans le processus de déblocage du phosphore du sol.

Conslusion

Ces premiers résultats nous amènent à conclure que pour le maintien de la fertilité de ces sols ferrallitiques et l'amélioration de leur capacité de production, l'apport de matière organique par le fumier de parc paraît indispensable. Une dose minimale de 5 t ha^{-1} est conseillée, mais une dose plus forte ne peut être que bénéfique.

Aspects de la fertilisation dans les sols acides des rizières des Hauts-Plateaux malgaches

H. NABHAN et H. RAKOTOMANANA*

Résumé

   Le Programme Engrais Malagasy/FAO, en collaboration avec le FOFIFA et l'IRRI, a entrepris un programme de recherche appliqué sur le riz inondé en vue d'améliorer les recommandations de fertilisation suivant les paramètres du sol et les zones agro-écologiques. En 1987, 82 essais ont été mis en place comprenant quatre niveaux d'azote et phosphore (0, 40, 80, 120 kg ha⁻¹) avec un contrôle (0-0-0) et une parcelle 80-80-80 + 20 S kg ha⁻¹. Les essais sont prévus pour quatre ans.

   Cinq types de sols ont peu être identifiés : sols hydromorphes minéraux à gley, sols hydromorphes à amphigley, sols hydromorphes minéraux sableux, sols hydromorphes organiques et sols tourbeux. Après deux saisons d'essais, il a été confirmé que la matière organique joue un rôle important dans l'efficacité de la fertilisation de la rizière et doit être considérée comme un facteur dans la distinction des types de sol pour l'affinage des recommandations des engrais.

   Les corrélations suivantes ont été observées :
-    *une corrélation négative entre la réponse à N et matière organique (MO), Fe⁺², Fe DTPA, saturation en Al, N total;*
-    *une corrélation positive entre la réponse à N et le pH-eau, S assimilable, argile et limon;*
-    *une corrélation négative entre la réponse à P et le pH du sol;*
-    *une corrélation positive entre la réponse à P et la MO, Al échangeable et S assimilable.*

*   Respectivement : Coordonnateur de Projet PEM/FAO, FAO Antananarivo, BP 3971, Madagascar; et Directeur National du Programme Engrais Malagasy, PEM Nanisana, PB 1028, Antananarivo, Madagascar.

Le rendement de paddy du témoin 0-0-0 dans quatre vallées ayant une concentration élevée de Fe++ (moyenne de 171 ppm) est nettement inférieur (75%) au rendement dans les vallées avec faible concentration de Fe++ (moyenne de 57 ppm). Les indices de productivité de N et P dans les quatre vallées à haute concentration en Fe++ sont relativement bas. L'apport de 120 kg de P_2O_5 ha^{-1} a un effet dépressif sur l'absorption de Fe, un aspect qui mérite une investigation plus profonde.

Le rendement moyen du paddy avec l'apport de dose optimale d'azote et phosphore (100 kg N, 84 kg P_2O_5 ha^{-1}) est supérieur à 5,5 t ha^{-1}, avec un rapport valeur-coût de 4,7 pour N et 6,5 pour P. Une stratification des recommandations des engrais a été établie en fonction des teneurs en matière organique et de la concentration de Fe++ dans les sols.

Abstract

Some aspects of the fertilization of acid soils in ricefields on the high plateaux of Madagascar

The Madagascar/FAO Fertilizer Programme, with collaboration from FOFIFA and IRRI, has undertaken a research programme on flooded rice with the aim of gaining a better understanding of fertilizer requirements in relation to soil features and agroecological zones. In 1987, 82 trials were set up, using form levels of nitrogen and of phosphorus (0,40,80, and 120 kg ha^{-1}), with a control (0-0-0) and one plot with 80-80-80 + 20 S kg ha^{-1}. The trials will last for four years.

Five types of soils have been identified: gleyic mineral hydromorphic soils, amphigleyic hydromorphic soils, sandy hydromorphic soils, organic hydromorphic soils, and peat soils.

After two seasons, it has been confirmed that organic matter plays an important role in a fertilizer's efficiency on ricefields, and should be considered as one of the factors in distinguishing different types of soil when arriving at a more precise definition of fertilizer requirements.

The following correlations have been established:
-   a negative correlation between the response to N and organic matter (OM), Fe++ Fe DTPA, Al saturations, and total N;
-   a positive correlation between the response to N and the water pH, assimilable S, clay, and silt;
-   a negative correlation between the response to P and the soil pH;
-   a positive correlation between the response to P and the OM, exchangeable Al, and assimilable S.

The paddy yield from the control plots (nil NPK) in the four valleys which had a high concentration of Fe++ (averaging 171 ppm) is considerably less (75%) than the yield in the valleys with a low concentration of Fe++ (averaging 57 ppm). The productivity indexes of N and P in the four valleys with a high concentration of Fe++ are relatively low. The application of 120 kg of P_2O_5 ha^{-1} reduces the absorption of Fe, which should be a matter for more detailed investigation.

The average paddy yield applying the optimum slope of nitrogen and phosphorus (100 kg N, and 84 kg P_2O_5 ha^{-1}) is more than 5.5 t ha, and has a cost-benefit ratio of 4.7 for N and 6.5 for P. A classification of fertilizer requirements has been made with respect to the organic-matter contents and the concentration of Fe^{++} in the soils.

Introduction

La population mondiale est estimée à 5,1 milliards en 1988, dont 500 millions sont mal nourris. L'estimation de la population en l'an 2000 est de 6,1 milliards, ce qui signifie qu'il y aura un milliard de plus à nourrir d'ici 10 ans. Par conséquent, la production vivrière devra augmenter d'environ 40%. Pour 93 pays en voie de développement, il est estimé (FAO, 1987), que 63% de l'augmentation nécessaire de la production agricole doivent provenir de l'augmentation de rendement de la superficie déjà cultivée.

Mais il nous faut produire plus d'aliments sans détruire la fertilité du sol parce que si cette fertilité diminue, il sera difficile de nourrir les générations actuelles et futures.

Toutes les mesures possibles pour le maintien de la fertilité des sols, y compris l'apport d'intrants agricoles appropriés, sont indispensables pour assurer l'alimentation de la population. En général, l'utilisation rationnelle des engrais peut augmenter la production agricole d'environ 50%.

En résumé, les sols acides (upland) couvrent environ 50% de la zone tropicale. Selon les deux systèmes de classification de sol (Soil Survey Staff, 1975; FAO, 1974) deux types de sols acides existent : oxisols "ferralsols" et ultisols "acrisols". En général, ces sols ont les caractéristiques suivantes : pauvre en fertilité, pauvre en saturation basique (Ca, Mg), faible capacité de rétention d'eau, déficit en phosphore, teneur élevée/toxicité en Al, Fe et autres oligo-éléments.

L'approche reconnue pour le développement et l'amélioration de la productivité de ces sols acides consiste dans la :
- maintien de la couverture végétale,
- maintien/augmentation de la teneur en matière organique, et
- maintien/augmentation de la fertilité par le système de jachère, combiné avec l'application de chaux et des engrais.

En examinant le cas de Madagascar, dans ce contexte, on peut constater les aspects suivants :
- population (en 1987) de 11 millions avec une superficie cultivée d'environ 2,3 millions ha, dont 1,2 million ha de culture de riz;
- une production totale de riz qui varie entre 2 et 2,3 millions t par an (moyenne 2,16 millions t de 1980 à 1988), avec un rendement moyen de 1,855 t par hectare (1986 à 1988);
- en général et avec approximation, il y a une exportation annuelle par les cultures d'environ 100 kg N - P_2O_5 - K_2O ha^{-1} de riz contre un apport de 30 kg N - P_2O_5 - K_2O ha^{-1} environ par le fumier organique (moyenne 2 t ha^{-1} de teneur 0,63% N, 0,33% P_2O_5, 0,54% K_2O) et 5 kg N - P_2O_5 - K_2O ha^{-1} par l'apport des engrais, soit un déficit annuel d'environ 65 'unités fertilisantes' par hectare.

Cela indique qu'une diminution de la productivité et une détérioration de la fertilité du sol des rizières, avec le système d'exploitation (riz--riz) est en cours. Des mesures adéquates de correction en matière de fertilisation doivent être envisagées.

Cependant, la potentialité d'une production/rendement élevé de riz sur les Hauts-Plateaux Malgaches existe avec le maintien de la fertilité, l'apport rationnel d'engrais et les pratiques culturales améliorées (figure 8).

Le Programme Engrais Malagasy a envisagé et entrepris un travail de recherche appliquée pour examiner et quantifier les aspects de réponse du riz aux engrais, en relation avec les paramètres du sol et les zones agro-écologiques.

Méthodes

En collaboration avec le FOFIFA et l'IRRI, le Programme Engrais Malagasy a mis en place, depuis 1987/88, un réseau d'essais de fertilisation sur riz irrigué/inondé sur les Hauts-Plateaux. Au total, 82 essais de 18 parcelles ont été installés sur les champs des agriculteurs dans sept vallées (cinq dans la région d'Antananarivo et deux dans la région de Fianarantsoa).

Le traitement "4 x 4 NP factoriel" comprend 4 niveaux d'azote et phosphore (0, 40, 80, 120 kg ha^{-1}) avec un cotrôle (0-0-0) et une parcelle avec 80-80-60 + 20 S kg ha^{-1}.

La localisation des vallées, l'emplacement des sites par vallée et le plan d'essai sont présentés dans les figures 1, 2, 3.

Ces essais sont prévus pour trois ans, avec l'application des engrais; en 4ème année, l'observation de rendement sera faite (sans application d'engrais) pour mesurer les arrières effets de P.

L'étude et l'analyse détaillée des sols des sites expérimentaux ont été faites en collaboration avec le CIRAD, Montpellier (France). L'analyse foliaire des plantes est réalisée par l'IRRI (Philippines). L'analyse statistique et les traitements des données de rendement/sol/plante sont faits par un consultant de la FAO. La réalisation de ce travail est faite par l'équipe nationale et internationale du Programme Engrais Malagasy (PEM).

Caractéristiques des sols de rizières de bas-fonds des Hauts-Plateaux malgaches

Sept vallées ont été choisies pour l'étude/classement pédologique des sols. Dans chaque vallée, les échantillons de surface (0 à 20 cm) de 12 sites ont été prélevés pour l'analyse détaillée des sols, soit au total 82 échantillons. En plus 2-3 profils pédologiques complets ont été prélevés et analysés pour chaque vallée.

Il est à noter que les sols tourbeux de la bordure Est et les sols formés sur des matériaux d'origine volcanique (Miarinarivo et Antsirabe) ne sont pas représentés. Cependant, l'échantillonage effectué représente une bonne majorité des rizières des Hauts-Plateaux.

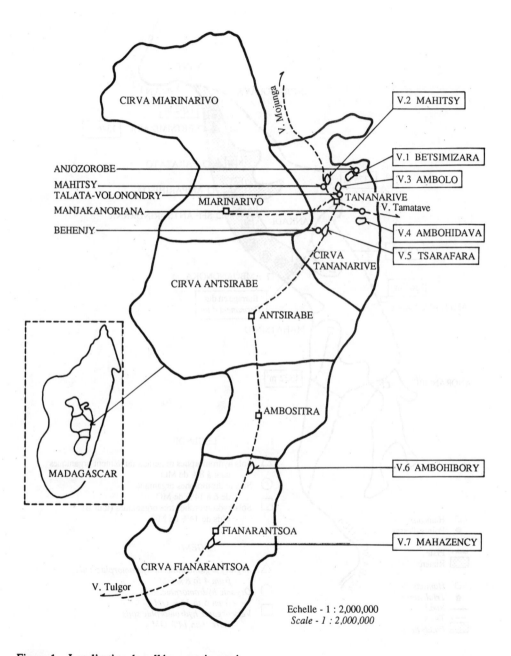

Figure 1. Localisation de vallées : essais sur riz.

Location of rice trials in valley areas.

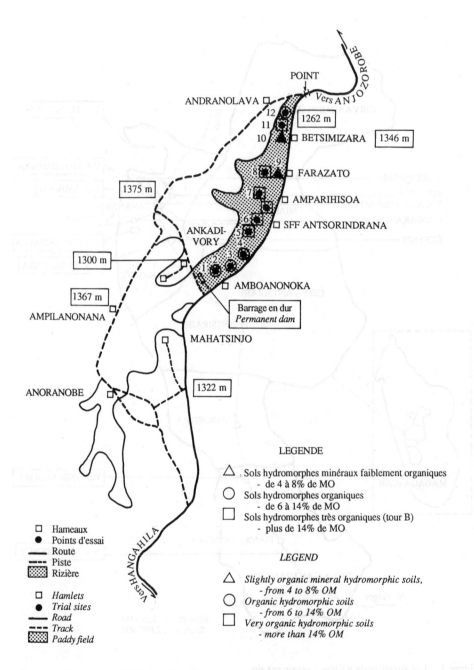

Figure 2.   Vallée Betsimizara (zone d'Anjozorobe).

Betsimizara Valley (Anjozorobe area).

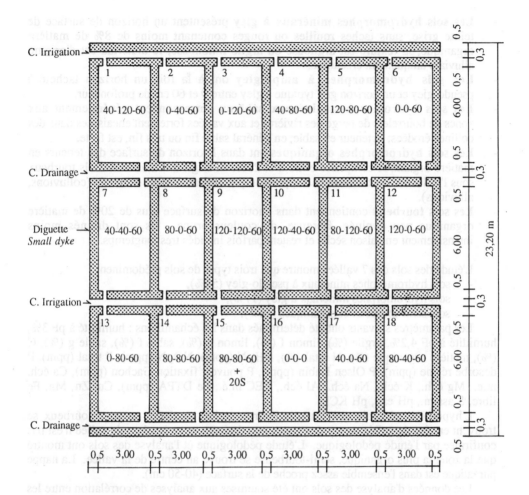

Figure 3.   Plan de l'essai.

Layout of the trial.

Les caractéristiques principales des sols de rizière sont les suivantes :
- **Les sols hydromorphes minéraux à pseudo-gley** présentent en surface des taches rouilles ou rouges (réoxydation du fer). Cependant, en saison des pluies, dès que les rizières sont submergées par l'eau de pluie provenant du bassin versant, tous ces sols hydromorphes ont un aspect comparable : l'horizon de surface devient gris foncé ou gris brun.

- **Les sols hydromorphes minéraux à gley** présentent un horizon de surface de teinte grise, sans taches rouilles ou rouges contenant moins de 8% de matière organique; la texture est argileuse ou argilo-limoneuse, le matériau est d'origine alluviale uniquement cristalline.
- **Les sols hydromorphes à amphigley** ont à la fois un horizon tacheté à pseudo-gley et un horizon gris typique à gley entre 0 et 60 cm de profondeur.
- **Les sols hydromorphes minéraux sableux** correspondent généralement aux zones de bourrelets de berge des rivières et aux vallées fortement encaissées dans des collines érodées; la teneur en sable, en général sable fin ou très fin, est forte.
- **Les sols hydromorphes organiques** ont dans l'horizon de surface des teneurs en matière organique comprises entre 8 et 20%. Il s'agit d'évolution de sols tourbeux sous l'influence du drainage et d'apports importants de matière minérale (colluvions, alluvions).
- **Les sols tourbeux** contiennent dans l'horizon de surface plus de 20% de matière organique peu évoluée, fibreuse, de couleur brune. Ils sont drainés moins intensivement en saison sèche et restent parfois inondés très longtemps.

L'étude des sols des 7 vallées montre que trois types de sols prédominent :
- sols hydromorphes minéraux à pseudo-gley (54%),
- sols hydromorphes minéraux à gley (17%),
- sols semi-tourbeux et organiques (16%).

Les paramètres suivants ont été déterminés dans les échantillons : humidité à pF 3%, humidité à pF 4,2%, argile (%), limon f (%), limon g (%), sable f (%), sable g (%), C (%), matière organique (%), N total (%), NH_4^-N (ppm), NO_3^-N (ppm), P total (ppm), P désorbé résine (ppm), P Olsen Dabin (ppm), P pouvoir fixation Gachon (ppm), Ca éch. m.e., Mg éch., K éch., Na éch., Al éch., CEC m.é., Fe DTPA (ppm), Cu, Zn, Mn, Fe libre, S assim., pH eau, pH KCl.

L'hypothèse générale que les sols hydromorphes tourbeux ou semi-tourbeux se trouvent en tête de vallée ou sur des zones non drainées et en bordure de vallée n'est pas confirmée par l'étude pédologique. L'étude pédologique et l'analyse des sols ont montré que la zone la plus organique, semi-tourbeuse se trouve au centre de la vallée. La nappe phréatique est dans l'ensemble assez proche de la surface (40-50 cm).

Les données d'analyse des sols ont été soumises aux analyses de corrélation entre les différents paramètres, ainsi qu'à l'analyse factorielle en composantes principales. Parmi les différents éléments analysés, cinq caractéristiques (sur lesquelles s'appuie la variabilité de 82 sites) ont été retenues pour le groupement de sols des rizières :
- argile + limon
- matière organique
- pH eau
- phosphore désorbé résine
- taux de saturation en Al

Les trois premières caractéristiques sont les données dominantes pouvant faire l'objet d'une estimation sur le terrain.

En résumé, l'ensemble de l'étude montre quatre groupes de sols. Les caractéristiques de ces quatre groupes de sols sont présentés en tableau 1.

Tableau 1.   Caractéristiques des quatre principaux groupes de sols.

Groupes Caractéristiques	(1) Sols hydromorphes minéraux		(2) Sols hydromorphes faiblement organiques		(3) Sols hydromorphes organiques	(4) Sols hydromorphes très organiques/ semi-tourbeux
Pourcentage de sites	43%		38%		10%	9%
Matière organique (%)	<4		4-8		8-14	>14
Argile + limon (%)	<50	>50	<50	>50	>50	>50

Valeurs analytiques moyennes

pF 4,2%	17,4	-	21,5	-	26,4	34,5
PF 3%	31,6	-	33,9	-	41,3	50,3
Argile + limon (%)	38,2	60,3	39,7	61,3	62,0	72,3
Matière organique (%)	2,7	3,2	6,5	5,1	10,5	21,0
NH_4-N (ppm)	12,2	-	16,2	-	14,1	35,9
P désorbé résine (ppm)	4,7	5,1	8,0	3,4	6,4	11,6
Pouvoir fixateur de P (ppm)	565	838	1070	855	1171	1992
pH eau	5,5	5,7	5,4	5,3	4,8	4,9
F DTPA (ppm)	170	-	306	-	334	330
Taux saturation Al échangeable (%)	12,9	14,3	25,9	14,0	44,7	49,6
S assimilable (ppm)	2,3	2,8	3,2	2,9	7,6	17,2
Zn DTPA (ppm	1,2	-	0,5	-	0,5	0,3

Les principales corrélations observées entre les paramètres de sols sont présentées en tableau 2.

Tableau 2.   Principales coorélations observées.

	Argile	Limon fin	MO†	N	P désorbé résine	Pouvoir fixateur de P
MO	0,331	0,513				
N	0,361	0,522	0,978			
P désorbé résine		0,469				
Pouvoir fixateur P		0,648	0,819	0,804	0,579	
CEC	0,562	0,493				
Al échangeable			0,665			0,659
pH eau			-0,552			-0,418
Fe DTPA			0,542			0,396

†   Matrère organique.

Corrélation entre le paramètre de sol et la réponse à N et P

Plusieurs approches/méthodes ont été utilisées pour examiner les relations entre les 36 différents paramètres de sols d'une part, et entre les paramètres de sols et la réponse à l'application d'azote et phosphore d'autre part. Quelques 30 logiciels ont été développés et utilisés pour l'étude (tableau 3).

Tableau 3. Logiciels développés/utilisés pour l'étude.

FACT	COMENU programme pour "orthogonal polynomial ANOVA"
FACT2AM	ANOVA independant pour 2 factoriels + 2 additionnels
FP1DF	FPDATA programme pour analyse économique sans EO
FP1TF	FPDATA programme pour courbes, économie et EO "Optimum Economique"
FPCORFIG	FPDATA programme pour corrélation
FPSISTRA	Stratification des données par site
GETSET1	Pour utiliser avec MADGRAPH
GFERTF	Version de GENFERT2 pour Madagascar
MADFIX1	Split composantes des données de rendement et composant individuel
MADFIX10	Split données des sols par vallée
MADFIX11	Pour corrélation avec MULGF
MADFIX12	
MADFIX13	
MADFIX14	
MADFIX15	
MADFIX16	
MADFIX17	
MADFIX18	Change de format des données de YCOMIN
MADFIX20	
MADFIX23	
MADFIX3	
MADFIX4	Pour RVC
MADFIX5	Pour bénéfice, RVC et risque
MADFIX6	
MADFIX7	
MADFIX8	
MADFIX9	Formats données des sols
MADG1	Version de MADGRAPH pour composantes de rendement
MADGRAPH	Graphiques
MADIN	Pour entrer données de rendement (N par vallée)
MADINS	Pour entrer données de rendement (S par vallée)
MADSOLIN	Pour entrer les paramètres des sols
MULGF	Régression multiple
TRANSFOM	Transformation des données (arcsine)
MADFIX19	Pour 8 EOs
YCOMIN	Pour entrer les composantes de rendement

Les coorélations obtenues par le logiciel MULGF sont présentées en tableau 4.

Tableau 4.   Corrélations entre les paramètres de sols et la préponse à N et P_2O_5 (1988/89).

Paramètres	Réponse à N	Réponse à P	Matière organique
MO	-***	+***	
Pouvoir fixateur de P			+*
Fe^{+2}	-***	n.s.	n.s.
Fe DTPA	-***	n.s.	n.s.
Al échangeable	n.s.	+***	
Al sat.	-***	+**	+*
N total	-***	n.s.	+***
NH$_4$-N	n.s.	+*	+*
NO$_3$-N	n.s.	+*	n.s.
pH eau	+**	-***	
S assim	+***	+***	
Argile	+***	n.s.	n.s.
Limon G	+***	n.s.	+***

* = 90%; ** = 95%; *** = 99%; n.s. = non significatives.

En examinant la réponse à N et P après 2 saisons d'essais, il a été confirmé que la matière organique (MO) joue un rôle important en ce qui concerne l'aspect de la fertilisation de la rizière. Elle doit être considérée comme un facteur important dans la distinction des types de sol pour l'affinage des recommandations des engrais.

Une forte corrélation négative a été trouvée entre la MO et la réponse à N, et une forte corrélation positive entre la MO et la réponse à P.

Les autres corrélations significatives qui ont été observées sont les suivantes :
-    une corrélation positive entre la MO et le limon.
-    une forte corrélation négative entre l'Al et le Fe^{+2}.
-    une forte corrélation négative entre la réponse à N et le degré de saturation en Al.
-    une corrélation positive entre la réponse à N et le pH eau (échantillon).
-    une corrélation négative entre la réponse à P et le pH eau (échantillon).
-    une forte corrélation positive entre la réponse à P et le S assimilable.
-    une faible corrélation positive entre l'argile, le limon et la réponse à N.

Effets de fer sur le rendement du riz et la réponse à N et P

Les sols bien aérés ont des caractéristiques de "Redox potential (RP)" d'environ + 400 à + 700 mv, mais les sols submergés ont un RP de -250 à -300 mv seulement (Patrice et Reddy, 1977). La réduction d'oxyhydroxide ferrique et la relation de RP (Eh) avec le pH et Fe peut être représentée comme suit :

$$Fe\,(OH)_3 + 3H^+ + e^- \rightleftharpoons Fe^{+2} + 3\,H_2O$$

$$Eh = 1,057 + 0,059\,pFe^{+2} - 0,177\,pH \qquad \text{(équation Nernst)}$$

Réduction	Redox potential, mv
$NO_3^- - N_2$ ⎤	+280 à +220
$Mn^{+4} - Mn^{+2}$ ⎦	
$Fe^{+3} - Fe^{+2}$	+180 à +150
$SO_4^{-2} - S^{-2}$	-120 à -180

La réduction de Fe^{+3} à Fe^{+2} prend place dans les conditions submergées (métabolisme anaérobique par les bactéries) de rizières et peut produire des effets toxiques sur le riz ('bronzing') quand la concentration du Fe dans le plantes de riz est supérieure à 300 ppm Fe (Tanaka et Yoshida, 1970).

L'absorption de Fe est également influencée par d'autres cations. Les compétitions entre l'absorption de Fe et Mn, Zn, Cu, K, Ca et Mg; la diminution d'absorption de Fe en présence de pH élevé et une haute concentration de phosphore dans le sol ont été observées (Lingle et al., 1963). La matière organique, dans les conditions submergées peut accélérer la réduction de Fe^{+3}, et par conséquent la concentration de Fe^{+2} dans le sol peut augmenter de traces à 128 ppm pendant une période de 2 mois (Tisdale et Nelson, 1986).

Le problème de haute concentration du Fe^{+2} dans les sols de rizière à Madagascar est connu, mais il y a peu de travail en ce qui concerne la quantification des effets de Fe sur le rendement et la réponse du riz à l'application de phosphore et d'azote.

Les résultats des essais PEM/FAO sur cet aspect sont présentés dans les tableaux 5, 6, 7, 8 et figures 4, 5.

Tableau 5.   Résultats de pH et Fe^{++} (mesures sur le terrain 1988/89).

Vallée/Zone	pH		Fe^{++} (ppm)		Moyenne
	Valeurs extrêmes	Moyenne	Valeurs extrêmes	Moyenne	Fe DTPA† (ppm)
1. Tsarafara/Behenjy	4,12-6,34	5,40	5-100	20	94
2. Ambohibory/Fianar (est)	5,30-7,10	6,49	50-100	75	270
3. Mahazengy/Fianar (ouest)	6,40-7,50	7,00	25-100	77	242
4. Mahitsy/Mahitsy	5,15-6,78	6,40	25-250	153	191
5. Betsimizara/Anjozorobe	5,28-6,27	6,04	100-250	167	301
6. Ambolo/Talata	6,11-6,61	6,44	50-250	170	311
7. Ambohidava/Manjakandriana	6,14-6,37	6,29	100-250	195	364

†   Mesures sur les échantillons des sols.

Tableau 6.   Relation entre le Fe^{++}, rendement du témoin, et la réponse aux engrais (IP).

Zone	Fe^+ (ppm)	Rendement 0-0-0	M.O. (%)	pH	IP§		
					NPK	N	P
Behenjy	20	2834	2,91	5,40	12,66	27,98	24,20
Fianarantsoa (est)	75	3783	6,98†	6,49	13,84	23,35	32,98
Fianarantsoa (ouest)	77	3283	3,01	7,00	15,25	30,60	30,85
(Moyenne) A	(57)	(3300)			(13,92)	(27,31)	(29,85)
Mahitsy	153	3398	3,80	6,40	10,46	22,11	14,45
Anjozorobe	167	2365	14,74	6,00	11,05	8,21‡	32,48
Talata	170	2274	5,63	6,44	9,67	16,83	14,38
Manjakandriana	195	1921	7,58	6,29	11,61	9,71	32,55
(Moyenne) B	(171)	(2409)			(10,69)	(14,22)	(23,65)
B/A	300%	75%			77%	52%	79%

†   Y compris 2 sites avec 15 et 27% de M.O.
‡   6 sites avec > 15% M.O.
§   IP = Indice de productivité.

Tableau 7.   La teneur en Fe† dans la paille de riz en relation avec le Fe^{++} dans le sol.

Zone	Fe^{++} sol (ppm)	Fe plante (ppm)
Behenjy	20	175
Fianarantsoa (est)	75	‡
Fianarantsoa (ouest)	77	‡
Mahitsy	153	240
Anjozorobe	167	319
Talata	170	320
Manjakandriana	195	553

†   Moyenne par vallée, parcelle témoin.
‡   Possibilité de contamination lors de l'analyse de fer.

Tableau 8.   Effets de N et P sur l'absorption de Fe (vallée d'Ambohidava/Manjakadriana).

Traitement			Teneur de Fe dans la plante (ppm)	
N	P₂O₅	K₂O		
0	- 0 -	0	553 ⎤	
0	- 120 -	60	360 ⎟	effets de NPK (-28%)
120	- 120 -	60	397 ⎦	
80	- 0 -	60	465 ⎤	effets de P (-22%)
80	- 120 -	60	365 ⎦	

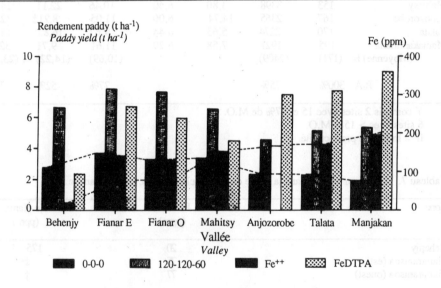

Figure 4. Relation entre le rendement du paddy et la concentration en fer du sol (Fe⁺⁺, Fe DTPA).

Relationship between paddy yield and soil iron concentration (Fe⁺⁺, Fe DTPA).

- On peut constater qu'il y a plus de Fe DTPA dans les groupes des sols hydromorphes organiques et très organiques que dans les sols minéraux et faiblement organiques (tableau 1).

   Dans les sept vallées étudiées on trouve des relations positives entre le Fe DTPA (mesuré dans les échantillons de sols) et le Fe^{+2} (mesuré dans les parcelles des rizières). En prenant la moyenne par vallée, la concentration de Fe^{+2} varie entre 20 ppm Fe^{+2} dans la zone de Behenjy et 195 ppm dans la zone de Manjakandriana.

   Le rendement de paddy du témoin (0-0-0) dans les quatres vallées ayant une concentration élevée de Fe^{+2} (moyenne 171 ppm) est nettement inférieur (75%) au rendement dans les vallées avec faible concentration de Fe^{+2} (moyenne 57 ppm).

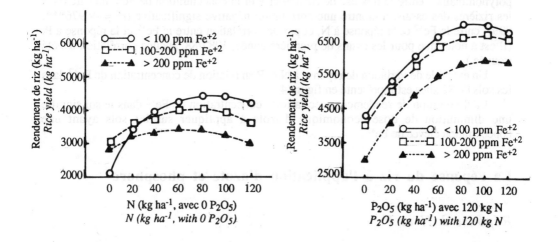

Figure 5.   Relation entre le niveau de Fe^{+2} dans le sol et la réponse à l'application d'azote et phosphore.

Relationship between the level of Fe^{+2} in the soil and the response to nitrogen and phosphorus application.

- Parallèlement, les 'indices de productivité' pour NPK, N et P sont aussi bas dans les quatre vallées, avec haute concentration de Fe^{+2}. Cela signifie que la haute concentration de Fe^{+2} dans les rizières diminue le rendement et l'efficacité des engrais (N et P).

    On peut supposer, par conséquent, que certaines pratiques doivent être envisagées quand la limite de Fe^{+2} dans les rizières dépasse 100 ppm afin d'assurer un bon rendement.

- On constate aussi une relation positive entre le Fe^{+2} dans les sols de rizières et la teneur en Fe dans les plantes (avec exception des deux vallées de Fianarantsoa, tableau 7).

- Dans la zone de Manjakandriana (par exemple) avec une concentration très élevée de Fe^{+2} dans le sol, on peut observer que l'apport d'engrais, notamment le phosphore, a diminué la concentration de Fe dans la plante (tableau 8). La moyenne de teneur en Fe dans la plante, par vallée, a diminué de 553 ppm, sans engrais, à 465 ppm avec l'apport de N et K, et a été de 365 ppm avec l'apport de NPK. L'apport de 120 kg P_2O_5 ha^{-1} a un effet dépressif sur l'absorption de Fe (environ 22% de réduction de teneur). Cet aspect mérite une investigation plus profonde.

En considérant la situation des 82 sites séparément, une étude détaillée de corrélation (régression) avec le logiciel "MULGF", estimée par les "coéfficients orthogonaux polynominaux", entre la réponse de riz à N et P et la concentration de Fe^{+2} mesuré dans les rizières des essais, a montré une corrélation négative significative (r^2 = -9,976***, 99%) entre le Fe^{+2} et la réponse à N, et pas de corrélation entre le Fe^{+2} et la réponse à P. (Il est à noter que pour les essais de première année, 1987/88, aucune corrélation n'a été trouvée.)

Un exemple de variation de réponse à N et P en relation de concentration de Fe^{+2} dans les sols de 82 sites est représenté en figure 4.

La diminution de rendement avec la forte concentration de Fe^{+2} dans le sol signifie une diminution de dose économique d'azote à appliquer sur les sols ayant une concentration élevée de Fe^{+2}

La réponse du riz à l'application d'azote et phosphore

Réponse à N et P

Le rendement de riz par site (et par traitement), ainsi que les composantes de rendement, ont été analysés par ordinateur (IBM) en utilisant plusieurs techniques et logiciels (principalement FP DATA) afin d'établir les équations de réponse et de déterminer les recommandations économiques des engrais.

Des exemples de courbes de réponse à N et P par vallée et pour l'ensemble des sept vallées sont présentés en figures 6 et 7.

Un exemple d'équation de réponse (quadratique) à N et P, à deux niveaux, pour la vallée d'Ambolo est comme suit :

Courbes de N
Avec 40 P   Y = 3721,2778 + 1082,0531 * X   -130,3672 * X * X
Avec 80 P   Y = 3812,8882 + 1015,4430 * X   -140,6480 * X * X

Courbes de P
Avec 40 N   Y = 4523,7529 + 72,2965 * X   -1,2637 * X * X
Avec 80 N   Y = 4674,2046 + 786,6211 * X   -184,1211 * X * X

Le besoin en engrais "economic optimum" par vallée est présenté dans le tableau 9. Ces doses/recommandations sont établies à partir de courbes de réponse en tenant compte du prix du paddy et des unités fertilisantes en 1988/89 (riz = FMG 200 kg^{-1}, N = FMG 761 kg^{-1}, P_2O_5 = FMG 783 kg^{-1}).

On peut constater la variation de besoin en N et P_2O_5 entre les vallées ainsi qu'entre les vallées du nord et les vallées du sud (Ambohibory et Mahazengy).

Il est à noter que la moyenne générale des besoins en engrais pour les sept vallées en 1988/89, est de 100 kg N ha^{-1} et 84 kg P_2O_5 ha^{-1}, comparée avec 120 kg N ha^{-1} et 90 kg P_2O_5 ha^{-1} en 1987/88 (avec le prix du riz = FMG 135 kg^{-1}, N = FMG 652 kg^{-1}, P_2O_5 = FMG 720 kg^{-1}).

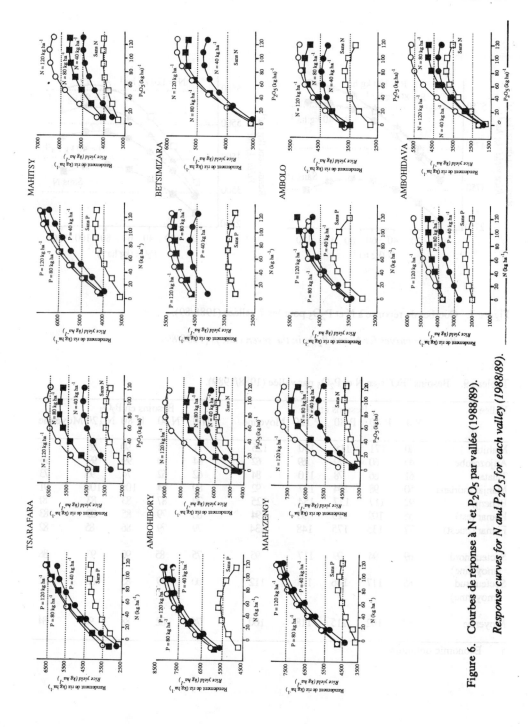

Figure 6. Courbes de réponse à N et P₂O₅ par vallée (1988/89).

Response curves for N and P₂O₅ for each valley (1988/89).

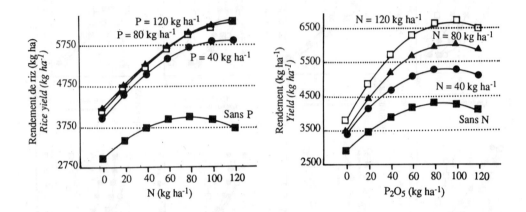

Figure 7.   Courbes de réponse à N et P$_2$O$_5$ pour les 7 vallées (1988/89).

Response curves for N and P$_2$O$_5$ in the seven valleys (1988/89).

Tableau 9.   Besoins "EO"† en N et P$_2$O$_5$ par vallée (1988/1989).

Vallée	Besoins en N					Besoins en P$_2$O$_5$				
	P 0	P 40	P 80	P 120	Moyenne	N 0	N 40	N 80	N 120	Moyenne
Mahitsy	80	171	118	130	125	71	94	90	86	85
Anjozorobe	40	60	79	69	62	84	90	99	99	93
Talata	61	86	78	110	84	62	62	74	93	73
Manjakandriana	50	98	94	127	92	80	84	100	100	91
Behenjy	92	117	137	154	125	70	87	83	83	81
Fianar (est)	55	102	123	95	94	80	79	85	89	83
Fianar (ouest)	77	135	175	148	134	79	79	86	85	82
Secteur nord (moyenne)	69	94	99	117	95	75	85	90	93	86
Secteur sud (moyenne)	65	117	146	121	112	80	79	85	86	82
Moyenne	67	102	115	118	100	76	82	88	90	84

†   Economic optimum.

En résumé, les données sur le rendement, le 'rapport valeur coût' (RVC), l'indice de productivité (IP) et les risques sont présentés dans le tableau 10.

Tableau 10. Le rendement et les données économiques avec l'apport optimal de N et P.

		Moyenne pour les 7 vallées (1988/89)								
	Dose optimale (kg ha^{-1})	Rendement (kg ha^{-1})	Augmentation du rendement (%)	RVC	IP	Bénéfice net (FMG ha^{-1})	\multicolumn{4}{c|}{Risque (%) (RVC)}			
							<1	1-2	2-4	>4
N	100	5632	32	4,7	18	288.032	1	5	34	60
P$_2$O$_5$	84	5568	38	6,5	25	360.703	-	6	23	71

Effets de variation des prix de paddy et des engrais

En utilisant les courbes de réponse à N et P, les effets de variations de prix de paddy et des engrais sur la dose optimale à recommander sont présentés dans le tableau 11. On peut constater que la dose à recommander n'est pas trop sensible à l'augmentation du prix des engrais. Néanmoins, un équilibre doit être maintenu entre le prix des engrais et le prix des produits agricoles.

Tableau 11. Effets de variation de prix sur la dose optimale de N et de P.

	\multicolumn{2}{c}{Prix FMG kg^{-1}}		\multicolumn{2}{c}{Dose optimale (kg ha^{-1})}	
Riz	N	P$_2$O$_5$	N	P$_2$O$_5$
200	761	783	100	84
200	989	1018 (+30%)	97	82
200	1218	1253 (+60%)	93	80
200	1522	1566 (+100%)	87	76
320 (+60%)	761	783	106	88

Réponse à K et S

La comparaison entre les deux traitements 0-0-0 et 0-0-60, montre une réponse localisée à l'application de potasse. Sur 78 sites des essais, 47 sites (soit 60%) ont montré une réponse positive à K, 29 sites ont montré une réponse négative. L'augmentation de rendement moyen de riz par vallée par l'application de 60 kg K$_2$O ha^{-1} varie entre 96 et 815 kg ha^{-1}, avec une moyenne générale d'environ 266 kg ha^{-1}.

La comparaison entre les deux traitements, 80-80-60 et 80-80-60+20 kg ha^{-1}, montre une légère réponse positive (en moyenne 102 kg de paddy ha^{-1}) sur 42 sites et une réponse négative (-50 kg de paddy ha^{-1}) sur 36 sites des essais.

Arrières effets de P

Dans les cinq vallées du secteur nord, la parcelle n°15 qui a reçu le traitement 80-80-60+20S la 1ère année, a été divisée en 2 parcelles (split) avec des traitements 80-0-60 et 80-120-60 pour la 2ème année, pour avoir une indication des arrières effets de P. Les résultats sont présentés dans le tableau 12.

Tableau 12.   Les arrières effets de P sur le rendement de riz.

Traitement/Vallées		Betsimizara	Mahitsy	Ambolo	Ambohidava	Tsarafara	Moyenne
		------------------------- Rendement moyen (kg ha^{-1}) -------------------------					
87/88	80-80-60	4157	4432	5538	4432	4915	4695
	80-80-60+20	4127	4461	5464	4461	4986	4700
88/89	80-0-60† A	2758	4324	3799	1804	4009	3339
	80-0-60‡ "split" B	5288	5213	4906	3994	6137	5108
Augmentation en %, B/A		92%	21%	29%	121%	53%	53%
Augmentat°+, diminut°- de rendement (1989/1988)		+26%	+23%	-13%	-16%	+17%	+8%

† Traitement en 2ème année et en 1ère année.
‡ Traitement en 2ème année et 80-80-60 en 1ère année.

Les résultats de 1ère année montrent qu'il n'y avait pratiquement pas de réponse au soufre (rendement moyen de 4700 contre 4695 kg paddy ha^{-1} pour les traitements 80-80-60+20S contre 80-80-60). Cela signifie que l'augmentation de rendement entre le traitement (B) et (A) peut être attribuée aux arrières effets de 80 unités de P_2O_5 appliquées en 1ère année (1987/88), ainsi qu'à la variation de rendement d'une année à l'autre. Mais on constate une augmentation moyenne de rendement, entre le traitement (B) et (A), nettement supérieure à l'augmentation de rendement de l'année 1989 par rapport au rendement en 1988. Avec une approximation, l'augmentation moyenne de rendement due à l'arrière effet de P peut être considérée de l'ordre de 45%, soit 1500 kg paddy ha^{-1}. (Il est à noter que 80 unités P_2O_5 appliquées en 2ème année ont augmenté le rendement en 1989 d'environ 2080 kg ha^{-1}, soit 57%; moyenne de sept vallées, traitement 80-80-60 contre 80-0-60.)

Stratification de résultats en fonction des types de sol

En fonction de la matière organique

La distribution des sites des essais, dans les sept vallées, en fonction de la teneur en matière organique est présentée dans le tableau 13.

Tableau 13.   Distribution des sites des essais en fonction de la matière organique.

Vallée (zone)	Moyenne MO (%)	MO <5%	MO >5%
Tsarafara (Behenjy)	2,91	12†	0
Mahazengy (Fianar, ouest)	3,00	12	0
Mahitsy (Mahitsy)	3,80	7	3
Ambolo (Talata)	5,63	2	10
Ambohibory (Fianar, est)	6,98	6	6
Ambohidaya (Manjakandr)	7,58	0	12
Betsimizara (Anjozorobe)	14,75	0	12

†   Sites.

En tenant compte des corrélations/régressions entre la matière organique et la réponse à N et P, les recommandations des engrais (rendement et prix de 1988/89) peuvent être considérées comme indiquées dans le tableau 14.

Tableau 14.   Recommandation des engrais pour le riz irrigué selon la teneur en matière organique.

Conditions	Sol	MO (%)	N	P_2O_5	K_2O
				- - - - - - - - - - - - - kg ha^{-1} - - - - - - - - - - - - -	
Favorables	Minéral	< 5	125	80	†
	Organique	> 5	80	85	
Moins favorables	Minéral	< 5	80	55	‡
	Organique	> 5	52	57	

†   Estimation à 40 kg K_2O ha^{-1}
‡   Estimation à 30 kg K_2O ha^{-1}

Dans les conditions favorables, la recommandation pour les *sols minéraux* (<5% MO) peut être de 125 kg N et 80 kg P_2O_5 ha^{-1}, et pour les *sols organiques* de 80 kg N et 85 kg P_2O_5 ha^{-1}.

Ces essais ont été réalisés dans les conditions favorables. Un ajustement de recommandation pour des conditions moins favorables chez les agriculteurs doit être faite (variétés, pratiques culturales, date de repiquage, situation d'eau, utilisation de produits phytosanitaires). Par conséquent, la recommandation pour les conditions moins favorables peut être de l'ordre de 80, 55 kg N, P_2O_5 ha^{-1} (sol minéral) et 52, 57 kg N, P_2O_5 ha^{-1} (sol organique).

En fonction de la concentration de Fe^{+2} du sol

Compte tenu des corrélations/régressions observées entre le Fe^{+2}, mesuré dans les parcelles d'essais, et la réponse à N et P, la stratification des recommandations des engrais est présentée dans le tableau 15 (rendement et prix de 1988/89).

Tableau 15.   Stratification de recommandation en fonction de la concentration de Fe^{+2} dans le sol.

Fe^{+2} (ppm)	Azote (kg ha^{-1})					Phosphore (kg ha^{-1})				
	P 0	P 40	P 80	P 120	Moyenne	N 0	N 40	N 80	N 120	Moyenne
< 100	81	120	127	144	118	74	79	85	87	82
101-200	61	93	102	113	92	75	82	85	90	83
> 200	49	92	120	93	88	79	86	96	93	89

Ces résultats montrent qu'une diminution de la dose de N de l'ordre de 25% et une augmentation de la dose de P_2O_5 d'environ 8% peuvent être envisagées sur les sols ayant une concentration élevée en Fe^{+2} (>200 ppm) par rapport aux sols faibles en Fe^{+2} (<100 ppm).

Par exemple, on peut recommander 118 kg N ha^{-1} sur le sol faible en Fe^{+2} (<100 ppm) et 88 kg N ha^{-1} pour le sol ayant une concentration de Fe^{+2} supérieure à 200 ppm.

Effets des amendements des sols acides

Si l'utilisation d'amendements (par exemple la dolomie) pour la correction du pH bas des sols et l'amélioration de l'efficacité des engrais est plus connue sur les sols ferrallitiques de tanety "upland", l'amendement peut être utilisé sur les rizières.

Les résultats de quelques essais/démonstrations sur le riz irrigué réalisés par le PEM en 1981/1982 sont présentés en tableau 16.

Il est difficile de tirer des conclusions de ce travail à cause de l'absence de données sur les caractéristiques des sols où les essais ont été installés. Néanmoins, on peut constater la variation d'effets de la dolomie selon les zones. En général, on peut obtenir une augmentation d'environ 430 kg de paddy par hectare, soit 14%, par l'application de 500 kg dolomie ha^{-1}.

Tableau 16.   Effets de la dolomie et des engrais sur le rendement de riz irrigué.

Zone	Nbre de sites	Rendement de riz (kg ha⁻¹)			% augmentation†	
		$N-P_2O_5-K_2O$ 0-0-0	$N-P_2O_5-K_2O$ 30-30--30	$N-P_2O_5-K_2O$+Dolomie 30-30-30+500	kg ha⁻¹	(%)
Fianar	14	2215	2734	3218	484	18
Ambositra	2	944	1325	2804	1479	112
Somalac PC 15	3	3483	3750	4317	567	15
Somalc PC 23	16	2647	3241	3295	54	2
Somalac (nord)	18	3578	4286	4844	558	13
Antananarivo	21	1976	2919	3351	432	15
Antsirabe	25	2370	3452	3674	222	6
Moramanga	16	1825	2243	2400	157	7
Ambatondrazaka	20	2388	3300	3423	123	4
Amparafaravola	10	3414	3745	3965	220	6
(Total)	(145)					
Moyenne		2484	3100	3529	429	14

†   Due à l'application de la dolomie (500 kg ha⁻¹).

L'identification des zones avec un pH bas, les méthodes, les doses et l'époque d'application de la dolomie méritent des études ultérieures plus approfondies.

Thèmes de recherche future

La figure 8 indique les moyens d'assumer un meilleur rendement de riz sur les sols hydromorphes à Madagascar. Compte tenu des résultats préliminaires sur les aspects de fertilisation des rizières (figure 8) obtenus par le PEM, et afin de maintenir/améliorer la fertilité et la productivité de ces sols, les thèmes de recherche suivants méritent l'attention :

- Quantification des contraintes du faible rendement du riz, par des essais et des enquêtes, dans les différentes zones de Madagascar : aspects de toxicité de Fe, pratiques culturales, l'interaction eau/fertilisation, variété, etc.
- Techniques appropriées pour la gestion des sols concernant : maintenance de teneur adéquate de matière organique dans les sols pauvres en matière organique et diminution de la teneur (en MO) dans les sols tourbeux, capacité de rétention d'eau, méthodes de labour, drainage/assèchement des sols de rizières en contre-saison, introduction des légumineuses dans la rotation.
- Les effets des amendements (chaux, dolomie) sur la productivité des sols acides : interaction dolomie/engrais/matière organique, méthodes, dose, époque et intervalle d'application.

- Affinage des recommandations de fertilisation selon les types de sols et les zones agro-écologiques, méthodes appropriées pour la diffusion des recommandations aux agriculteurs.

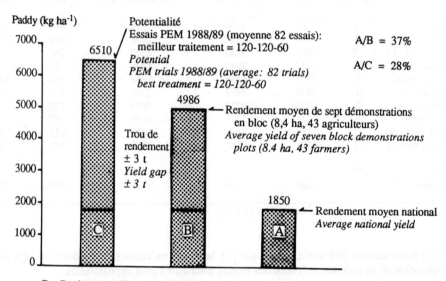

Figure 8.   Potentialité de la production/rendement de riz à Madagascar.

Production/yield potential for rice in Madagascar.

Bibliographie

COPE, F. 1989. Rice response to fertilizers in Madagascar : 1987/88 and 1988/89 trails. AGLF/ FAO consultancy reports, GCPF/MAG/060/NOR. Rome: FAO.

DELANNOY, M. 1987. *Sols de rizières.* AGLF/FAO rapport de mission. Rome : FAO.

FAO (Food and Agriculture Organization of the United Nations). 1974. *FAO/Unesco soil map of the world, 1:5 000 000.* Vol 1, Legend. Paris: Unesco.

FAO 1987. *Agriculture: toward 2000.* FAO conference document. Rome: FAO.

FAO 1986. *Efficient fertilizer use in acid upland soils of the humid tropics*. AGLF Bulletin no. 10. Rome: FAO.

LINGLE, J.C., TIFFIN, L.O. and BROWN, J.C. 1963. Iron-uptake transport of soybeans as influenced by other cations. *Plant physiology* 38: 71-76.

PATRICE, W. and REDDY, C. 1977. Chemical changes in rice soil. In: *Soil and rice* (Symoposium, IRRI, 1977).

ROCHE, P. 1988. *Sols de rizières des Hautes Plateaux de Madagascar*, CIRAD et AGLF. Rome: FAO.

SOIL SURVEY STAFF. 1975. *Soil taxonomy: A basic system of soil classification for making and interpreting soil surveys*. Soil Conservation service, U.S. Department of Agriculture. Agriculture Handbook no. 436. Washington, DC: Government Printing Office.

TANAKA, A. and YOSHIDA, S. 1970. *Nutritional disorders of rice plants in Asia*. International Rice Research Institute. Technical Bulletin no. 10. Los Baños, Philippines: IRRI.

TISDALE, S. and NELSON, W. 1966. *Soil fertility and fertilizers*, 327-328. London: Macmillan and Co.

Sélections des espèces et des variétés pour l'adaptation aux sols acides

RAKOTONDRAMANANA, R. A. RANDRIANTSALAMA, B.E. RAKOTOARISOA, A. RAVELOSON et E. TORSKENAES[*]

Résumé

Les sols acides représentent la majeure partie du Vakinankaratra et de l'Ankaratra. Mis à part les sols d'origine volcanique récente, la plupart des sols ont été formés sur socle cristallin ou sur alluvions anciennes lessivées, avec des pH moyens très bas et une carence en phosphore très marquée. Le seuil de toxicité à l'aluminium échangeable est atteint pour plusieurs plantes. La plupart des espèces ou variétés ne supportent pas les conditions de sols acides et exigent des amendements préalables en dolomie, phosphore et fumier, souvent à très fortes doses.

Des essais de longue durée menés à FIFAMANOR, mettant en rotation le maïs, la pomme de terre et le blé ou triticale, ont montré que l'arrière-effet de 2 t ha^{-1} de dolomie et celui de 200 kg ha^{-1} de P$_2$O$_5$ disparaissent en sixième campagne. L'analyse du rapport valeur des produits/coût des intrants montre que la rentabilité de la dolomie est la plus élevée sur volcanisme ancien que sur les autres substrats, et que celle du phosphore est très faible à cause de son prix élevé.

Les deux principaux critères de sélection pour le blé et le triticale sur les Hautes Terres malgaches sont l'adaptation aux sols acides et la résistance à la rouille noire (Puccinia graminis f. sp. tritici). Après un criblage des diférentes lignées pour ces deux caractères, plus de la moitié ont été éliminée en une saison. Plusieurs lignées de blé et de triticale associant ces deux caractères ont été identifiées. La rentabilité du chaulage calculé sur le blé seul est marginal; il en est de même pour le fumier; par conséquent, il faut toujours considérer les amendements du blé dans un système de cultures.

---

[*] Respectivement : Chef du Département Recherche, Section Recherche (blé), Section Recherche (pommes de terre), Section Recherche (plantes fourragères), et Expatrié NORAD (recherche plantes fourragères), FIFAMANOR, BP 198, Antsirabe 110, Madagascar.

Le surpâturage et les feux de brousse successifs ont été considérés comme étant des facteurs importants de dégradations des sols; ces deux problèmes persistent à cause du faible développement du pâturage artificiel. L'acidité des sols affecte beaucoup les rendements des espèces fourragères; le Pennisetum purpureum cv. Kizozi est le plus tolérant et les légumineuses fourragères et les espèces tempérées sont les plus sensibles. Avec une espèce assez sensible comme le Chloris gayana, la rentabilité des amendements et de la fertilisation minérale peut être très élevée avec un pH de l'ordre de 5,0.

Des différences importantes existent entre les clones et les variétés de pomme de terre sur leurs réponse aux sols acides. A pH très bas (de l'ordre de 4,2) le nombre de plantes lévées peut être réduit de moitié et la végétation des plantes est très faible. Les clones sélectionnés à FIFAMANOR à partir de graines botaniques sont plus performants que les clones introduits.

Abstract

The selection of species and varieties for adaptation to acid soils

Most of Vakinankaratra and Ankaratra areas in Madagascar are covered by acid soils. Apart from soils of recent volcanic origin, most of the soils are formed on a crystalline substratum or on ancient leached alluvion deposits, and have a very low average pH, a very pronounced phosphorus deficiency, and for a number of crops the level of exchangeable aluminium toxicity is barely acceptable. Most species or varieties of plant cannot tolerate acid soil conditions, and require prior cultivation, sometimes, intensive amendment of dolomite, phosphorus and manure.

Long-term trials at FIFAMANOR have been established with a maize-potato-wheat/ triticale rotation, and have shown that the after effects of 2 t ha^{-1} of dolomite and 200 kg ha^{-1} of P_2O_5 disappear in the sixth season. The value/cost ratio analysis shows that the highest profits from dolomite are achieved on old volcanic soil, and that phosphorus gives very poor returns because it is so expensive.

The two main criteria used in the selection of wheat and triticale in the Malagasy highlands are their adaptability to acid soils and their resistance to stem rust. After testing the lines for these two characteristics, more than a half are eliminated in a single season.

Several varieties of wheat and triticale with both of these characteristics have been identified. The profitability of liming wheat is marginal based on one season, as it is with manure. Consequently the question of using amendments with wheat always needs to be taken into account in the frame of a cropping system.

Overstocking and successive bush fires are considered to be the main factors contributing to soil degradation. These two problems persist because of the poor development of artificial pasture. Soil acidity affects the yield of many fodder crops considerably. *Pennisetum purpureum* cv. Kizozi is the most acid-tolerant species, and legumes and temperate species are the most acid-sensitive. With a rather susceptible species like *Chloris gayana*, and a pH of about 5.0, the profitability of the amendments and the mineral fertilizers can be very high.

There are important differences between the clones and the varieties of potato in their reaction to acid soils. With a very low pH (about 4.2), the number of plants emerging may be reduced by half, and the growth of the plants is very poor. The clones selected by FIFAMANOR from botanical seeds have performed better than introduced clones.

Introduction

Les sols des Hautes Terres malagaches sont pour la plupart formés sur socle cristallin, à l'excéption des sols d'origine volcanique. L'origine même des sols (gneiss et granite), la pluviométrie élevée et souvent torrentielle, la disparition progressive de la couverture végétale à cause des feux de brousse successifs et du surpâturage, l'insuffisance des amendements organiques et calcaires dans les exploitations, les fortes pentes des collines sont parmi les facteurs importants qui ont favorisé l'érosion des sols et le lessivage des éléments minéraux. Les lessivages successifs des éléments, en particulier du calcium et du magnésium, ont entrainé une acidification du milieu. Dans les sols minéraux des régions tropicales humides avec du pH inférieur à 5,0, l'aluminium échangeable peut être élevé, et en dessous du pH 5,5, le manganèse devient soluble et peut avoir un effet toxique sur les plantes (Webster *et al.*, 1980). Les différentes espèces végétales et variétés réagissent différemment à la teneur en aluminium échangeable du sol. Kamprath (1970) a trouvé par exemple, que la croissance du coton diminue avec 10% de saturation en aluminium échangeable de la capacité d'échange, celle du soja avec 20% et celle du maïs avec 45%.

Au cours des années, des efforts ont été entrepris à FIFAMANOR pour sélectionner les différentes espèces et variétés contre l'acidité du sol et pour trouver des moyens pour redresser les pH des sols fortement lessivés.

Le présent article a pour objet de faire une synthèse de ces travaux pour pouvoir formuler des recommandations. L'essentiel des travaux a été fait dans le Vakinankaratra avec les principales cultures du projet, à savoir le blé, le triticale, la pomme de terre et les fourrages.

Le milieu physique du Vakinankaratra

La région du Vakinankaratra comprend cinq 'fivondronana' (équivalent du district), à savoir Antsirabe I, Antsirabe II, Detafo, Antanifotsy et Faratsiho. le climat est du type tropical d'altitude, avec six mois de saison chaude et humide (octobre à avril) et six mois de saison fraîche et sèche (mai à octobre). L'altitude varie de 1400 m à 3200 m. La pluviosité moyenne varie de 1300 mm à 2000 mm sur les sommets (Raunet, 1981). La plus grosse partie des pluies tombe du 15 octobre au 15 avril (tableau 1).

C'est en décembre que l'intensité pluviométrique est la plus forte (tableau 2), ce qui provoque de fortes érosions et par conséquent l'ensablement des rivières et des bas-fonds.

La végétation naturelle est une savane herbeuse à *Aristida*, *Heteropogon*, *Hyparrhenia*, *Helichrysum*, *Ctenium* et *Trachypogon*. Les quelques lambeaux de forêts

Tableau 1. Pluviosité moyenne de quelques stations dans la région de Vakinankaratra (en mm par an).

	Altitude	J	F	M	A	M	J	J	A	S	O	N	D	Par an
Betafo	1402	316	247	248	49	30	12	14	13	22	72	154	290	1467
Antsirabe	1506	293	241	218	77	30	12	17	15	23	77	158	268	1429
Ambatolampy	1555	331	291	279	82	42	16	22	22	23	60	178	279	1625
Antanifotsy	1560	297	157	182	38	14	16	26	27	36	40	125	303	1261
Ambohiman-droso	1600	300	151	180	40	12	15	27	27	37	39	120	312	1260
Ambohibary	1658	288	266	226	92	38	24	28	17	31	74	174	287	1545
Faratsiho	1750	399	336	313	109	32	13	19	14	28	101	195	352	1911
Soanindrariny	1800	300	160	182	40	18	20	31	30	40	45	128	310	1304
Manjakatompo Stat° forest.	1806	362	322	386	106	55	28	30	40	40	96	211	329	2005
Antsapandrano Stat° forest.	1844	298	254	272	92	34	22	25	26	32	72	196	366	1689
Nanokely	2100	351	290	272	94	28	13	16	13	30	92	182	293	1674

Source : Raunet (1981).

Tableau 2. Maxima pluviométrique en 24 h sur 22 ans (1936-1958).

	Altitude en m.	J	F	M	A	M	J	J	A	S	O	N	D	Par an
Betafo	1402	104	135	111	35	60	10	23	18	33	55	80	68	135
Antsirabe	1506	117	100	106	97	45	27	32	55	50	70	72	103	117
Ambatolampy	1555	145	166	103	87	60	17	30	40	79	50	61	87	166
Ambohibary	1658											101		101
Faratsiho	1750	92	178	95	38	45	39	43	22	60	57	59	85	178
Manjakatompo	1806	130	114	122	57	66	28	38	33	100	58	86	119	130
Nanokely	2000	102	106	105	65	44	21	20	18	40	54	71	70	106

Source : Raunet (1981).

qui restent sont à base d'eucalyptus, d'*Acacia decurrens* (mimosa) ou quelques reboisements à base de *Pinus* spp. Cette végétation et ces quelques lambeaux de forêts sont en partie brûlés presque tous les ans en fin de saison sèche, surtout en septembre, octobre et novembre.

En début de pluie, les jeunes pousses subissent une forte densité de troupeaux de zébus, ralentissant ainsi la couverture végétable. Durant la saison sèche on peut avoir le gel à plusieurs endroits; les mois les plus gélifs sont juillet, août et septembre.

On distingue différents types de sols dans le Vakinankaratra (Raunet, 1981) :
- les sols ferrallitiques rouges formés sur socle cristallin qui occupent la majeure partie de la région, en particulier à l'est de la route nationale (RN) 7;

- les sols ferrallitiques "rouges" ou "bruns" formés sur roches volcaniques ancienes (basaltes ou trachytes) qu'on rencontre dans le massif de l'Ankaratra;
- les sols développés sur roches volcaniques récentes;
- les sols "chocolat" sur roches basiques qui résultent des émissions de scories : Antsirabe, Betafo, Tritriva et Vananinkarena;
- les andosols peu différenciés sur les projections volcaniques très récentes (holocène) qui occupent le Vakinankaratra occidental (région de Betafo et Tritriva) et qui réprésentent les sols les plus riches de la région;
- les alluvions volcano-lacustres anciennes des bassins d'Antsirabe, de Betampona, et Antanifotsy-Ambohimandroso (bordure de l'Onive).

Les sols les plus cultivés sont les sols sur roches volcaniques récentes (sols chocolat et andosols), et les moins cultivés sont les sols développés sur socle cristallin et sur une bonne partie des sédiments volcano-lacustres anciens.

La figure 1 montre une carte des pH des sols dans les principales régions du Vakinankaratra (Rakotondramanana et Randriantsalama, 1987). Il faut noter que ces pH sont mesurés sur des sols déja cultivés destinés à recevoir les essais multilocaux de blé de saison pluviale en 1984, 1985 et 1986. On note ainsi que seuls les sols d'origine volcanique des environs de Betafo, Tritriva et la partie ouest d'Antsirabe ont un pH supérieur à 5,5 (sols "chocolat" et les andosols peu différenciés), seuil en dessous duquel le chaulage est généralement recommandé.

Des échantillons de sols ferrallitiques prélevés sur dépôt volcano-lacustre d'Andranomanelatra et de sols volcaniques récents de Betafo et Tritriva analysés à l'Université d'Oslo donnent les principales conclusions suivantes (FIFAMANOR, 1982) :
- les sols ferrallitiques ont un pH faible (4,8), un taux de saturation en aluminium élevé (23,2%), une teneur en calcium et magnésium faible (1,3 cmol kg^{-1} et 0,4 cmol kg^{-1}), une faible teneur en phosphore, une teneur en manganèse élevée (75 ppm), et une faible concentration en matière organique;
- les sols sur roches volcaniques récentes ont un pH faiblement acide (6,2), pas d'aluminium échangeable; les concentration en phosphore et en calcium sont élevés mais les teneurs en bore sont très faibles (0,087 ppm).

Arrière-action du phosphore et de la dolomie

Compte tenu des carences marquées en phosphore et en calcium sur la plupart des sols, plusieurs essais de longue durée ont été mis en place par FIFAMANOR pour étudier l'arrière-action du phosphore (appliqué sous forme de scorie Thomas) et celle de la dolomie sur différents types de sols. Il s'agit d'essais factoriels avec trois niveaux de dolomie (0, 1 t ha^{-1} et 2 t ha^{-1}) et trois niveaux de phosphore (0, 100 kg ha^{-1} et 200 kg ha^{-1}) et deux répétitions par site. Le phosphore et la dolomie ont été appliqués seulement à la première culture, et par la suite on n'a appliqué que de l'azote et de la potasse. Le maïs, la pomme de terre et le blé ou triticale se sont succédés sur la parcelle pendant quatre ans, soit au total six cycles de culture, car la pomme de terre et le blé sont cultivés en une saison pluviale.

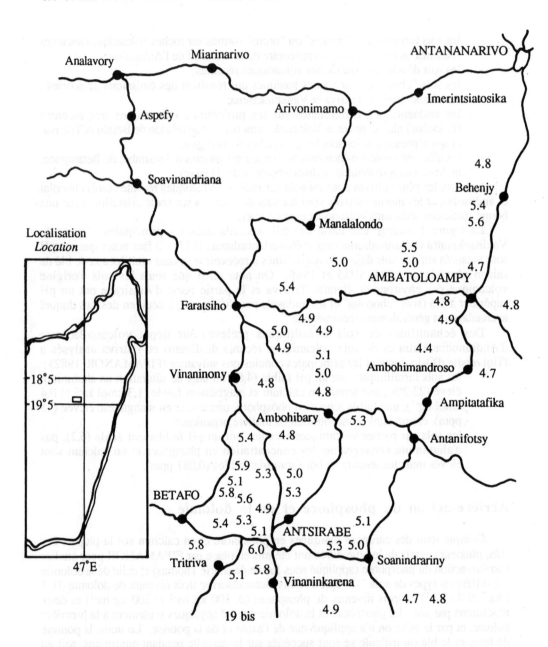

Figure 1.   pH des sols dans la région de Vakinankaratra (Rakotondramanana *et al.*, 1987

pH of soils in the Vakinankaratra district (Rakotondramanana et al., *1987).*

On note ainsi que l'effet du phosphore et celui de la dolomie sont les plus marqués sur sol ferrallitique formé sur socle (tableau 3) et sur volcanisme ancien (tableau 4) que

Tableau 3. Arrière action du phosphore (scorie Thomas) et de la dolomie sur sol ferrallittique formé sur socle précambrien (moyennes de 4 sites).†

Périodes d'essais	Cultures	Doses de phosphore (kg ha⁻¹)	Doses de dolomie (t ha⁻¹)			Récolte moyenne (t ha⁻¹)
			0	1	2	
Octobre		0	0,7	1,4	2,5	1,5
1979		100	2,3	2,6	3,5	2,8
à	Maïs	200	3,0	3,4	4,3	3,5
juin		Moyenne	2,0	2,4	3,4	
1980						
Octobre		0	7,6	12,3	12,3	10,7
1980	Pomme	100	11,6	15,8	16,4	14,6
à	de	200	15,0	17,9	19,0	17,3
février	terre	Moyenne	11,4	15,3	15,9	
1981						
Février		0	0,8	1,0	1,0	0,9
1981		100	1,0	1,1	1,2	1,1
à	Blé	200	1,2	1,4	1,6	1,4
juin		Moyenne	1,0	1,1	1,2	
1981						
Septembre		0	0,4	0,6	1,0	0,6
1981		100	0,8	1,0	1,2	1,1
à	Maïs	200	1,1	1,3	2,0	1,4
mai		Moyenne	0,7	0,9	1,4	
1982						
Octobre		0	8,1	8,7	10,5	9,1
1982	Pomme	100	9,1	11,8	11,7	10,8
à	de	200	10,8	10,4	13,8	11,6
février	terre	Moyenne	9,3	10,3	12,0	
1983						
Février		0	0,9	0,6	0,9	0,8
1983		100	0,9	0,9	0,9	0,9
à	Triticale	200	1,0	1,0	1,0	1,0
juin		Moyenne	0,9	0,8	0,9	
1983						

† Localités : Antanifotsy, Ambohimiarivo, Soanindrariny et Sahanivotry Est.

Tableau 4. Arrière action du phosphore (scorie Thomas) et de la dolomie sur sol formé sur roches volcaniques anciennes (moyennes de 3 sites).†

Périodes d'essais	Cultures	Doses de phosphore (kg ha^{-1})	Doses de dolomie (t ha^{-1})			Récolte moyenne (t ha^{-1})
			0	1	2	
Octobre		0	2,3	3,2	3,2	2,9
1979		100	2,5	4,3	5,7	4,1
à	Maïs	200	3,6	6,8	6,8	5,7
juin		Moyenne	2,8	4,7	5,2	
1980						
Octobre		0	10,1	11,2	13,1	11,4
1980	Pomme	100	12,1	12,9	13,9	12,9
à	de	200	13,3	13,6	14,7	13,8
février	terre	Moyenne	11,8	12,5	13,9	
1981						
Février		0	0,4	0,4	0,5	0,4
1981		100	0,5	1,0	1,2	0,9
à	Blé	200	0,5	1,1	1,4	1,0
juin		Moyenne	0,5	0,8	1,0	
1981						
Septembre		0	2,1	3,0	3,4	2,8
1981		100	2,0	3,5	3,5	3,0
à	Maïs	200	2,6	3,6	4,1	3,4
mai		Moyenne	2,2	3,3	3,6	
1982						
Octobre		0	7,2	12,2	13,0	10,8
1982	Pomme	100	10,2	12,5	14,3	12,3
à	de	200	10,7	13,0	16,5	13,4
février	terre	Moyenne	9,3	12,5	14,6	
1983						
Février		0	0,9	1,1	1,2	1,0
1983		100	1,1	1,4	1,5	1,3
à	Triticale	200	1,1	1,5	1,6	1,4
juin		Moyenne	1,0	1,3	1,4	
1983						

†   Localités : Faratsiho, Ampitatafika et Tsaramody/Ambohibary.

sur sol sur roches volcaniques recentes. Sur sol formé sur roches volcaniques récentes, la dolomie et le phosphore ont des effets significatifs en première et en deuxième année (tableau 5). Dans tous les cas le phosphore et la dolomie ont des effets

Tableau 5. Arrière action du phosphore (scorie Thomas) et de la dolomie sur sol formé sur roches volcaniques récentes (moyennes de 4 sites).†

Périodes d'essais	Cultures	Doses de phosphore (kg ha⁻¹)	Doses de dolomie (t ha⁻¹)			Récolte moyenne (t ha⁻¹)
			0	1	2	
Octobre		0	0,9	2,9	3,2	2,3
1979		100	2,5	2,9	3,5	2,9
à	Maïs	200	2,6	4,0	4,1	3,5
juin		Moyenne	2,0	3,2	3,6	
1980						
Octobre		0	11,8	11,4	12,00	11,7
1980	Pomme	100	12,4	15,6	18,6	13,8
à	de	200	14,2	16,7	14,4	15,1
février	terre	Moyenne	12,8	14,5	13,3	
1981						
Février		0	1,2	1,2	1,7	1,3
1981		100	1,7	1,6	1,9	1,7
à	Blé	200	1,5	1,6	1,7	1,6
juin		Moyenne	1,4	1,4	1,7	
1981						
Septembre		0	3,0	3,3	3,8	3,3
1981		100	3,5	3,5	3,8	3,6
à	Maïs	200	3,8	3,9	4,2	3,9
mai		Moyenne	3,4	3,5	3,9	
1982						
Octobre		0	19,6	20,5	20,9	20,3
1982	Pomme	100	19,9	22,5	21,7	21,3
à	de	200	20,7	23,6	22,7	22,3
février	terre	Moyenne	20,0	22,2	21,7	
1983						
Février		0	1,9	1,6	2,0	1,8
1983		100	1,9	2,1	2,0	2,0
à	Triticale	200	2,1	2,1	2,1	2,1
juin		Moyenne	1,9	1,9	2,0	
1983						

†   Localités : Andranomafana-Betafo, Ampamelomana-Betafo et Tritriva.

significatifs dès la première année. Il n'y a pas eu d'interactions significatives entre le phosphore et la dolomie dans ces essais. Eriksen *et al.* (1984) ont trouvé une interaction significative entre la méthode d'application du phosphore et de la chaux (CaCO₃) avec des essais en phytotron conduits sur des sols ferrallitiques de la région du Vakinankaratra : lorsque la chaux est appliquée, il n'y a pas de différence significative entre la localisation du phosphore et l'incorporation au sol, mais en l'absence de chaux, le phosphore est mieux absorbé par le blé quand il est localisé autour des plantes (tableau 6).

Tableau 6.   Interaction variété-chaulage-P pour du blé cultivé sur sol ferrallitique.[†]

Chaux -------------->	Romany		PF 71131	
	$0 \, t \, ha^{-1}$	$4 \, t \, ha^{-1}$	$0 \, t \, ha^{-1}$	$4 \, t \, ha^{-1}$
P 0 $kg \, ha^{-1}$	0,2	1,8	0,4	3,8
P 48 $kg \, ha^{-1}$ mélangé au sol	0,02	3,4	2,5	5,8
P 48 $kg \, ha^{-1}$ localisé	1,6	3,3	5,7	5,9

[†]   Rendement en grain en $g \, pot^{-1}$ du blé.
Source: Eriksen et Njos (1984).

L'analyse du rapport valeur des produits/coût de production (tableau 7) avec les prix 1989 de la dolomie et des produits agricoles et un taux d'interêt de 20% par an montre que $1 \, t \, ha^{-1}$ et même $2 \, t \, ha^{-1}$ de dolomie sont rentables en fertilisation de fond sur sol forme sur roches volcaniques anciennes et sur sol ferrallitique formé sur socle cristallin. Sur sol forme sur roches volcaniques récentes la fertilisation de fond en dolomie n'est rentable qu'à la dose de $1 \, t \, ha^{-1}$. La rentabilité de la fertilisation de fond avec du phosphore ne semble être possible que sur sol ferrallitique à $100 \, kg \, ha^{-1}$; ceci est dû au coût très élevé du phosphore à l'importation.

Tableau 7.   Rapport valeur/coût avec les essais de longue durée de dolomie et de phosphore.

Type de sols	Dolomie		P_2O_5	
	$1 \, t \, ha^{-1}$	$2 \, t \, ha^{-1}$	$100 \, kg \, ha^{-1}$	$200 \, kg \, ha^{-1}$
Sols volcaniques récentes	3,2	1,9	1,5	1,1
Sols volcaniques anciens	6,0	4,4	1,8	1,6
Sol ferrallitique (socle)	3,2	3,1	2,1	1,8

Base de calcul (prix 1989) :
- Dolomie   = ... FMG   110 kg^{-1}
- P_2O_5 = ... FMG   2.618 kg^{-1} *
- Maïs   = ... FMG   200 kg^{-1}
- Blé   = ... FMG   350 kg^{-1}
- Triticale   = ... FMG   315 kg^{-1}
- Pomme de terre   = ... FMG   100 kg^{-1}
- *   (Phosphate bicalcique 38% P_2O_5)   = ... FMG   995 kg^{-1}
- Interêt du capital : 20% par an, 4 ans

En raison du coût de la fumure de redressement en phosphore et de la dolomie, il est nécessaire de sélectioner en même temps des espèces et des variétés résistantes à l'acidité du sol, et de mieux valoriser les ressources locales tel que le fumier.

Sélections des variétés de blé et de triticale résistantes à l'acidité du sol

Le début de la sélection des variétés de blé résistantes l'acidité du sol est conduit grâce à la sélection en navette ("shuttle breeding") entre le CIMMYT et les stations brésiliennes (Kohli et Rajaram, 1988). Les scientifiques brésiliens ont identifié des variétés comme Preludio, Carazinbo (1956 et 1957) et un peu plus tard (1960) la variété IAS 20 comme étant tolérantes à la toxicité aluminique (Kohli and Rajaram, 1988). En 1980 la variété Alondra a été vulgarisée au Brésil et a été reconnue comme étant performante même sur sol acide. En fait, il a été reconnu que cette variété était capable d'extraire le phosphore dans le sol même à de faible quantité. Alondra a été ainsi introduite dans le pedigree de plusieurs cultivars, dont le Thornbird (BR14) qui a été reconnue comme étant tolérante à la toxicité aluminique. Le CIMMYT a utilisé des parents résistants à la toxicité aluminique dans les hybridations successives et a distribué des lignées en ségrégation à des pays riches en sols acides comme Madagascar, la Zambie, le Rwanda, le Cameroun et l'Ecuador.

A FIFAMANOR, toutes les introductions sont systématiquement testées par rapport à l'acidité du sol (pH 4,0 à 4,5) et par rapport à la rouille noire. Une méthode de sélection combinée à ces deux critères de sélection a été décrite (Rakotondramanana et Randriantsalama, 1987). Les sources d'introductions ont été variées pour permettre l'obtention de variétés avec une base génétique assez large. En une saison, plus de la moitié des lignées sont rejetées pour non adaptation au sol acide et/ou à la sensibilité à la rouille noire (tableau 8). Les variétés retenues des criblages de départ rentrent dans les essais préliminaires en station où l'inoculation artificielle à la rouille noire continue. Les lignées retenues sont mises en essais avec quatre répétitions sur plusieurs sites et sur différents types de sols; ces essais à répétitions qui durent deux ans, suivis par des essais multilocaux que confirment ou non la tolérance des variétés aux sols acides.

A FIFAMANOR il a été démontré que le triticale est plus performant que le blé sur sols acides (Rakotondramanana, 1984, 1986) et qu'il existe une corrélation significative entre le pH du sol et le rendement, (Rakotondramanana et Randriantsalama, 1987). Il a été démontré également que les lignées provenant des régions à sols acides comme la Zambie et le Kenya sont plus performantes sur les Hautes-Terres malgaches que les lignées provenant d'autres origines (Rakotondramanana et Randriantsalama, 1987; Rakotondramanana *et al.*, 1989). La résistance à la rouille noire (*Puccinia graminis* f. sp. *tritici*) est un autre critère important de sélection, et par conséquent, pour être retenue une variété doit combiner ces deux caractères principaux en dehors de qualités de rendement et des autres caractères telles que les qualités boulangères.

Le tableau 9 montre l'effet de 500 kg ha^{-1} de dolomie sur des variétés de blé et de triticale. On note que les variétés CNT 7 et PF 70354.1, qui sont des variétés brésiliennes, ne repondent pas à la dolomie. Le rapport valeur des produits/coût de production montre que l'apport de dolomie n'est pas très rentable lorsqu'elle est calculée sur le blé uniquement, mais l'on sait qu'elle a un arrière-effet sur la culture suivante.

Tableau 8. Nombre de lignées, origine, pourcentage retenu durant les deux saisons 1988.

"Nursery"	Origine	Lignées SP 88	Lignées retenus SA 1988	% retenus SA 1988	Lignées retenus pour 1989
EAWPC 1st	KEN	85	26	30,5	15
EAWPC 2nd	KEN	76	30	39,4	20
FAWPC 3rd	KFN	113	50	42,3	20
OPC 7th	KFN	110	42	38,1	20
SIN 12th	KEN	244	80	32,7	30
ZAM-Wheat	ZAM	95	35	36,8	18
SNACWYT 11th	KEN	101	26	25,7	5
HCWSN 3rd	THA	-	59	-	10
HCWSN 4th	THA	-	104	-	32
ZAM-TCL	ZAM	95	39	41,0	20
F6 Modified	MEX	134	15	11,0	8
ITSN 11	MEX	132	31	23,4	10
F3 AS/WARM	MEX	105	49	48,5	40
MASA 79	MEX	127	-	-	40

SP 88 = Saison pluviale 88   KEN = Kenya   ZAM = Zambie
SA 88 = Sous arrosage 88   THA = Thailande   MEX = Mexico

Tableau 9. Effet de la dolomie sur différentes variétés de blé et de triticale (moyenne de 24 essais multilocaux, saison pluviale 1985, FIFAMANOR).

Variétés		Doses de dolomie		
		0 kg ha⁻¹	500 kg ha⁻¹	Moyenne
Puppy/beagle	Triticale	2147	2572	2369 a
IRA/DRIRA	Triticale	1702	2024	1863 b
PAT 7219//KAL/BB	Blé	1320	1481	1400 c
CNT 7	Blé	1404	1418	1411 c
PF 70.354.1	Blé	1701	1761	1731 b
Moyenne		1654 b	1855 a	
Rapport valeur/coût†			1,3	

†   Dolomie = FMG 110 per kilogramme.
    a -   Effets variétés   :   significatif à $P = 0,01$
    b -   Effets dolomie   :   significatif à $P = 0,05$
    c -   Interaction dolomie-variétés   :   significatif à $P = 0,01$

Tableau 9. (suite)

Interaction variétés-dolomie				kg ha^{-1}
Puppy/beagle	x	500	kg ha^{-1} de dolomie	2592 a
Puppy/beagle	x	0	kg ha^{-1} de dolomie	2147 b
IRA/DRIRA	x	500	kg ha^{-1} de dolomie	2024 b
PF 70.354.1	x	500	kg ha^{-1} de dolomie	1761 c
IRA/DRIRA	x	0	kg ha^{-1} de dolomie	1702 c
PF 70.354.1	x	0	kg ha^{-1} de dolomie	1701 c
P A T	x	500	kg ha^{-1} de dolomie	1481 d
CNT 7	x	500	kg ha^{-1} de dolomie	1413 d
CNT 7	x	0	kg ha^{-1} de dolomie	1404 d
P A T	x	0	kg ha^{-1} de dolomie	1320 d

L'apport de fumier a un effet significatif sur le rendement (tableau 10), mais calculé par rapport au prix généralement pratiqué (FMG 10 000 t^{-1}), il n'est pas rentable, mais il a l'avantage d'être disponible dans la plupart des exploitations agricoles. Il a été démontré avec des essais multilocaux de blé pluvial en contre-saison qu'en doublant la dose de fumier, on peut réduire de moitié celle de la dolomie (tableau 11). Toutefois, il ne sera pas possible de supprimer la dolomie sous peine de voir diminuer le pH et d'aboutir à des carences marquées en calcium, magnésium et phosphore.

Tableau 10. Effets du fumier sur variétés de blé et triticale : résultats en kg ha^{-1} (moyenne de 10 essais multilocaux, saison pluviale 1987, FIFAMANOR).

Variétés	Doses de fumier		
	0 t ha^{-1}	5 t ha^{-1}	Moyenne
1. Jules 87	1687	1680	1683 b
2. Daniel 88	1585	1872	1728 b
3. Fifa 74	1490	1788	1639 b
4. Egil 87	1635	1789	1712 b
5. TCL BULK 50MA	1274	1333	1303 c
6. Puppy/beagle	2253	2440	2346 a
Moyenne	1654 a	1817 b	
Rapport valeur/coût†		1,1	

† Fumier = FMG 10 000 per tonne
- Effet du fumier : significatif à P = 0,13
- Effet variétés : significatif à P = 0,001
- Interaction variétés-dose du fumier : non significatif

Tableau 11.   Combinaisons dolomie-fumier sur blé de saison pluviale et de contre-saison.

Moyennes de 20 essais de saison pluviale 1988				Moyennes de 25 essais de contre-saison 1988			
Variétés	F1	F2	Moyenne	Variétés	F1	F2	Moyenne
1. Daniel 87	1503	1648	1575 a	1. Daniel 87	2584	2599	2592 a
2. Daniel 88	1472	1497	1484 b	2. Honoré 87	2455	2342	2398 bc
3. Corinne 87	1431	1457	1444 b	3. Daniel 88	2327	2321	2324 cd
4. IKBAL 87	1501	1330	1416 b	4. IKBAL 87	2340	2408	2374 bc
5. Andry 87	1581	1634	1607 a	5. Corinne 87	2397	2554	2476 ab
6. Tsara 87	1517	1381	1449 b	6. Veery 88	2553	2213	2235 d
Moyennes	1501 a	1491 a		Moyennes	2443 a	2407 a	

F1 = 20 t ha^{-1} fumier + 0,5 t ha^{-1} dolomie, F2 = 10 t ha^{-1} fumier + 1,0 t ha^{-1} dolomie
Source : FIFAMANOR (1988).

Les variétés de blé et de triticale mises en circulation actuellement figurent au tableau 12 avec leurs principales caracteristiques. La variété Veery 88, bien que sensible au sol acide a été multipliée pour les sols volcaniques où elle donne de très bonnes performances.

Tableau 12.   Les principales variétés de blé et de triticale diffusées, avec leurs principales caracté-ristiques.

Rubriques	Rouille noire	Rendement	Cycle	Verse	Sols acides
Variétés de saison pluviale					
Blé					
- Daniel 87	MS	Très B	Moyen	S	Très bon
- Daniel 88	MR	Bon	Moyen	R	Très bon
- IKBAL 87	MR	Bon	Moyen	R	Bon
- Egil 87	MR	Bon	Moy. tardif	R	Bon
- Andry 87	R	Bon	Précoce	Moy.	Bon
Triticale					
- Puppy/beagle resel.	R	Bon	Moyen	R	Très bon
- Merino bulk 87	R	Bon	Moyen	R	Très bon
- RAM Bulk 87	R	Bon	Tardif	R	Très bon
Variétés de contre-saison					
Blé					
- Daniel 87	MS	Très bon	Moyen	S	Très bon
- Bozy 87	MR	Bon	Tardif	R	Bon
- Honoré 87	R	Moy. à TB	Moyen	R	Mauvais
- Andry 87	R	Bon	Précoce	Moy.	Bon
Triticale					
- Puppy/beagle resel.	R	Bon	Moyen	R	Très bon
- Merino bulk 87	R	Bon	Moyen	R	Très bon

Sélections des espèces fourrageres à l'acidité des sols

La production de fourrages est une activité importante du fait de l'accroissement régulier du cheptel laitier; les femelles laitières (races améliorées et métisses) sont passées de 13 100 en 1985 à 15 700 en 1988, et la production collectée par les différents transformateurs de lait est passée de 2,8 millions à 4,8 millions de litres de lait par an dans la seule région de Vakinankaratra durant la même période. La superficie de cultures fourragères mise en place n'est passée que de 1140 ha à 1875 ha durant la même période dans la même région (FIFAMANOR, 1988), c'est à dire que la majeure partie des animaux vivent sur les pâturages naturels qui sont de mauvaise qualité. On pourrait alors supposer que l'augmentation de la production serait plus importante si la mise en place de pâturage artificielle était plus importante. Les principales raisons du faible développement des cultures fourragères sont : la priorité donnée par les paysans aux cultures vivrières par rapport aux cultures fourragères, le coût initial élevé des installations de pâturages à cause de la pauvreté des sols, et l'insuffisance de terrains disponibles. Pour résoudre en partie ces problèmes, il y a lieu de sélectioner les espèces les plus tolérantes aux sols acides.

L'acidité des sols affecte beaucoup les rendements des espèces fourragères; ceci est reflété par l'effet de la dolomie et du phosphore. On note qu'à Soanindrariny, aucune plante n'a poussé sans application de phosphore; le phosphore a été retrogradé à cause du pH trop bas. La dolomie donne un effet significatif jusqu'à 3 t ha^{-1}. A Vinaninony, par contre, le phosphore n'a pas eu d'effet significatif, mais la dolomie a eu un effet significatif jusqu'à 3 t ha^{-1}. Calculés sur cinq coupes, 3 t ha^{-1} de dolomie s'avèrent rentables avec les prix 1989.

Des pâturages associant du *Pennisetum clandestinum* à du tréfle plantés sur des sols volcaniques anciens de Vinaninony (pH 4,9) et sur les sols ferrallitiques de Soanindrariny (pH 4,4) (tableau 13).

Tableau 13. Effet de la dolomie et du phosphore sur espèces fourragères : rendements en matière sèche du *Pennisetum clandestinum* (Kikuyu) plus trèfle en kg ha^{-1}.

	Vinaninony : volcanisme ancien (moyenne de 5 coupes 1986 et 1987: pH du sol au départ = 4,9)					**Soanindrariny** : sol ferrallitique (une coupe 1986: pH du sol au départ = 4,4)				
Dolomie (t ha^{-1})	Doses de P$_2$O$_5$ (kg ha^{-1})				RVC	Doses de P$_2$O$_5$ (kg ha^{-1})				RVC
	0	80	160	Moyenne		0	80	160	Moyenne	
0	1629	2105	2245	1993 b	-	0	1563	2837	1467 b	-
1,5	2073	2542	2515	2377 a	4,4	0	1905	3235	1714 b	0,7
3,0	2356	2541	2863	2587 a	3,4	0	3386	4462	2616 a	1,6
Moyenne	2019 a	2396 a	2541 a			0c	2285 b	3511 a		
RVC	-	-	-			-	4,9	3,7	-	

† RVC = rapport valeur/coût

Le *Pennisetum purpureum* cv. Kizozi est une grande graminée reconnue comme étant la plus tolérante aux sols acides; c'est d'ailleurs l'espèce la plus cultivée. On l'utilise surtout en vert et accessoirement en ensillage. Il doit être coupé jeune car la qualité de fourrage diminue très vite avec l'âge; la tendance est d'ailleurs la même avec les espèces similaires tel le *Setaria sphacealata* et la *Panicum maximum* (Webster et Wilson, 1980). Même sur cette espèce, l'effet de la dolomie est très significatif lorsque le sol est très acide (tableau 14). Avec une espèce plus sensible à l'acidité du sol comme le *Chloris gayana*, l'effet de la dolomie et du fumier est très significatif et très rentable, même sur un sol à pH assez élevé (tableau 15). Le chloris est très répandu car il est plus apprecié pour la fabrication de foin. (On note sur cet essai la rentabilité très élevée des fumures azotées et NPK.)

Tableau 14. Effet de la dolomie, du fumier et des éléments majeurs : rendements en matière sèche du *Pennisetum purpureum* cv. Kizozi.

Sahanivotry Est : sol ferrallitique

(moyenne de 5 coupes entre 1988 et 1989 : pH du sol au départ = 4,2)

	O	F	D	F + D	Moyenne	RVC†
O	1338	1861	1837	1956	1748 b	-
N	1616	1587	2124	2049	1844 b	1,2
P	2019	1874	1999	2103	1999 a	0,5
K	1757	1632	2127	2017	1883 b	1,2
NPK	2106	2366	2290	2724	2372 a	2,0
Moyenne	1767 c	1864 bc	2075 ab	2170 a		
RVC	-	0,9	2,7	1,8		

† RVC : rapport valeur/coût

F   =   20 t ha^{-1} du fumier
D   =   2 t ha^{-1} de dolomie
N   =   46 kg ha^{-1} (sous forme d'urée)
P   =   88 kg ha^{-1} P_2O_5 (sous forme superphosphate)
K   =   60 kg ha^{-1} K_2O (sous forme KCl)
NPK =   46-88-60 (sous forme 11-22-16)

Avec les espèces tempérées comme le radis et le navet (tableau 16), l'effet de la dolomie est très significatif; en effet ce sont ces espèces (radis, navet, raygrass) et les légumineuses fourragères (trèfle, luzerne, etc.) qui demandent un important amendement calco-magnésien de redressement du sol.

Tableau 15. Effet de la dolomie, du fumier et des éléments majeurs : rendements en matière sèche kg ha⁻¹ du *Chloris gayana*.

Ambatomena : sol ferrallitique
(une coupe, 1988 : pH du sol au départ = 5,0)

	O	F	D	F + D	Moyenne	RVC†
O	2062	3371	2459	3669	2890 c	-
N	2499	5434	2776	5374	4021 b	14,8
P	2181	3887	2677	5097	3460 bc	1,1
K	2201	3570	2578	4958	3327 c	4,0
NPK	4660	7318	4760	7378	6029 a	10,3
Moyenne	2721 b	4716 a	3050 b	5295 a		
RVC	-	4,6	0,7	2,8		

† RVC : rapport valeur/coût

N = 46 kg ha⁻¹ (sous forme d'urée)
P = 88 kg ha⁻¹ P_2O_5 (sous forme superphosphate)
K = 60 kg ha⁻¹ K_2O (sous forme KCl)
NPK = 400 kg ha⁻¹ (sous forme 11-22-16)

Tableau 16. Effet de la dolomie et du fumier sur espèces tempérées : rendement en feuilles, (matière sèche kg ha⁻¹) du radis et du navet (année 1988).

Kianjasoa : sol ferrallitique
(pH du sol au départ = 4,5)

	Radis			RVC	Navet			RVC
	0	10 t ha⁻¹ fumier	Moyenne		0	10 t ha⁻¹ fumier	Moyenne	
0	353	512	432 b	-	637	928	782 b	-
1 t ha⁻¹ de dolomie	670	952	811 a	1,8	1383	1037	1210 a	2,0
Moyenne	512 a	732 a			1010 a	983 a		
RVC	-	-			-	-		

† RVC : rapport valeur/coût

Sélections des variétés de pomme de terre à l'acidité des sols

Comme il a été constaté dans les essais de longue durée (tableaux 3 à 5), la pomme de terre répond aussi aux chaulage; cette réponse de la pomme de terre au chaulage peut être due à une carence marquée du sol en calcium et en magnésium, mais elle peut aussi résulter de l'acidité du sol.

Il est important de noter la différence de réaction des variétés de pomme de terre à l'acidité du sol. Cette différence est d'autant plus marquée que le sol est plus acide (tableau 17). La variété Boda a été reconnue comme une variété à très bonne performance dans les sols volcaniques et le sol chaulé de la station Mimosa. Son comportement est en revanche médiocre sur les sols de faible fertilité et acides. A la suite de ces observations, nous avons entrepris la sélection de clones tolerants aux sols acides. Comme les expé rimentations présentés ont été mises eu place en 1989, les rendements en tubercules ne sont pas encore disponibles. Le tableau 17 présenté donc uniquement le nombre de plantes lévées et le pourcentage de recouvrement du sol en cours de végétation. Les rendements sont en général fonctions de ces deux critères.

Tableau 17. Effet de la dolomie sur différentes variétés de pomme de terre : nombre de plantes levées (sur 20).

Var. Pdt	Tsiafajavona pH :5,6					Ambohimdroso pH : 4,2				
	O	F	D	F + D	Moyenne	O	F	D	F + D	Moyenne
Garana	14	18	16	18	16,5 ab	4	19	20	20	15,4 a
Kinga	20	16	18	20	18.5 a	2	18	20	19	14,5 a
Boda	12	19	13	19	15,5 b	0	0	8	6	3,6 c
800934	19	17	17	20	18,3 a	1	19	20	20	14,7 a
377851.11	16	17	17	19	17,2 ab	0	4	16	8	6,9 b
377835.12	17	17	14	18	15,7 b	0	7	14	13	8,2 b
Moyenne	15,9 a	17,3 a	15,7 a	18,9 a		1,1 c	11,1 b	15,9 a	14,2 a	

F = 20 t ha^{-1} fumier
D = 2 t ha^{-1} dolomie

Les tableaux 18 et 19 montrent les comportements de 15 clones de pomme de terre sur deux sols de pH différents. A Antsira où le sol est à pH 5,3, il n'y pas eu de différence significative entre le nombre de plantes lévées, mais par contre les différences entre les clones sur le pourcentage de couverture du sol au 38 ème jour ont été significatives. A Andranomanelatra où le pH est plus bas (pH 4,6), le nombre de plantes lévées avec le clone Boda est significativement inférieur à celui des autres clones.

A Ampitatafika où le pH est encore plus faible (tableau 20), la différence entre les clones sur la lévée est hautement significative. Presque la moitié des plantes n'ont pas levé quand la dolomie et le fumier n'ont pas été appliquées, et les plantes levées sont très peu développées (taux de recouvrement du sol de 4,20%).

Tableau 18. Nombre de plantes lévées et recouvrement du sol avec différentes variétés/ou clone de pomme de terre (Andranomanelatra, 1989 : pH = 4,6).

N°	Variétés/ clones	Nombre de plantes lévées (sur 8)			Recouvrement du sol au 40è jour en pourcent		
		0	Fumier 20 t ha⁻¹ + dolomie 1 t ha⁻¹	Moyennes	0	Fumier 20 t ha⁻¹ + dolomie 1 t ha⁻¹	Moyennes
1	HK.86.26.1	8,00	8,00	8,00 a	6,50	40,00	23,25 bc
2	HK.86.27.1	8,00	7,50	7,75 a	7,50	40,00	23,75 bc
3	HK.86.33.1	8,00	8,00	8,00 a	9,00	41,50	25,25 ab
4	HK.86.35.1	6,50	8,00	7,25 a	4,00	37,50	20,75 bc
5	HK.86.35.2	8,00	8,00	8,00 a	8,00	36,50	22,25 bc
6	HK.86.36.1	8,00	8,00	8,00 a	10,00	50,00	30,00 a
7	HK.86.238.1	6,50	8,00	7,25 a	5,50	31,50	18,50 cd
8	382124.3	8,00	8,00	8,00 a	5,00	39,00	22,00 bc
9	382135.1	7,50	8,00	7,75 a	5,50	52,50	29,00 a
10	382143.1	8,00	8,00	8,00 a	8,00	39,00	23,50 bc
11	382169.4	8,00	8,00	8,00 a	5,00	40,00	22,50 bc
12	800927	7,50	7,00	7,25 a	3,50	38,00	20,75 bc
13	Boda	1,50	6,00	3,75 b	1,50	28,00	14,75 d
14	Kinga	7,50	8,00	7,15 a	4,00	26,50	20,25 bc
15	Garana	7,50	8,00	7,75 a	3,00	35,00	19,00 cd
Moyennes		7,20 a	7,80 a		5,70 b	39,00 a	

Tableau 19. Nombre de plantes lévées et recouvrement du sol avec différentes variétés/ou clones de pomme de terre (Antsira, 1989 : pH = 5,3).

N°	Variétés/ clones	Nombre de plantes lévées (sur 8)			Recouvrement du sol au 40è jour en pourcent		
		0	Fumier 20 t ha⁻¹ + dolomie 1 t ha⁻¹	Moyennes	0	Fumier 20 t ha⁻¹ + dolomie 1 t ha⁻¹	Moyennes
1	HK.86.26.1	8,00	8,00	8,00 a	9,00	32,00	20,5 abcd
2	HK.86.27.1	8,00	7,50	7,80 a	13,50	35,00	24,3 ab
3	HK.86.33.1	8,00	8,00	8,00 a	12,00	31,50	21,8 abcd
4	HK.86.35.1	8,00	7,80	7,80 a	10,00	35,00	22,5 abcd
5	HK.86.35.2	8,00	8,00	8,00 a	11,00	34,00	22,5 abcd
6	HK.86.36.1	8,00	8,00	8,00 a	14,00	35,00	24,5 ab
7	HK.86.238.1	8,00	6,00	7,00 a	8,50	27,50	18,0 cd
8	382124.3	8,00	6,50	7,30 a	13,50	26,00	19,8 abcd
9	382135.1	7,50	8,00	7,80 a	10,00	34,00	22,0 abcd
10	382143.1	8,00	8,00	8,00 a	13,50	33,00	22,3 abc

Tableau 19. (suite)

Nº	Variétés/ clones	Nombre de plantes lévées (sur 8)			Recouvrement du sol au 40è jour en pourcent		
		0	Fumier 20 t ha^{-1} + dolomie 1 t ha^{-1}	Moyennes	0	Fumier 20 t ha^{-1} + dolomie 1 t ha^{-1}	Moyennes
11	382169.4	8,00	8,00	8,00 a	15,50	35,50	25,3 a
12	800927	7,00	6,00	6,50 a	9,00	25,00	17,0 d
13	Boda	7,50	6,50	7,00 a	8,00	26,50	17,0 d
14	Kinga	8,00	7,50	7,80 a	8,00	37,50	22,8 abcd
15	Garana	8,00	8,00	8,00 a	7,00	30,00	18,5 bcd
Moyennes		7,80 a	7,40 a		10,80 b	31,30 a	

Tableau 20. Nombre de plantes lévées et recouvrement du sol avec différentes variétés ou clones de pomme de terre (Ampitatafika, 1987 : pH = 4,2).

Nº	Variétés/ clones	Nombre de plantes lévées (sur 8)			Recouvrement du sol au 40è jour en pourcent		
		0	Fumier 20 t ha^{-1} + dolomie 1 t ha^{-1}	Moyennes	0	Fumier 20 t ha^{-1} + dolomie 1 t ha^{-1}	Moyennes
1	HK.86.26.1	6,50	8,00	7,30 ab	5,00	36,50	20,8 b
2	HK.86.27.1	5,50	7,00	6,30 abc	4,00	35,50	19,8 b
3	HK.86.33.1	2,50	8,00	5,30 bc	2,00	37,50	19,8 b
4	HK.86.35.1	3,00	7,00	5,00 c	2,00	35,00	18,5 b
5	HK.86.35.2	3,00	7,00	5,00 c	12,50	35,00	18,5 b
6	HK.86.36.1	5,00	8,00	6,50 abc	3,50	36,50	20,0 b
7	HK.86.238.1	6,00	6,00	6,00 abc	4,00	35,50	19,8 b
8	382124.3	6,00	6,00	6,00 abc	6,50	31,50	19,0 b
9	382135.1	1,50	8,00	4,30 cd	3,50	29,00	16,3 b
10	382143.1	7,00	7,50	7,30 ab	4,00	26,50	15,3 b
11	382169.4	4,50	8,00	6,30 abc	5,50	34,00	19,8 b
12	800927	3,50	7,00	5,30 bc	5,00	34,00	19,5 b
13	Boda	0,00	6,00	3,00 d	0,00	9,00	4,5 c
14	Kinga	3,50	7,50	5,50 bc	2,50	25,50	14,0 b
15	Garana	3,00	7,50	5,20 bc	2,50	31,10	16,8 b
Moyennes		4,40 b	7,30 a		4,20 b	32,30 a	

Si on se réfère au témoin local Garana, celui-ci a entièrement levé à Antsira et Andranomanelatra, mais la moitié des plantes n'ont pas levé à Ampitatafika, c'est à dire que plusieurs clones testés sont plus performants que le témoin local au point de vue tolérance aux sols acides. Tous les clones qui se développent bien sur sols acides (tableau 20) sont des sélections de FIFAMANOR produites localement à partir de graines botaniques (n° 1 à 12). La variété Kinga se comporte à peu près de la même manière que la variété Garana. Le clone Boda n'a pas du tout levé à Ampitatafika.

Conclusion

Pour pouvoir mettre en valeur les sols de tanety (collines) fortement lessivés il faut des applications de doses fortes de dolomie de phosphore et de fumier. Ces pratiques ne sont pas toujours à la portée des paysans à faibles revenus. On estime que c'est à cause de son prix élevé que la dolomie a été très peu utilisée à Madagascar (5 à 6000 t seulement par an), alors qu'elle est disponible localement. La technique d'élevage utilisée par la plupart des paysans ne leur permet pas la production de fumier en quantité et les qualité.

Il y a lieu, alors, de sélectionner en même temps les variétés et les espèces tolérantes aux sols acides. L'élevage amélioré, en particulier l'élevage de vache laitière, qui exige l'enlèvement quotidien de la litière, permet une production de fumier en quantité avec un nombre restreint d'animaux; ceci réduit le surpâturage et permet ainsi la couverture végétale des sols. FIFAMANOR s'oriente actuellement de plus en plus vers l'agroforesterie qui inclue les arbustes fourragers dans les exploitations agricoles. Des essais sont en cours pour trouver les espèces appropriées; c'est une recherche multidisciplinaire qui associe l'agronome, le zootechnicien, le forestier, l'économiste, voire même le sociologue.

Bibliographie

ERIKSEN, A.B. et NJØS, A. 1984. Recherche en phytotron avec les sols du Vakinankaratra. In: *Le blé et la pomme de terre à Madagascar, productions et contraintes*, ed. Rakotondramanana. Madagascar: FIFAMANOR.

FIFAMANOR (Fiompiana, Fambolena Malagasy Norveziana). 1979 à 1989. *Rapport des essais*. Madagascar: FIFAMANOR.

FIFAMANOR. 1988. *Rapport des activités 1988*, 24-30. Madagascar: FIFAMANOR.

KAMPRATH, E.J. 1970. Exchangeable aluminium as a criterion for liming leached mineral soils. *Soil Science Society of America Proceedings* 34:252-254.

KOHLI, M.M. and RAJARAM, S. eds. 1988. *Wheat breeding for acid soils: review of Brazilian/ CIMMYT collaboration*, 1974-1986. International Maize and Wheat Improvement Center. Mexico: CIMMYT.

RAKOTONDRAMANANA. 1984. Avenir du triticale en complément du blé. In: *Le blé et la pomme de terre à Madagascar, productions et contraintes*, ed. Rakotondramanana. Madagascar: FIFAMANOR.

ROKOTONDRAMANA. 1986. Performance of triticale in Madagascar. In: *Proceedings of the International Triticale Symposium*. Australian Institute of Agricultural Sciences. Sydney: AIAS.

RAKOTONDRAMANANA and RANDRIANTSALAMA, R.A. 1987. Breeding wheat and triticale for stem rust resistance and adaptation to acid soils. In: *Fifth Regional Wheat Workshop for Eastern, Central and Southern Africa and Indian Ocean*, ed. M. Van Ginkel and D.G. Tanner. International Maize and Wheat Imporvement Center. Mexico: CIMMYT.

RAKOTODRAMANANA, RANDRIANTSALAMA, R.A. and RAKOTONIRAINY, H.J. 1989. Evaluation of nurseries for stem rust resistance and aluminium tolerance. Paper submitted to the Sixth Regional Wheat Workshop, CIMMYT, Addis Ababa, 2-6 October 1989. Mimeo. 9p.

RAUNET, M. 1981. Le milieu physique de la région volcanique Ankaratra-Vakinankaratra-Itasy. Institut de Recherches AgronomiquesTropicales. IRAT/MDRRA - Projet blé. Paris: IRAT.

WEBSTER, C.C. and WILSON, P.N. 1980. *Agriculture in the tropics*, 2d ed. Harlow, Essex: Longman Group Ltd.

Section 3: Soil organic matter and its management

The role of organic inputs and soil organic matter for nutrient cycling in tropical soils

E.C.M. FERNANDES and P.A. SANCHEZ*

Abstract

It is acknowledged that soil organic matter plays a key role in soil fertility and ecosystem productivity, but its precise contributions are unclear despite decades of research. In the tropics, increasing populations and migrations to marginal lands exacerbate the need for sustainable alternatives to shifting cultivation. The most promising alternatives, low-input crop production and agroforestry, rely more on organic inputs than on chemical fertilizers for nutrient supply than conventional farming systems practiced in nonmarginal areas. The traditional estimates of total soil organic carbon, however, offer little in the way of management parameters for optimizing nutrient cycling and sustainable productivity. Most research on soil organic matter has been conducted in temperate regions dominated by soils with a permanent charge, and much remains to be elucidated on SOM dynamics in the tropics, where soil mineralogy, climate, and vegetation are markedly different from the glaciated temperate areas.

*   Tropical Soils Program, North Carolina State University, Department of Soil Science, PO Box 7619, Raleigh, NC 27695-7619, USA.

Introduction

The intensity and duration of weathering in parts of the tropics have resulted in vast areas dominated by acid, infertile soils which are generally classified as Oxisols and Ultisols. These two soil orders occupy approximately 1600 million ha, or 43% of the tropics (Sanchez, Gichuru and Katz, 1982). The main soil constraints in the humid and subhumid tropics are chemical rather than physical. In terms of area, the most frequent constraints are: low nutrient reserves (64%), Al toxicity (56%), high P fixation by Fe (37%), and low effective cation-exchange capacity (11%) (Sanchez, Nicholaides and Couto,1982).

Natural forest ecosystems thrive under such fertility constraints because of well-established nutrient-cycling mechanisms (Nye and Greenland, 1960; Golley *et al.*, 1975; Vitousek, 1984; Jordan, 1985; Vitousek and Sanford, 1986) and negligible exports of biomass. A large proportion of the total nutrient stocks are in organic form and exist primarily in the biomass, litter, and SOM. Soil organic-matter pools are especially important reservoirs of N, P, and S.

Given these soil constraints, it is evident that the management of organic inputs and soil organic matter is crucial for sustainable soil productivity in the tropics. Allison (1973) summarized the potential benefits of soil organic matter to soil productivity as follows: soil organic matter is (i) a source of inorganic nutrients to plants, (ii) a substrate for microorganisms, (iii) an ion-exchange material, (iv) a factor in soil aggregation and root development, and consequently (v) a factor in soil and water conservation.

We will discuss the various management aspects of organic inputs and SOM and their relevance to nutrient cycling and sustainable productivity in acid soils of the tropics.

Organic inputs and soil organic matter in tropical agroecosystems

Numerous studies in both temperate and tropical regions have shown that in addition to adding nutrients, organic inputs can improve soil physical properties and help maintain soil organic-matter content, all of which positively affect nutrient cycling and plant productivity (Allison, 1973; Larson *et al.*, 1978; Lal and Kang, 1982). While these favourable effects of SOM have been known for some time, no consistent correlation has been found between total SOM and existing or potential soil productivity (Goh, 1980; Greenland, 1986; Sanchez and Miller, 1986; Swift, 1984). In most cases, the use of Walkley-Black total C and Kjeldhal total N are at best crude indicators of potential plant growth, and are not easily used as management parameters for soil fertility or productivity.

In much of the literature, the term "organic matter" is used indiscriminately to refer to organic inputs and/or soil organic matter. Organic inputs to the soil consist of the above- and the belowground litter, crop residues, mulches, green manures, and animal manures. Some of these inputs are grown and recycled on the site, while others are brought in from other sites, and thus constitute an external source of nutrients and organic carbon. Soil organic matter on the other hand, is the result of the partial or complete transformation in the soil of these organic inputs, and is located below the soil surface. In order to better evaluate the management and effect of these organic materials on nutrient cycling and sustainable soil productivity, Sanchez et al. (1989) proposed the use of "organic inputs" and "soil organic matter (SOM)" rather than the general term "organic matter".

Organic inputs as a source of nutrients

One of the functions of organic inputs is to supply nutrients to crops. Nutrient release and availability depends on the rate of decomposition of these inputs, which is influenced in part by temperature, moisture, soil texture, and mineralogy, as well as the quantity, quality, placement, and time of application. Much of the knowledge about the processes of decomposition in agroecosystems has been obtained from soils dominated by permanent-charge materials. Such soils are usually well supplied with bases, but limited by N, and have microbial populations dominated by bacteria. It is likely that decomposition of organic inputs and subsequent nutrient availability will operate differently in acid, infertile Oxisols and Ultisols with variable-charge minerals, often high in phosphorus-fixation capacity, and with microbial populations presumably dominated by fungi.

Studies using uniformly ^{14}C-labelled organic residues reveal a rapid initial stage of decomposition, during which about two-thirds of the plant ^{14}C is lost as $^{14}CO_2$, followed by a much slower rate of release over a long period of time (Figure 1). According to Swift (1984, 1985), successful management of organic inputs should aim to synchronize nutrient release with the crop's nutrient uptake pattern. Asynchrony between the decomposition of organic inputs and crop nutrient demands may be the rule in most tropical agroecosystems, as illustrated in Figure 2 from Anderson and Swift (1983).

The diverse range of tropical agroecosystems, such as shifting cultivation, managed fallows, cover/pasture species in tree plantations, and various agroforestry systems, offer plenty of scope for the management of organic inputs.

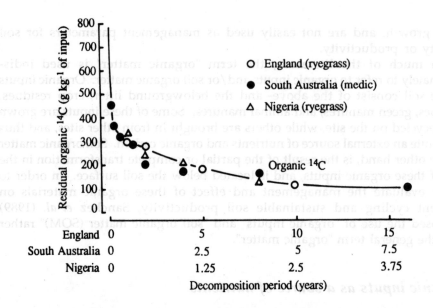

Figure 1. Decomposition of ^{14}C-labelled plant materials in different climates.
Reproduced from Ladd and Amato (1985) with permission from the Instituten
voor Bodemvruchtbaarheid, the Netherlands, and IITA, Nigeria.

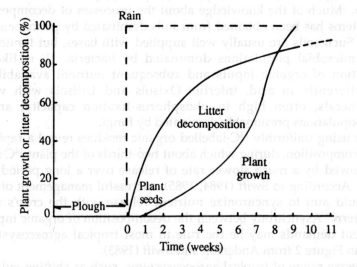

Figure 2. Theoretical illustration of asynchrony between nutrient release from
decomposing litter and nutrient demand by a crop receiving incorporated
litter (Anderson and Swift, 1983).

Included in these systems are hundreds of plant species (Fernandes and Nair, 1987) not found in temperate agroecosystems and likely to differ in both nutrient concentrations and the presence of secondary compounds (McVey *et al.*, 1978; Muller *et al.*, 1987). These secondary compounds may affect decomposition and nutrient release from organic inputs. Both the quantity and quality of organic inputs will be important in determining the nutrient availability pattern via decomposition of these inputs.

Organic-input quantity

The amount of nutrients added via organic inputs depends on the biomass and nutrient concentrations in the tissues, which in turn will vary with soil properties, climate, and the production system in which the organic inputs were grown. Table 1 provides estimates of the potential quantities of aboveground organic inputs in various agroecosystems. In general, carbon and nutrients from aboveground organic inputs in fertilized and/or systems on high base status soils equal or exceed that of natural systems. This is due to increased plant productivity on the more fertile sites, resulting in greater amounts of residues.

While there is a severe lack of reliable data on the biomass and nutrient content of aboveground organic inputs as related to soil data in tropical agroecosystems, even less is known about the role of roots as organic inputs and their contributions to SOM. Sanchez *et al.* (1989) estimated that root production in tropical agroecosystems is similar in magnitude to that estimated for tropical forests. Estimates of carbon and nutrient inputs to soil via root turnover, sloughing or exudation, however, are not available.

In an Amazon forest on an Oxisol in Venezuela, Cuevas (1983) found that annual fine-root production in the first 10 cm of soil and the root mat above the mineral soil was 8 t ha^{-1} yr^{-1}. Nutrient fluxes in litterfall for the same forest amounted to 121 kg N ha^{-1} yr^{-1} and 2 kg P ha^{-1} yr^{-1} (Cuevas and Medina, 1986), indicating that fine-root production and turnover in forests on very infertile soils may be more important for nutrient cycling than litterfall. Root turnover may be even faster in those agroforestry systems where trees and cover crops are pruned or grazed. In pot studies to determine the effect of shoot removal on fine-root dynamics at Yurimaguas (Peru), decreases in live fine-root biomass were detected three to four days after shoot removal, and new fine roots began growing eight to ten days after shoot removal (Fernandes, unpublished data).

Table 1. Dry-matter and nutrient inputs via litterfall/prunings in various production systems in the humid tropics.

Systems	Dry matter t ha^{-1} yr^{-1}	N	P	K	Ca	Mg	Source[†]
				kg ha^{-1} yr^{-1}			
Fertile soils							
Rainforest	10.5	162	9	41	171	37	1
High-input cultivation[‡]	9.3	139	15	98	52	23	2
Alley cropping							
L. leucocephala	8.1	276	23	122	126	31	3
Erythrina sp.	8.1	198	25	247	111	26	3
Shade systems							
Cacao/mixed shade	8.4	52	4	38	89	26	4
Cacao/*Erythrina*[‡]	6.0	81	14	17	142	42	5
Infertile soils							
Rainforest Oxisol/Ultisol	8.8	108	3	22	53	17	1
Low-input cultivation Ultisol	6.0	77	12	188	27	12	2
Alley-cropping Palaeudult							
Inga edulis	5.6	136	10	52	31	8	6
Shade systems							
Erythrina sp. Dystropept	11.8-18.4	170-238	14-24	119-138	84-222	27-56	7

[†] 1.   Vitousek and Sanford (1987)
   2.   Sanchez *et al.* (1989)
   3.   Salazar *et al.* (unpublished)
   4.   Boyer (1973)
   5.   FAO (1985)
   6.   Szott (1987)
   7.   Russo and Budowski (1986)
[‡] Fertilized.
Source: Szott *et al.* (1990).

Organic input quality

The efficiency of nutrient transfer from organic inputs to crops might be managed by varying the quality or the timing of application of organic inputs (Swift, 1985). Our definition of quality is based on the lignin:nitrogen ratio proposed by Mellilo *et al.* (1982). Inputs with "high quality" (a low L/N ratio) decay rapidly, and yield proportionally small quantities of humified material. Conversely, "low-quality" inputs are high in lignin and decompose slowly. The quality and nutrient contents of organic inputs in natural ecosystems may reflect the inherent fertility of the sites. For example, Vitousek (1984) and Vitousek and Sanford (1986) found that litter in tropical forests on infertile Oxisols and Ultisols was lower in P concentrations than the litter of more fertile tropical soils.

Other factors, such as polyphenolic and nutrient concentrations, can also affect the quality or decomposability of the plant material (Swift *et al.*, 1979). Phenolic polymers, found in both organic inputs and SOM are an important group of materials that inhibit microorganisms and enzymes, and are often involved in protecting other materials from biological degradation by a tanning process (Benoit and Starkey, 1968; Ladd and Butler, 1975). Recent work suggests that nitrogen release from leguminous organic inputs (low C/N, low L/N) is better correlated with polyphenolic content than with lignin and nitrogen content (Vallis and Jones, 1973; Palm, 1988).

Little is known about the quality of root litter from agricultural crops, perennial crops, and tropical forest species. In general, small-diameter roots, such as those produced by food crops, have comparatively low L/N ratios and would be expected to decompose and release nutrients rapidly, whereas decomposition and nutrient release from larger, more lignified root litter would be slower (Amato *et al.*, 1987). There are indications, however, that indices such as the nonstructural carbohydrates:nitrogen ratio may describe root-litter quality better than the L/N ratio (Berg *et al.*, 1987).

The effect of the quality of organic inputs on nutrient release is mostly discussed in terms of nitrogen release and availability, with little mention of phosphorus. This is probably due to N limitation in many temperate ecosystems and the difficulty of measuring phosphorus availability and mineralization, particularly on high P-fixing soils (Sanchez *et al.*, 1989). Two factors related to organic input quality and P release from these inputs will be especially important in the tropics: (i) acid, P-deficient, or P-fixing soils are abundant; and (ii) crop residues are usually low in P, as most of this element is concentrated in the grain and hence removed by harvest. It will hence be important to consider C/P or N/P ratios of organic inputs, and the interactions between the organic material and mineral soil.

The incorporation of organic inputs into the soil has been shown to temporarily reduce aluminium in soil solution (Davelouis, 1989) and hence Al toxicity to plants. The process involved is believed to be a complexation of aluminium in soil solution by organic acids (Hue *et al.*, 1986). Develouis (1989) found that high-quality organic inputs (rice straw, kudzu, and cowpea) were more effective in reducing Al in soil solution than low-quality inputs (*Inga edulis*). Another factor to consider is the reduction of aluminium saturation by the calcium and magnesium released from the organic inputs (Wade and Sanchez, 1983).

Effect of organic inputs on soil physical properties

In addition to managing nutrient-release patterns, organic inputs can also be used to influence soil physical properties, with a view to maximizing nutrient cycling and soil productivity. Organic inputs may be used as a surface mulch or incorporated into the soil.

Mulching with organic residues has been found to improve soil structure and porosity, reduce rainfall impact and runoff velocity, and increase the infiltration rate, water-retention capacity, and biological activity. Surface mulching was found to increase feeding and burrowing of earthworms, which then alters bulk density, macroporosity, and water-retention and transmission characteristics (Rusek, 1986). For most soils, runoff and erosion decrease exponentially with increasing mulch rate (Lal, 1976). The reduced loss of nutrients from soil via decreased soil erosion is likely to significantly enhance nutrient cycling and soil productivity (Young, 1987).

The effects of organic inputs on soil physical properties depend on the residence time or quality and placement of the material. Thus organic inputs that decompose slowly (low quality) will have a greater and more durable effect on soil physical properties than high-quality organic residues. The importance of mulching with organic inputs will be especially important on slopes (Young, 1987) and those sandy soils which are subject to surface crusting by raindrop impact (Chase et al., 1987).

Incorporation of organic inputs into the soil generally results in soils that are easier to work, are less susceptible to drought, and have a higher capacity to retain nutrients. The conversion of organic inputs into SOM via microbial decomposition results in the production of mucilages and cementing agents that bind primary particles into compound units called peds or aggregates. The resistance of these inputs to the slaking action of water is crucial to the structural integrity of the soil. Tisdall and Oades (1982) also showed that aggregate stability depends on the quality of organic inputs. Aggregation increased with lower-quality inputs in poorly structured Alfisols of Australia.

In many tropical soils, this improvement in soil physical conditions may far outweigh the value of organic inputs as nutrient sources (Sanchez et al., 1989). Although many studies have shown the existence of a positive correlation between SOM and aggregate stability and improved physical properties, the exact mechanisms are unknown, and present a barrier to the optimal management of organic inputs.

Organic inputs and soil organic matter

In addition to supplying nutrients and improving soil physical properties, organic inputs also lead to the formation and maintainance of soil organic matter. In fact, many of the effects of organic inputs already discussed (such as nutrient availability patterns and aggregation) are indirect via the formation of soil organic matter. Given the absence of a correlation between total SOM and crop growth or soil productivity, many researchers suggest the division of total soil organic matter into fractions or pools of different turnover times (Jenkinson and Rayner, 1977; van Veen and Paul, 1981; Jenkinson et al., 1987; Parton et al., 1987).

The framework of the Century model (Figure 3; Parton et al., 1987) is a convenient tool for the discussion of the concepts of SOM dynamics. Organic inputs are divided into a structural pool (turnover time 1 to 5 years) and a metabolic pool (turnover time 0.01 to 1 year) based on their L/N ratios (Mellilo

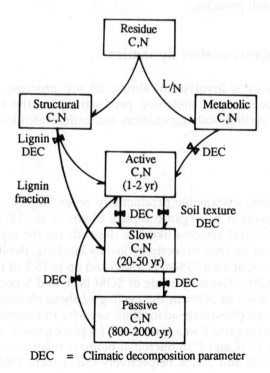

Figure 3. Flow diagram for the Century SOM model.

et al., 1982). The model describes transformations of the structural and metabolic pools of plant residue into three main SOM pools - active, slow, and passive. The active pool consists of live microbes and microbial products, and has a turnover time of the order of 1 to 2 years. The slow pool is more resistant to decomposition due to its chemical nature and/or physical protection and has a turnover time of 20 to 40 years. The passive SOM is chemically recalcitrant, physically protected or nutrient-depleted, and has a turnover time of 200 to 1500 years.

Although a major problem is that there are currently no methods for quantifying these different SOM pools, integrated research efforts are currently under way on this topic. From a management perspective, it is hypothesized that the active SOM pool could play a major role in nutrient release, and perhaps be estimated by microbial biomass; the slow SOM pool may play a major role in the aggregation of macroaggregates, while the passive SOM pool may be largely inert as a nutrient release source but may play a major role in the binding of primary soil particles.

Effects of soil organic-matter dynamics

Soil organic matter is involved in almost all soil processes. Three areas where SOM is important for sustained productivity in the tropics can be identified: nutrient cycling, soil aggregation, and cation retention.

Nutrient cycling

In natural systems, sustainable productivity relies largely on the efficient and rapid decomposition of organic inputs (Swift *et al.*, 1979). Sustained production in agricultural systems likewise depends on the replenishment of soil nutrients removed by crop harvests, or lost by leaching, denitrification, and soil erosion. In general, at least 95% of N and S and up to 75% of the P in surface soils are found in SOM. The magnitude of SOM N, P and S pools emphasizes the potential importance of SOM in the cycling of these elements. Of special significance is the high phosphate-adsorption capacity in Oxisols and Andepts and the importance of organic P as a source of P for plant growth on these soils.

The proportions of N and P in microbial biomass relative to total SOM are in the order of 0.5-15% and 3-20% respectively (Stevenson, 1986). Despite its relatively small size, the importance of this pool to nutrient cycling becomes even greater if we consider the rapid turnover of microbial biomass. Mineralization of N occurs when N-rich organic materials are used as energy substrates by soil microbes. In the course of decomposition, N-C covalent bonds are broken and CO_2 and N released (McGill and Cole, 1981). Most organic P, on the other hand, is linked to C by an ester (C-O-P) bond. Hence its rate of mineralization from SOM, relative to C or N, may not be in strict proportion to the decomposition of SOM (McGill and Cole, 1981).

Clay mineralogy can also be expected to influence the relative importance of physical protection and chemical recalcitrance, and will influence the turnover time of SOM and hence the rate of nutrient release and total loss of C from a soil. While soils dominated by variable charge are extensive in the tropics, most work on the relationship between clays and SOM has been limited to permanent-charge soils. The dynamics of SOM pools could be markedly different in soils with variable-charge mineralogies. For example, kaolinite absorbes less humic material than montmorillonite at comparable pH and solution concentration because of less surface area and exchange capacity. Sesquioxides, oxide-coated layer silicates, and allophane may have high specific surfaces, depending on the extent of amorphism of the material, and hence more associated SOM than crystalline oxides. In soils with allophanic mineralogy, organic materials can be protected by having pores too small for microbes to enter (Wada, 1980). Allophane also forms stable complexes with humid substances, increasing the chemical recalcitrance of the organic complexes. Laboratory studies, on the other hand, have shown that fertilization with P leads to increased soil respiration and N mineralization (Fox, 1980), indicating nutrient availability as a factor. These factors are apt to result in greater mineral-SOM affinity and lower SOM decomposition rates in soils dominated by variable-charge minerals.

It is apparent that mineralization of SOM constitutes a rate-limiting step in the supply of nutrients, and especially N and P, to plants. Disruption or displacement of this cycling may lead to nutrient loss and fertility decline. Lack of synchrony between mineralization and plant demand creates conditions which promote these losses (Swift and Sanchez, 1984). Therefore management of organic inputs and SOM to promote improved synchrony between mineralization and plant N and P uptake is of primary importance. Little information exists on the effects of SOM dynamics on the availability of other nutrients.

Assuming the various SOM pools can be manipulated by varying tillage practice, quality, and placement of organic inputs, Sanchez et al. (1989) suggested the following scenario to optimize nutrient availability and cycling in tropical agroecosystems. For short-cycle food crops with high demands for nutrients over a short time, applications of high-quality organic inputs could provide nutrients more in synchrony with plant demands than nutrients from inorganic fertilizers, whose availability often exceeds plant demand. Systems involving tree crops on the other hand, have lower nutrient demands over similar time periods than do short-cycle crops, and the use of inorganic fertilizers or high-quality organic inputs will probably result in large losses of nutrients. The buildup of a slow SOM pool by additions of low-quality organic residues would release lower levels of nutrients than inorganic fertilizers, but these would be more in synchrony with the nutrient demand pattern of trees.

Soil structure

Good soil structure is important for the conservation, and eventually the recycling, of soil nutrients via the minimization of soil erodibility. Organic inputs and SOM are central to the formation and stability of soil structure, although soil texture, mineralogy, and biological activities are also important.

Tillage practices can influence the maintainance of SOM and the relative proportions of the various SOM pools through their effects on aggregation. This assumes that SOM associated with micro- and macroaggregates corresponds to passive and slow pools respectively (Tisdall and Oades, 1982; Tiessen et al., 1984c; Elliott, 1986). Increases in microbial activity and reserves of organic N in no-till systems result from surface placement of residues, better soil moisture status for microbial activity, and accumulation of SOM (Ayanba et al., 1976; Doran, 1987).

Improper land clearing or tillage reduces the mean diameter of soil aggregates (Allegre and Cassel, 1986), resulting in increases in bulk density and decreases in aeration, hydraulic conductivity, and increased erodibility (Lal and Greenland, 1979; Lal et al., 1986). It is likely that the loss of SOM following land clearing and cultivation (Seubert et al., 1977) traditionally attributed to a new and lower equilibrium between inputs and losses (Greenland and Nye, 1959) is also related to decreases in the active and slow SOM pools which are exposed to microbial attack by disruption of macroaggregates (Elliott, 1986). There is a lack of data on the short- and long-term effects of tillage and placement of organic inputs on SOM nutrient reserves in tropical soils (Duxbury et al., 1989).

Cation retention

The ability of the soil to retain cations is an important component of nutrient cycling because it ameliorates leaching losses and regulates nutrient concentrations in the soil solution. Both soil minerals and SOM contribute to CEC in soils; but as soils are weathered, their CEC drops due to changes in mineralogy from 2:1-type layer aluminosilicate minerals to kaolinite and amorphous oxides of Fe and Al.

Soil organic matter is a source of cation-exchange capacity (CEC) because negative charges are created by the dissociation of its carboxyl and hydroxyl groups. Consequently all SOM-CEC is pH-dependent. Cation-exchange capacity is commonly measured using extractants buffered at pH 7 or 8.2. Such methods grossly overestimate the CEC of acid soils because of the pH-dependent charge of SOM and clay minerals such as kaolinite, iron oxides, and allophane. Measurements of the effective cation-exchange capacity (ECEC) at the soil's actual pH are therefore more relevant (Sanchez, 1976).

Levels of ECEC are low (less than 5 cmol$^+$ kg^{-1} of soil) in many soils of the tropics, with SOM a major source of ECEC in soils with siliceous, kaolinitic, or oxidic mineralogies. The net negative charge of many A horizons of oxidic or allophanic soils is due to SOM. As SOM decreases with depth, subsoils may approach net zero charge (in Acrustox and Acrorthox), or occasionally show net positive charge (Sanchez, 1976). Given the large areas of highly weathered soils in the tropics, maintainance of SOM to provide CEC will be crucial for optimum conservation and cycling of nutrient cations.

Lopes and Cox (1977) in a survey of 518 topsoils of the Brazilian Cerrado found no relation between ECEC and SOM contents between 1 and 4% when the soil pH was less than 5.0. In contrast, they observed a sharp increase in ECEC with increasing SOM in soils with a pH above 5.5 (Figure 4). Is the poor rela-

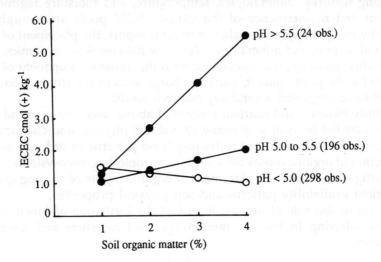

Figure 4. Differential relationship between SOM and effective cation-exchange capacity (ECEC) according to topsoil pH in soils of the Brazilian Cerrado (Lopes and Cox, 1977).

tionship between SOM and ECEC in these kaolinitic and oxidic soils at low pH values due to the possibility that these soils have mainly slow and passive SOM pools? Alternatively, is active SOM highly correlated to ECEC? If so, this would allow considerable scope for optimizing CEC by management of organic inputs.

The higher the ECEC of a soil, the more cations it can retain. This is good if the main exchangeable cations are plant nutrients. This is not the case, however, in acid soils where Al is likely to occupy the exchange sites at the expense of nutrient cations. Thus an increase of ECEC in these soils without a decrease in the original Al-saturation value, may not be desirable. A good point, however, is that the complexing of Al by SOM leads to a reduction in exchangeable Al (Davelouis, 1989) which is toxic to the roots of most plant species. Improved root growth may have more significance to plant growth in those agroecosystems subject to periods of drought stress. In general, there are positive and negative effects of SOM on CEC but a reduction in SOM levels has detrimental consequences for nutrient cycling and soil structure.

Conclusions and research priorities

A key challenge is to understand the relative role of above- or belowground inputs on the different SOM pools in different agroecosystems, and on soils with contrasting textures, mineralogies, temperature, and moisture regimes. The formation and maintainance of the various SOM pools are thought to be affected by the quantity and quality of organic inputs, the placement of organic inputs, soil texture, and mineralogy. Very few data on SOM dynamics, organic input quality, quantity, and interaction with the mineral component of soil are available for the predominant, variable-charge soils of the tropics. Sanchez *et al.* (1989) have suggested several key research needs:

- quantify biomass and nutrient content of above- and belowground organic inputs on the basis of well-characterized soil physical and chemical data.
- Obtain criteria that predict nutrient-release patterns in order to aid in the selection of organic inputs for a variety of tropical agroecosystems.
- Investigate the effects of placement of organic inputs of varying quality on nutrient availability patterns and soil physical properties.
- Determine the role of organic inputs in the formation of functional SOM pools differing in texture, mineralogy, and moisture and temperature regimes.

References

ALEGRE, J.C. and CASSEL, D.K. 1986. Effect of land-clearing methods and post-clearing management on aggregate stability and organic carbon content of a soil in the humid tropics. *Soil Science* 142: 289-295.

ALLISON, F.E. 1973. *Organic matter and its role in crop production*. Amsterdam: Elsevier. 637p.

AMATO, M., LADD, J.N., ELLINGTON, A., FORD, G., MAHONEY, J.E., TAYLOR, A.C. and WALSGOT, D. 1987. Decomposition of plant material in Australian soils. IV. Decomposition *in situ* of ^{14}C and ^{15}N-labelled legume and wheat materials in a range of southern Australian soils. *Australian Journal of Soil Research* 25: 95-105.

ANDERSON, J.M. and SWIFT, M.J. 1983. Decomposition in tropical forests. In: *Tropical rainforests: ecology and management*, ed. L.C. Sutton, T.C. Whitmore and A.C. Chadwick, 287-309. London: Blackwell.

AYANABA, A., TUCKWELL, S.B. and JENKINSON, D.S. 1976. The effects of clearing and cropping on the organic reserves and biomass of tropical forest soils. *Soil Biology and Biochemistry* 8: 519-525.

BENOIT, R.E. and STARKEY, R.L. 1968. Inhibition of decomposition of cellulose and some other carbohydrates by tannin. *Soil Science* 105: 291-296.

BERG, B., MULLER, M. and WESSEN, B. 1987. Decomposition of red clover (Trifolium pratense) roots. *Soil Biology and Biochemistry* 19: 589-593.

BOYER, J. 1973. Cycles de matière organique et des éléments minéraux dans une cacaoyère Camerounaise. *Café Cacao Thé* 17 : 2-23.

CUEVAS, E. 1983. Crecimiento de raices finas y su relacion con los procesos de descomposicion de materia organica y liberacion de nutrientes en bosques del Alto Rio Negro en el Territorio Federal Amazonas. Ph.Sc. dissertation. Instituto Venezolano de Investigaciones Cientificas, Caracas, Venezuela.

CUEVAS, E. and MEDINA, E. 1986. Nutrient dynamics within Amazonian forest ecosystems: I. Nutrient flux in fine litter fall and efficiency of nutrient utilization. *Oecologia* (Berlin) 68: 466-472.

DAVELOUIS, J.R. 1989. Green manure applications to minimize aluminium toxicity in the peruvian Amazon. Ph.D. dissertation. Department of Soil Science, North Carolina State University, Raleigh, NC.

DORAN, J.W. 1987. Microbial biomass and mineralizeable nitrogen distributions in no-tillage and plowed soils. *Biology and Fertility of Soils* 5: 68-75.

DUXBURY, J.M., SCOTT SMITH, M. and DURAN, J.W. 1989. Organic matter as a source and a sink of plant nutrients. In: *Dynamics of soil organic matter in tropical ecosystems*, ed. D.C. Coleman, J.M. Oades and G. Uehara. Honolulu: University of Hawaii Press.

ELLIOTT, E.T. 1986. Aggregate structure and carbon, nitrogen, and phosphorus in native and cultivated soils. *Soil Science Society of America Journal* 50: 627-633.

FAO (Food and Agriculture Organization of the United Nations). 1985. *Yearbook for agriculture for 1984*. Rome: FAO.

FERNANDES, E.C.M. and NAIR, P.K.R. 1987. An evaluation of the structure and function of tropical homegardens. *Agricultural Systems* 21: 279-310.

FOX, R.L. 1980. Soils with variable charge: agronomic and fertility aspects. In: *Soils with variable charge*, ed. B.K.G. Theng, 195-224. New Zealand Society of Soil Science. Lower Hutt: NZSSS.

GOH, K.M. 1980. Dynamics and stability of organic matter. In: *Soils with variable charge*, ed. B.K.B. Theng, 373-393. New Zealand Society of Soil Science. Lower Hutt: NZSSS.

GOLLEY, F.B., McGINNIS, J.I., CLEMENTS, R.G., CHILD, G.I. and DUEVER, M.J. 1975. *Mineral cycling in a tropical moist forest ecosystem*. Athens: University of Georgia Press.

GREENLAND, D.F. 1986. Soil organic matter in relation to crop nutrition and management. Paper presented at the International Conference on the Management and Fertilization of Upland Soils, Nanjing, China. September 7, 1987.

GREENLAND, D.J. and NYE, P.H. 1959. Increases in carbon and nitrogen contents of tropical soils under natural fallows. *Journal of Soil Science* 10: 284-175.

HUE, N.V., CRADDOCK, G. and ADAMS, F. 1986. Effects of organic acids on aluminium toxicity in subsoils. *Soil Science Society of America Journal* 50: 28-34.

JENKINSON, D.S., HART, P.B.S., RAYNER, J.H. and PARRY, L.C. 1987. Modelling the turnover of soil organic matter in long-term experiments at Rothampstead. *INTECOL Bulletin* 15: 1-8.

JENKINSON, D.S. and RAYNER, J.H. 1977. The turnover of soil organic matter in some of the Rothampstead classical experiments. *Soil Science* 123: 298-305.

JORDAN, C.F. 1985. Nutrient cycling in tropical forest ecosystems. Chichester, England: John Wiley and Sons.

LADD, J.N. and BUTLER, J.H.A. 1975. Humus-enzyme systems and synthetic organic polymer-enzyme analogs. In: *Soil biochemistry*, ed. E.A. Paul and A.D. McClaren, 143-194. New York: Marcel Dekker Inc.

LAL, R. and GREENLAND, D.J. 1979. *Soil physical properties and crop production in the tropics*. New York: John Wiley and Sons.

LAL, R. and KANG, B.J. 1982. Management of organic matter in soils of the tropics and subtropics. In: *Transactions of the 12th International Congress of Soil Science* (New Dehli), vol. IV, 152-178. Indian Society of Soil Science. New Delhi: ISSS.

LAL, R., SANCHEZ, P.A. and CUMMINGS, Jr., R.W. 1986. *Land clearing and development in the tropics*. Boston: A.A. Balkema. 450p.

LARSON, W.P., HOLT, R.F. and CARLSON, C.W. 1978. *Residues for soil conservation*, 1-16. American Society of Agronomy. Special Publication no. 31. Madison, Wisconsin: ASA.

LOPES, A.S. and COX, F.R. 1977. A survey of the fertility status of surface soils under "Cerrados" vegetation in Brazil. *Soil Science Society of America Journal* 41: 742-747.

McGILL, W.B. and COLE, C.V. 1981. Comparative aspects of organic C, N, S and P cycling through organic matter during pedogenesis. *Geoderma* 26: 267-286.

McVEY, D., WATERMANN, P.G., MBI, C.N., GARTLAN, J.S. and STRUHSAKER, T. 1978. Phenolic content of vegetation in two African rainforests: ecological interpretations. *Science* 202: 61-64.

MELILLO, J.M., ABER, J.D. and MURATORE, J.F. 1982. Nitrogen and lignin control of hardwood leaf litter decomposition dynamics. *Ecology* 63: 621-626.

MULLER, R.N., KALISE, P.J. and KIMMERER, T.W. 1987. Intraspecific variation in production of astringent phenolics over a vegetation-resource availability gradient. *Oecologia* (Berlin) 72: 211-215.

NYE, P.H. and GREENLAND, D.J. 1960. *The soil under shifting cultivation.* Commonwealth Bureau of Soils. Technical Communication no. 51. Slough, England: CBS.

OADES, J.M. 1989. An introduction to organic matter in mineral soils. In: *Minerals in soil environments*, 2d ed., ed. J.B. Dixonand, S.B. Weed, 89-159. Soil Science Society of America. Madison, WI: SSSA.

PALM, C.A. 1988. Mulch quality and nitrogen dynamics in an alley cropping system in the Peruvian Amazon. Ph.D. dissertation, North Carolina State University, Department of Soil Science, Raleigh, NC, USA.

PARTON, W.J., SCHIMEL, D.S., COLE, C.V. and OJIMA, D.S. 1987. Analysis of factors controlling soil organic matter levels in Great Plains grasslands. *Soil Science Society of America Journal* 51: 1173-1179.

RUSSEK, J. 1986. Soil microstructures: contributions on specific soil organisms. Faunal influences on soil structure. Department of Entomology, University of Alberta, Edmonton, Canada. *Questions Entomological* 21(4): 497-514.

RUSSO, R.O. and BUDOWSKI, G. 1986. Effect of pollarding frequency on biomass of *Erythrina poeppigiana* as a coffee shade tree. *Agroforestry Systems* 4: 145-162.

SANCHEZ, P.A. 1976. *Properties and management of soils in the tropics.* New York: John Wiley and Sons.

SANCHEZ, P.A. and BENITES, J.R. 1987. Low-input cropping for acid soils of the humid tropics. *Science* 128: 1521-1527.

SANCHEZ, P.A. and BUOL, S.W. 1975. Soils of the tropics and the word food crisis. *Science* 188: 598-603.

SANCHEZ, P.A. and MILLER, R.H. 1986. Organic matter and soil fertility management of acid soils in the tropics. In: *Transactions of the 13th International Congress of Soil Science* (Hamburg), vol 6, 609-625. International Soil Science Society. Wageningen, The Netherlands: ISSS.

SANCHEZ, P.A., GICHURU, M.P. and KATZ, L.B. 1982. Organic matter in major soils of the tropical and temperate regions. In: *Transactions of the 12th International Congress of Soil Science* (New Delhi), vol. I, 99-114. New Delhi: Indian Society of Soil Science.

SANCHEZ, P.A., NICHOLAIDES, J.J. and COUTO, W. 1982. Physical and chemical constraints to food production in the tropics. In: *Chemistry and the world food supplies: the new frontiers*, ed. G. Bixler and L.W. Shemilt, 89-106. International Union of Pure and Applied Chemistry and International Rice Research Institute. Los Baños, Philippines: IRRI.

SANCHEZ, P.A., PALM, C.A., SZOTT, L.T., CUEVAS, E. and LAL, R. 1989. Organic input management in tropical agroecosystems. In: *Dynamics of soil organic matter in tropical ecosystems*, ed. D.C. Coleman, J.M. Oades and G. Uehara, 125-152. Honolulu: University of Hawaii Press.

SEUBERT, C.E., SANCHEZ, P.A. and VALVERDE, C. 1977. Effects of land clearing methods on soil properties and crop performance in an Ultisol of the Amazon jungle of Peru. *Tropical Agriculture* (Trinidad) 54: 307-321.

STEVENSON, F.J. 1986. *Cycles of soil carbon, nitrogen, phosphors, sulfur micronutrients.* New York: John Wiley and Sons.

SWIFT, M.J. 1984. *Soil biological processes and tropical soil fertility: a proposal for a collaborative program of research.* International Union of Biological Sciences. Biology International Special Issue no. 5. Paris: IUBS. 38p.

SWIFT, M.J. 1985. *Tropical soil biology and fertility (TSBF): Planning for research.* International Union of Biological Science. Biology International Special Issue 9. Paris: IUBS. 24p.

SWIFT, M.J. and SANCHEZ, P.A. 1984. Biological management of tropical soil fertility for sustained productivity. *Nature and Resources* 10: 1-8.

SWIFT, M.J., HEAL, O.W. and ANDERSON, J.M. 1979. *Decomposition in terrestrial ecosystems.* Oxford, England: Blackwell Scientific.

SZOTT, L.T. 1987. Improving the productivity of shifting cultivation in the Amazon Basin of Peru through the use of leguminous vegetation. Ph.D. dissertation, North Carolina State University, Raleigh, NC.

SZOTT, L.T., FERNANDES, E.C.M. and SANCHEZ, P.A. 1990. Soil plant interactions in agroforestry systems. Paper presented at the International Meeting on Agroforestry Systems, Department of Forestry and Natural Resources, University of Edinburgh, 25-29 July 1989.

TATE, K.R. and THENG, B.K.G. 1980. Organic matter and its interactions with inorganic soil constituents. In: *Soils with variable charge,* ed. B.K.G. Theng, 225-249. New Zealand Society of Soil Science. Lower Hutt: NZSSS.

TIESSEN, H., STEWART, J.W.B. and COLE, C.V. 1984. Pathways of phosphorus transformations in soils of differing pedogenesis. *Soil Science Society of America Journal* 48: 853-858.

TISDALL, J.M. and OADES, J.M. 1979. Stabilization of soil aggregates by the root systems of ryegrass. *Journal of Soil Science* 17: 429-441.

TISDALL, J.M. and OADES, J.M. 1982. Organic matter and water stable aggregates in soils. *Journal of Soil Science* 33: 141-163.

VALLIS, I. and JONES, R.J. 1973. Net mineralization of nitrogen in leaves and leaf litter of *Desmodium intortum* and *Phaseolus artopurpureus* mixed with soil. *Soil Biology and Biochemistry* 5: 391-398.

VAN VEEN, J.A. and PAUL, E.A. 1981. Organic carbon dynamics in grassland soils. I. Background information and computer simulation. *Canadian Journal of Soil Science* 61: 185-201.

VITOUSEK, P.M. 1984. Litterfall, nutrient cycling and nutrient limitation in tropical forests. *Ecology* 65: 285-298.

VITOUSEK, P.M. and SANFORD, R.L. 1986. Nutrient cycling in moist tropical forests. *Annual Review of Ecology and Systematics* 17: 137-167.

WADA, K. 1980. Mineralogical characteristics of Andisols. In: *Soils with variable charge,* ed. B.K.G. Theng, 87-107. New Zealand Society of Soil Science. Lower Hutt: NZSSS.

WADE, M.K. and SANCHEZ, P.A. 1983. Mulching and green manure applications for continuous crop production in the Amazon Basin. *Agronomy Journal* 75: 39-45.
YOUNG, A. 1987. *The potential of agroforestry for soil conservation: II. Maintainance of fertility.* International Council for Research in Agroforestry. Nairobi, Kenya: ICRAF.

Le rôle de la matière organique dans les sols tropicaux : matière organique et pratique agricole en Côte d'Ivoire

GNAHOUA H. GODO[*]

Résumé

L'ordonnancement des itinéraires techniques tant en agriculture traditionnelle que moderne ne tient pas toujours compte de l'importance de la matière organique comme facteur essentiel de la fertilité du sol. Il en résulte généralement des baisses de productivité consécutives à un appauvrissement rapide des sols, et ce malgré des apports importants d'engrais minéraux.

Des essais de courte et longue durées réalisés respectivement dans le sud et le centre de la Côte d'Ivoire et comparant les effets de fumure minérale stricto sensu à ceux de fumure organique ou organo-minérale, montrent en substance que la fumure organique et notamment la fumure organo-minérale conduit à rendements meilleurs en culture d'igname que la fumure minérale, et qu'elle améliore certaines caractéristiques du sol cultivé telles que le pH, la somme des bases échangeables, le taux de saturation en bases et la teneur en matière organique du sol.

Abstract

The role of organic matter in tropical soils: organic matter and agricultural practices in Côte d'Ivoire

The organization of technical schedules, both for traditional and modern agriculture, does not take into account the importance of organic matter as an essential factor in soil fertility. Generally, by neglecting this factor, there is a progressive decline in

---

[*]   Laboratoire d'Agronomie, IIRSDA, BP. V 51, Abidjan, Côte d'Ivoire.

productivity, resulting from rapid soil degradation, which occurs even if significant amounts of mineral fertilizers are used.

Long- and short-term experiments, carried out respectively in the south and centre of Côte d'Ivoire, have been set up to compare the effects of purely mineral fertilization, with those of organic or organomineral inputs. Essentially, these experiments show that organic fertilizer, and especially organomineral fertilizer, gives rise to better yam yields than mineral fertilizer, and also improves some of the features of the cultivated soil, such as the pH, the exchangeable bases, the base saturation percentage, and the organic-matter contents of the soil.

Introduction

Sous les tropiques humides, les sols sont souvent caractérisisés, sur le plan physique, par un drainage excessif, et une charge gravillonnaire pouvant parfois occuper 60% du volume total d'un horizon donné. Il en résulte une faible capacité de rétention pour l'eau (Boa, 1989). Par ailleurs, ces sols sont sujets au ruissellement et à l'érosion dès lors qu'ils sont mis à nu.

Sur le plan chimique, les fortes teneurs en fer et aluminium, la prédominance d'argiles de type 1/1 et la faible teneur en matière organique entraînent un pH acide (souvent inférieur à 5,0), un taux d'aluminium échangeable élevé, une capacité d'échange de cations et un taux de saturation en bases très faibles.

A l'opposé, il est reconnu que la matière organique influe de façon prépondérante sur les propriétés physico-chimiques et microbiologiques des sols (Brady, 1974). En effet, sur le plan physique, la matière organique améliore la structure du sol, ce qui favorise l'aération, l'infiltration et surtout la capacité de rétention de l'eau. Ces conditions sont favorables à la vie de la plante. Sur le plan chimique, la matière organique augmente la capacité d'échange du complexe absorbant permettant ainsi de retenir dans le sol les éléments nutritifs provenant soit de l'altération *in situ*, soit des apports d'engrais. La matière organique (l'humus du sol) exerce un effet tampon sur l'acidité (Young, 1989). Enfin la matière organique est elle-même source d'azote, de soufre et de phosphore, et favorise la vie microbienne.

Vu sous cet angle, la matière organique est un facteur essentiel de la fertilité du sol en milieu tropical. Dans ce milieu, elle doit être appréhendée comme un facteur fondamental dans le processus de gestion des sols (Sanchez, 1976).

Cependant l'importance prise par la nutrition minérale des plantes ces vingt dernières années a quelque peu occulté le rôle de la matière organique dans la fertilité des sols. Cet état de faits, il faut le reconnaître, a eu une influence néfaste sur les pratiques agricoles moderne dans la plupart des pays tropicaux où l'on fonde trop d'espoirs (souvent déçus) sur l'utilisation exclusive des engrais minéraux.

Matière organique et pratique agricole en Côte d'Ivoire

Quelques observations

En milieu paysannal où l'agriculture itinérante est de mise, des cultures comme l'igname et la banane plantain sont invariablement pratiquées sur défriches de forêt ou de vieille jachère. La place obligatoire de la culture d'igname en tête de rotation ou d'assolement s'explique, selon les paysans, par le "maximum de fertilité de la terre" à ce moment précis. Le fait expérimental montre que pour une production maximum, l'igname exige un bon niveau de matière organique ainsi qu'une réserve adéquate d'élé ments nutritifs dans le sol (Coursey, 1968). Ces conditions organo-minérales sont géné ralement réunies dans le sol sous nouvelle défriche de forêt ou de vieille jachère.

Mais la pratique agricole paysanne ne favorise pas le maintien ou l'augmentation du stock organique du sol. En effet, les résidus de récolte ou de sarclage, facteurs d'amélioration ou de maintien du taux de matière organique du sol, sont habituellement considérés comme des "salissures" des champs et sont systématiquement exportés et brûlés hors des champs pour faire place aux cultures suivantes. Les effets dégradants de cette pratique culturale aussi bien sur le sol que sur les cultures sont particulièrement évidents dans les régions à forte pression démographique où l'agriculture itinérante est quasiment impossible. Sur les sols dénudés et pratiquement dépourvus de matière organique, les rendements baissent de plus de 50% dès la deuxième ou troisième année de culture continue (Godo et Yoro, 1985). L'exubérance des espèces végétales autour des cases, lieu de dépôt des ordures ménagères, devrait attirer cependant l'attention des paysans sur l'importance des résidus organiques en particulier, et de la matière organique en général dans la fertilité des sols cultivés.

En agriculture moderne, la mise en place d'exploitations agricoles s'accompagne de dé frichement motorisé (au bulldozer) de très grandes surfaces. Au cours de cette opération, la couche superficielle humifère du sol est systématiquement emportée. Les conséquences de cette pratique sont le ruisellement excessif et l'érosion (Roose, 1983), la péjoration de la plupart des propriétés physiques et chimiques du sol (Moreau, 1984) et surtout la baisse, à brève échéance, de la productivité en dépit de l'apport annuel de grandes quantités d'engrais minéraux. Ici également, l'incorporation des résidus de récolte n'est pas systématique. En effet lors d'une mission sur une exploitation de projet soja dans le nord-ouest de la Côte d'Ivoire, il a été constaté que pour faire place à une sole de maïs, les résidus de récolte du soja ont été exportés et brûlés hors de la parcelle (Diomande et Godo, 1982). On sait que cette exploitation a été par la suite abandonnée pour cause de baisse de rendement des cultures. Ainsi, les cas d'abandon de ces vastes étendues sont fréquents, surtout lorsque ces surfaces sont occupées par des cultures annuelles. L'ampleur et la fré quence de ce phénomène amènent à s'interroger sur l'efficacité et l'opportunité de ces grandes exploitations agricoles, dans la mesure où elles sont aussi éphémères que le cycle culture-jachère de l'agriculture traditionnelle itinérante et plus consommatrices de l'espace forestier (Godo, 1982).

Quelques faits expérimentaux

En Côte d'Ivoire, des résultats assez révélateurs de recherches menées sur l'effet des amendements organiques sur le rendement des cultures et les propriétés des sols cultivés sont significatifs. Il est donc paradoxal de voir que ces résultats n'ont pas toujours l'écho souhaité en milieu réel.

Expérimentation n°1 : Effet de fumure minérale et organo-minérale sur le rendement de l'igname

Cet essai de longue durée est réalisé sur la station IRAT de Bouaké dans le Centre de la Côte d'Ivoire, zone à pluviométrie irrégulière, bimodale de 1130 mm. L'essai teste plusieurs itinéraires techniques en vue de remplacer l'agriculture itinérante traditionnelle par une agriculture continue. Nous présentons ici une partie des résultats rapportés par Kalms et Chabalier (1981).

Le support édaphique de cet essai est un sol ferrallitique moyennement désaturé rajeuni. L'itinéraire technique choisi, comporte un labour attelé, 2 types de fumure (fumure minérale, NPK et organo-minérale, fumier de ferme + NPK) et 3 niveaux de fumure (F0, F1, et F2). Cet itinéraire technique est testé sur une rotation type Baoulé (igname-maïs/coton-riz pluvial) de trois ans. L'essai est mis en place sur trois blocs correspondant chacun à une sole de la rotation. Dans la fumure organo-minérale, le fumier de ferme est appliqué à l'igname et la fumure minérale NPK au maïs, coton et riz pluvial, de telle manière que l'igname bénéficie de l'effet résiduel de l'engrais minéral. Les types et niveaux de fumure sont rapportés dans le tableau 1.

Tableau 1. Types et niveaux de fumure en culture attelée.

	Fumure minérale NPK	Culture	Fumure organo-minérale Fumier + NPK
F$_0$	0-0-0	Iganme	0-0-0
	0-0-0	Maïs	0-0-0
	0-0-0	Coton	0-0-0
	0-0-0	Riz	0-0-0
F$_1$	40-30-60	Iganme	15 t ha^{-1} de fumier
	40-40-60	Maïs	20-40-40
	30-45-45	Coton	30-45-45
	33-40-40	Riz	33-40-40
F$_2$	80-60-120	Iganme	30 t ha^{-1} de fumier
	80-80-120	Maïs	40-80-80
	60-90-90	Coton	60-90-90
	66-80-80	Riz	66-80-80

Source : Kalms et Chabalier (1981).

Nous nous intéresserons uniquement aux rendements de la sole d'igname (figure 1.)

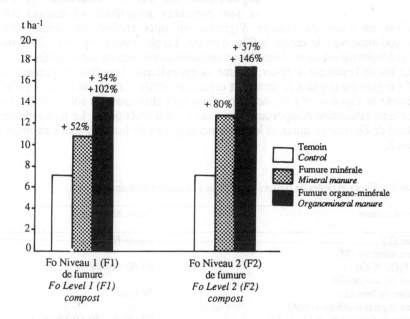

Figure 1. Rendements moyens pluriannuels (11 années) d'igname (t ha^{-1}) en fonction du type et du niveau de fumure.

Average yam yields (t ha^{-1}) over 11 years in relation to the type and level of amendments.

Au niveau F_1, la fumure minérale engendre un gain de rendement de 52% par rapport au témoin F_0 tandis que la fumure organo-minérale induit un gain de 102%. Le rendement moyen lié à la fumure organo-minérale est de 34% supérieur à celui lié à la fumure minérale simple.

Au niveau F_2, les gains sont respectivement de 80% pour la fumure minérale et 146% pour la fumure organo-minérale. Ici également, le rendement moyen lié à la fumure organo-minérale est de 37% supérieur à celui lié à la fumure minérale simple.

D'une façon générale, les rendements moyens enregistrés à chacun des niveaux de fumure permettent de dire que l'igname valorise la fumure, mais à un même niveau elle répond mieux à la fumure organo-minérale qu'à la fumure minérale simple. Il est aussi remarquable de constater que la fumure organo-minérale maintient un rendement de 14.5 t ha^{-1} et 17.5 t ha^{-1} sur 11 ans de culture continue respectivement aux niveaux F_1 et F_2 de fumure, soit au moins le double du rendement du témoin F_0.

Expérimentation n°2 : Effets des fumures minérale, organique et organo-minérale sur le rendement de l'igname et sur certaines propriétés chimiques du sol

C'est un essai de culture d'igname en pots réalisé au Centre ORSTOM d'Adiopodoumé dans le sud de la Côte d'Ivoire. La pluviométrie y est de 2000 mm, et le support édaphique est un sol ferrallitique appauvri modal issu de sables tertiaires.

Le but de l'essai est de tester, sur une saison culturale, les effets de fumures minérale (NPK), organique (fumier de ferme) et organo-minérale (fumier de ferme + NPK) sur le rendement de l'igname et certaines caractéristiques chimiques du sol. Il s'agit d'un essai monoplante randomisé comportant 4 traitements et 6 répétitions. Le matériel végétal est constitué de *Dioscorea alata*, et les traitements (doses de fumure) sont consignés dans le tableau 2.

Tableau 2. Types et doses de fumure appliquées à la culture d'igname en pots.

Types de fumure	Doses de fumure
Témoin (T)	Pas de fumure
Fumure minérale (M) (N; P_2O_5; K_2O)	80-60-120 kg ha^{-1}
Fumure organique (O) (fumier de ferme)	30 t ha^{-1}
Fumure organo-minérale (O-M) (fumier de ferme + N; P_2O_5; K_2O)	30 t ha^{-1} + 80-60-120 kg ha^{-1}

Les rendements induits par les différents traitements sont rapportés dans le tableau 3.

Tableau 3. Rendements (kg) moyens par pot en fonction des traitements.

Traitements	Rendement moyen par pot†	Gain de rendement par rapport à T	Gain de rendement par rapport à M
O-M	3,97 c	106%	38%
O	3,50 c	81%	22%
M	2,87 b	49%	-
T	1,93 a	-	-

† Différence significative (à 1% de probabilité) entre les valeurs n'étant pas suivies de la même lettre.
Source : de Saint-Amand (1985).

On constate, d'une manière générale, que les fumures minérale (M), organique (O) et organo-minérale (O-M) induisent des rendements supérieurs à celui du témoin (T). Mais les gains de rendement induits respectivement par la fumure organique (+81%) et la fumure organo-minérale (+106%) sont de loin supérieurs à ceux induits par la fumure minérale simple (+49%).

Par ailleurs, bien qu'il n'y ait pas de différence significative entre les traitements O et O-M, il est tout de même important de souligner que par rapport à la fumure minérale simple (M), la fumure organo-minérale entraîne une différence de rendement (38%) supérieure à celle de la fumure organique (22%).

Les effets des traitements sur certaines caractéristiques du sol sont consignés dans le tableau 4.

Tableau 4. Variations des valeurs de certaines caractéristiques du sol en fonction des traitements par rapport au sol de base en fin de cycle cultural.

	Sol de base	T	M	O	O-M
Somme des bases échangeables (cmol$^+$ kg^{-1})	1,38	1,23	1,59	1,94	1,94
CEC (cmol$^+$ kg^{-1})	5,77	4,77	4,93	4,43	4,98
Taux de saturation (%)	23,90	25,64	32,54	43,84	39,25
Matière organique (%)	2,2	2,6	2,65	2,8	2,95

La réaction du sol : d'une manière générale, il y a élévation du pH dans tous les traitements (T, M, O et O-M) par rapport au sol de base, mais cette hausse est plus marquée dans les sols de fumures organique et organo-minérale où la valeur du pH est égale ou supérieure à 5,0.

Le taux de matière organique : il y a une augmentation relative dans tous les traitements. Cette augmentation est en moyenne de 20% en T et M et 30% en O et O-M. Ainsi il y a un léger avantage dû aux fumures organique et surtout organo-minérale.

La somme des bases échangeables : cette somme passe de 1,38 cmol$^+$ kg^{-1} dans le sol de base à 1,23 dans le sol du traitement Témoin, puis augmente en M, O et O-M. Mais cette augmentation est plus accusée en O et O-M (où S avoisine 2 cmol$^+$ kg^{-1}) qu'en M (où S = 1,6 cmol$^+$ kg^{-1}).

La capacité d'échange des cations : sa valeur initiale (5,77 cmol$^+$ kg^{-1} dans le sol de base) est à la baisse dans tous les traitements. Cette baisse est de 1,0, 0,84, 1,34 et 0,79 cmol$^+$ kg^{-1} respectivement en T, M, O et O-M, soit en moyenne une baisse peu spectaculaire de 17% sur l'ensemble des traitements.

Le taux de saturation en bases : le taux est à la hausse dans tous les traitements mais cette hausse est plus prononcée dans les traitements organique et organo-minéral où la valeur moyenne est de 41,54 marquant ainsi une augmentation de 74, 62 et 28% respectivement par rapport au sol de base, au témoin (T) et au traitement de fumure minérale simple.

Bien que la tendance évolutive des caractéristiques du sol, notamment le taux de matière organique, le pH, la somme des bases et le taux de saturation, privilégie les traitements organique et organo-minéral, la durée du cycle de culture (1 année) est trop courte pour tirer des conclusions définitives. Cependant les enseignements qui se dégagent de ces résultats sont une claire indication de ce que pourrait être l'évolution de ces caractéristiques à long terme. Et le fait que les traitements de fumure organique et surtout organo-minérale induisent les meilleurs rendements, confirme à court terme (1 année) à Adiopodoumé, ce que l'essai de Bouaké montre à long terme (11 années). On peut donc supposer, en l'absence d'analyses de sol appropriées (essai de Bouaké), que le maintien du rendement de l'igname bien au delà de 10 t ha^{-1} sur 11 ans par la fumure organo-minérale serait dû aux effets améliorants du fumier sur les caractéristiques du sol et donc sur sa fertilité (Jones, 1971; Moukan et Tchato, 1987). En tout cas, au vu des résultats des deux essais, il apparaîtrait que les amendements organo-minéraux soient une pratique idéale, tout au moins dans le cas de l'igname, dans la double perspective de l'augmentation et du maintien du rendement de cette culture.

Conclusion

L'agriculture moderne en Côte d'Ivoire est caractérisée par la motorisation de presque toutes ses étapes - depuis le défrichement, le travail du sol jusqu'à la fertilisation et au semis. De toutes ces étapes, celle du défrichement mécanisée est la plus critique dans la mesure où il ampute au sol sa couche humifère.

Autant cette agriculture ne peut se passer des engrais minéraux, autant elle ne peut y recourir comme seul moyen d'assurer la reproductibilité des systèmes de production. De fait, la nature de nos sols ainsi que les nombreux cas d'abandon de sites d'exploitation agricole pour raison de baisse spectaculaire de productivité, amènent à repenser tout le processus de la pratique agricole moderne. Il faut en tout premier lieu que la méthode de défrichement mécanisé préserve au maximum la couche humidère et qu'ensuite les itinéraires techniques maintiennent, ou si possible augmentent le taux de matière organique dans le sol cultivé. Les résultats des essais rapportés ci-dessus montrent bien que la fumure minérale *sensu stricto* induit de moins bons résultats que la fumure organo-minérale. Ceci suggère que l'utilisation des divers types de résidus agricoles généralement disponibles (résidus de récolte, sciure de bois, fumier, plantes de couverture, etc.) seuls, mais surtout en association avec les engrais minéraux, permettrait une agriculture intensive et productive en Côte d'Ivoire, aussi bien dans le secteur moderne que paysannal.

Bradfield (1968) n'écrivait-il pas que "Les nations qui ont appris à judicieusement associer les engrais minéraux à la matière organique sont aujourd'hui les pays dont l'agriculture est la plus productive" - une idée à méditer.

Bibliographie

BOA, D. 1989. Caractérisation, propriétés hydrodynamiques, contraintes et potentialités agronomiques des sols gravillonnaires : cas de Boro-Borotou (Région de Touba, nord-ouest de la Côte d'Ivoire). Thèse de Docteur-Ingénieur, Université Nationale de Côte d'Ivoire, Abidjan. 126p.

BRADFIELD, R. 1968. The role of organic matter in soil management and the maintainance of soil fertility. In: *Study week on organic matter and soil fertility*, 107-121. Pontificia Academia Scientiarum no. 10. New York: John Wiley and Sons.

BRADY, N.C. 1974. *The nature and properties of soils*. 8th ed. New York: Macmillan Publishing Co.

COURSEY, D.G. 1967. *Yams*. Tropical Agriculture Series. London: Longmans.

DE SAINT-AMAND, J. 1985. Fertilisation de l'igname sur le sol de basse Côte d'Ivoire. Document ORSTOM. Centre d'Adiopodoumé. Abidjan, Côte d'Ivoire : ORSTOM. 102p.

DIOMANDE, M. et GODO, G. 1982. Rapport de mission BETPA. Document ORSTOM. Centre d'Adiopodoumé. Abidjan, Côte d'Ivoire : ORSTOM. 6p.

GODO, G. 1982. Réflexions sur l'utilisation de l'espace par les systèmes d'exploitation agricole en Côte d'Ivoire forestière. Nécessité d'une approche "agropédologique". Document ORSTOM. Centre d'Adiopodoumé. Abidjan, Côte d'Ivoire : ORSTOM. 8p.

GODO, G. et YORO, G. 1985. Recherche sur les systèmes de cultures à base de manioc en milieu paysannal dans le Sud-Est ivoirien (Bonoua-Adiaké). Deuxième phase : Résultats d'enquêtes et observations au champ. Document ORSTOM. Centre d'Adiopodoumé. Abidjan, Côte d'Ivoire : ORSTOM. 15p.

JONES, M.J. 1971. The maintainance of soil organic matter under continuous cropping at Samaru, Nigeria. *Journal of Agricultural Science* 77: 473-482.

KALMS, J.M. et CHABALIER, P.F. 1981. Bilan d'un essai agronomique de longue durée. "Systèmes culturaux de Bouaké". Synthèse des principaux résultats de 1967 à 1978 d'un test de différents systèmes culturaux. Institut des Savanes, Département des Cultures Vivrières. Bouaké, Côte d'Ivoire : IDESSA/DCV. 90p.

MOREAU, R. 1984. Evolution des sols sous différents modes de mise en culture en Côte d'Ivoire forestière et préforestière. *Cahiers ORSTOM, série Pédologie* 20(4) : 311-325.

MOUKAM, A. et TCHATO, E.D. 1987. Utilisation des résidus agricoles (parche de café) en vue de l'amélioration de la fertilité de certains sols ferrallitiques en culture continue. In: *IBSRAM Proceedings n°4*, 101-111. Bangkok : IBSRAM.

ROOSE, E. 1983. Ruissellement et érosion avant et après défrichement en fonction du type de culture en Afrique Occidentale. *Cahier ORSTOM, série Pédologie* 20 (4) : 327-339.

SANCHEZ, P.A. 1976. *Properties and management of soils in the tropics*. New York: John Wiley and Sons.

YOUNG, A. 1989. *Agroforestry for soil conservation*. Science and Practice of Agroforestry no. 4. Wallingford, UK: CAB International/Nairobi: ICRAF.

Maintenance and management of organic matter in tropical soils

ANTHONY S.R. JUO[*]

Abstract

Tropical soil ecosystems are characterized by continuously warm temperature and low clay activity. Published data indicate that the rates of decomposition of both fresh plant residues and humified soil organic matter were 3 to 5 times faster in the humid tropical environment than those under temperate conditions. The maintenance of adequate levels of soil organic matter may be achieved through (i) frequent recycling of plant and animal residues, (ii) judicial use of chemical fertilizers to offset nutrient losses due to crop removal and leaching, (iii) minimum disturbance of the soil surface, and (iv) integrating multipurpose trees and perennials into annual cropping systems.

Promising practices of organic-matter management, such as mulching, green manuring, composting, and tree and annual crop mixed farming are well known. Much research is needed to integrate these practices into the small-farm systems in the tropics.

Introduction

The role of organic matter in agriculture has turned a full circle. It began with mankind's discovery of the benefits of organic matter in shifting cultivation. This was followed by the use of green manure, compost, and

[*] Department of Soil and Crop Sciences, Texas A&M University, College Station, Texas, USA 77843.

farmyard manure on intensively cultivated small farms. The subsequent development of industrialized agriculture using chemical fertilizers and large farm machinery has, to a large extent, made the role of organic matter somewhat obscure. More recently, however, increased energy costs in agriculture, and environmental concerns for greenhouse-gas emission and groundwater pollution have all but put organic matter back in the centre of the agricultural arena.

The contribution of organic matter to soil productivity has long been recognized in traditional agriculture. Soil organic matter serves as a reservoir of plant nutrients, increases cation-exchange capacity, improves soil aggregation, regulates soil temperature, increases soil water content, and plays an important role in soil structure development. The beneficial effects of organic matter are generally more pronounced in soils with low-activity clays (i.e. kaolinite, halloysite) than those with high-activity clays (i.e. smectite, vermiculite).

Under similar soil and climatic conditions, a natural ecosystem generally maintains a higher soil organic-matter level at steady state than its cultivated or disturbed counterpart (agroecosystem). This is because a significant portion of the biomass in the agroecosystem is often removed for human and livestock consumption, and the regenerative cycles of soil organic matter and nutrients are thus disrupted. Moreover, conventional soil management practices such as ploughing and bare fallow generally accelerate biodegradation of organic matter in cultivated soils (Tate, 1987).

Soil organic matter also plays a significant role in regulating the global carbon budget. It has been estimated that organic matter in world soils retains approximately 300 billion metric tons of carbon (Bohn, 1976). Further loss of soil organic matter due to deforestation and improper management of farmlands may result in a significant increase in carbon dioxide emission into the atmosphere.

Most tropical upland ecosystems are characterized by a continuously warm climate, infertile soils, and a rapid rate of decomposition of organic matter. Thus the maintenance of adequate levels of soil organic matter in tropical agroecosystems rests on a delicate annual balance among production, decomposition, and accumulation. The role of organic matter in tropical agriculture has been reviewed by FAO (1975, 1977, 1980), Lathwell and Bouldin (1981), Wetselaar and Ganry (1982), Lal and Kang (1982), Greenland (1986), and Sanchez and Miller (1986). In general, these reviewers emphasized the importance of organic matter as a reservoir of plant nutrients, particularly N, P, and S. They also pointed out the lack of quantitative data dealing with the effects of organic matter on soil physical and biological properties.

This paper attempts to address selected topics of organic-matter research in the tropics with special reference to the management and maintenance of soil organic matter in soils dominating in low-activity clays.

Decomposition and accumulation

The term organic matter in a soil ecosystem generally refers to both the aboveground organic residues and the belowground organic matter. The aboveground organic matter includes plant and animal residues, young and old humic substances, soil macro- and microorganisms. The freshly added or partially decomposed plant residues and their humified turnover products, otherwise known as the young humic substances, constitute the labile organic-matter pool which controls the nutrient-supplying power of the soil ecosystem, particularly nitrogen.

The dynamics of organic matter in a soil ecosystem may be better understood by establishing simple models of productivity, decomposition, and accumulation in natural and disturbed ecosystems. It has been pointed out that tropical rainforests can produce five times as much organic matter and biomass as temperate forests; but the rate of decomposition in the tropics is also about five times greater than that under a temperate environment (Nye, 1961; Goh, 1980). As additions and losses of organic matter occur in the soil simultaneously, a steady state is attained when the rate of addition is equal to the rate of loss. The changes in soil organic matter which are usually expressed in terms of either organic carbon or nitrogen, follow approximately the first-order reaction, in which the rate of loss is a function of the total amount present.

$$dC/dt = -kC \qquad (1)$$

Upon integration, equation (1) becomes:

$$C(t) = C_o\, e^{-kt} \qquad (2)$$

where C is the amount of organic carbon per unit area in a fixed amount of soil, t is time, k is the decomposition constant, and C_o is the amount of C when $t = 0$.

On the other hand, accumulation of soil organic matter may occur as a result of the annual addition of fresh organic material to the system. Hence, the net change in "young" organic matter, C, may be described as in equation (3).

$$dC/dt = -kC + h\, C_i \qquad (3)$$

Upon integration, equation (3) becomes:

$$C = hC_i(1 - e^{-kt})/k \qquad (4)$$

at steady state,

$$C_{max} = hC_i/k \qquad (5)$$

where C_i is the addition of organic matter unit area per unit time and h is a constant.

According to Henin and Dupuis (1945), the net change of soil organic matter due to accumulation of young organic matter and the decomposition of old organic matter may be estimated by combining equations (2) and (4):

$$C = hC_i(1 - e^{-kt})/k - C_o e^{-kt} \qquad (6)$$

Or

$$C = C_{max} (1 - e^{-kt}) - C_o e^{-kt} \qquad (7)$$

The above equations assume that the same k value applies for both the young and the old organic-matter fractions. Moreover, long-term observations reported in temperate regions (Jenkinson, 1977) indicated that the decomposition constant, k, decreases with time. A rapid decomposition of plant and animal residues occurs during the initial phase, followed by a slower rate of decomposition. Therefore, the above simple equation is evidently not entirely satisfactory.

Based on knowledge of organic matter obtained primarily from temperate regions, several workers have developed more complex computer simulation models in which different decomposition rates for the various fractions of organic matter were treated separately as compartments (Jenkinson and Rayner, 1977; van Veen and Paul, 1981; van Veen et al., 1984; Parton et al., 1987, 1988). These workers further divide soil organic matter into several components, namely soil biomass (the most readily decomposable material), degradable plant material, resistant plant material, chemically resistant organic matter, and physically resistant organic matter. Although these complex simulation models help to advance our understanding of the properties and functions of the various fractions of organic matter in the soil ecosystem, calibration and validation of such complex models are extremely difficult because of the lack of quantitative knowledge on the physical, chemical, and biological processes of organic matter in tropical ecosystems. For practical purposes, the simple first-order treatment is still used by many researchers.

The rate of decomposition of both fresh and humified organic matter in soil ecosystems are determined by many factors, such as soil texture, mineralogy, temperature, moisture, aeration, microbial activity, and ecosystem disturbance. It has been shown that the rates of decomposition of both fresh plant materials and humified soil organic matter are four to five times faster under a tropical environment than those under temperate conditions (Jenkinson, 1981). Results from field investigations conducted at the International Institute of Tropical Agriculture (IITA) near Ibadan, Nigeria (7° 30'N, 3° 54'E, altitude 700 m), showed that the rate of decomposition of ryegrass residues was four times faster than that obtained at Rothamsted, England (Jenkinson and Ayanaba, 1977).

In another IITA experiment, Mueller-Harvey *et al.* (1985) reported that the half-lives of soil organic C, N, P, and S in the surface 10-cm layer of a Kaolinitic Alfisol newly cleared from secondary forest were 3.5, 3.3, 4.7, and 2.3 years respectively (Table 1). In this experiment, fresh forest litter was removed from the soil surface and the field was cropped with soybean and maize under no tillage. Subsequent observations on crop performance showed that acute symptoms of S deficiency in maize appeared during the second year of cropping, and N and P deficiencies occurred thereafter. While these experiments clearly illustrate the climatic effect on the rate of organic-matter degradation, other environmental factors such as soil texture, drainage condition, microbial activity, and the size of the labile organic-matter pool may also influence the rate and process of decomposition.

Table 1. Mineralization of organic C, N, and S in a Kaolinitic Alfisol after forest clearance in Ibadan, Nigeria.

Element	Initial level	After 22 months	Loss**	Half-life
	---------- mg kg^{-1} ----------		(%)	(yr)
C	14900	10600	29	3.5
N	1740	1190	32	3.3
S	163	92	44	2.3

** Significant at the 0.1% level.
Source: Mueller-Harvey *et al.*, 1985.

The relatively faster rate of decomposition under a tropical environment implies that high equilibrium levels of soil organic matter in tropical agro-ecosystems are difficult to attain. Thus in order to maintain a reasonably high level of organic-matter content in cropped soils comparable to that in the natural soil ecosystem, frequent additions of fresh plant residues and minimum disturbance of the soil surface would be required.

It should also be pointed out that the quantity and quality of soil organic-matter accumulation depend upon the type of organic matter used. For example, application of easily degradable materials (i.e. C/N ratio <30) would increase the labile nitrogen pool. The application of plant materials with a high C/N ratio (i.e. >100) favours humus formation, and therefore benefits soil structural development. What are the adequate levels of young and old organic matter for crop growth and for maintaining soil structural stability in tropical soils? What management practices should be used? Unfortunately, few long-term studies in tropical locations are designed to address these fundamental issues.

Effects on soil properties

The beneficial effects of organic matter on soil chemical and physical properties are well known. But experimental evidence in the tropics is scanty. Soil organic matter reduces phosphate fixation capacity of tropical soils high in Fe and Al oxides or sesquioxides. Uehara and Gilman (1981) compared the phosphate-sorption capacity of surface and subsurface soil samples as a function of sesquioxide content (Figure 1). These results imply that for high P-'fixing' soils (i.e. oxide-rich soils derived from volcanic and ferromagnesian rocks), management systems that are capable of accumulating and maintaining a greater amount of soil organic matter in the surface horizon would increase P availability from both organic and fertilizer sources.

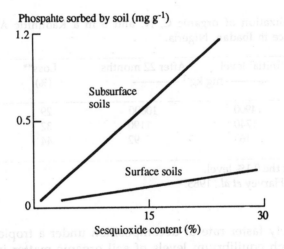

Figure 1. Effect of organic-matter content in reducing phosphate sorption on high sesquioxide soils (Uehara and Gilman, 1981).

For the widely spread kaolinitic and siliceous soils in tropical West Africa, low cation-exchange capacity is a major chemical constraint to soil management. In such soils, maintaining an adequate level of soil organic matter is crucial to crop growth (Ofori, 1973; Juo and Lal, 1977; Sanchez et al., 1983; Juo and Kang, 1987). Results from a long-term experiment conducted at Ibadan, Nigeria, have depicted a close correlation between soil organic carbon content and effective CEC under three management systems (Figure 2). These results also showed that the level of soil organic matter depends upon the amount of annual fresh biomass returned to the soil as mulch. The highest level of organic carbon occurred in the continuous maize plots, where crop residues were returned as mulch twice a year.

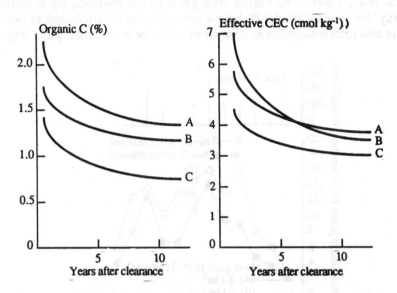

Figure 2. Changes in soil organic C and effective cation-exchange capacity (ECEC) of a Kaolinitic Alfisol in Ibadan, Nigeria, after forest clearance and as affected by three management schemes: (A) continuous maize, 2 crops per year, no-till and stover mulched; (B) maize/cassava intercropped, no-till and maize stover mulched; (C) continuous maize, 2 crops per year, no-till, maize stover removed (Juo and Kang, 1987).

The effects of organic matter on soil physical conditions are equally evident. Soil structure and aggregates in kaolinitic and sandy soils are less stable. Good water infiltration in soils under natural vegetation is primarily due to intensive root and soil faunal activities. It may take 10 to 30 years of forest fallow to build a good physical condition of the surface horizons of the soil in terms of soil structure, bulk density, and water infiltration. But it would take a relatively short period of time for the soil to lose its good physical condition under cultivation (Juo and Lal, 1977). It has been observed that bulk density of the surface 10 cm of a newly cleared forest Alfisol increased from 1.13 to 1.31 g cm^{-3} after the soil was cropped for 22 months (Mueller-Harvey et al., 1985).

The effect of organic matter on soil water retention in the subhumid tropics may be illustrated by comparing soil moisture contents under mulched and unmulched soils monitored throughout the two growing seasons at Ibadan (Juo, 1980; Lal *et al.*, 1984). The higher crop yield in the mulched plots during the unusually dry year (rainfall 20% below normal) may be attributed to better soil moisture and nitrate supplies as compared with the unmulched plots (Figure 3).

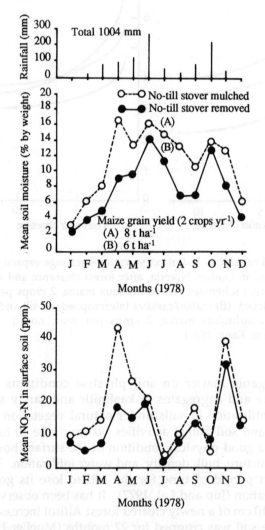

Figure 3. Effects of crop-residue mulch on soil moisture storage and nitrate content in the surface layer (0-15) of a Kaolinitic Alfisol in Ibadan, Nigeria (Juo, 1980).

For the strongly acidic soils (i.e. Ultisols and Oxisols) under natural vegetation, cation-exchange sites on clays and soil organic matter are predominantly Al-saturated (Juo, 1977; Bloom et al., 1979; Juo and Kamprath, 1979). Thus on land newly cleared from tropical rainforest, Al phytotoxicity and P, Ca and Mg deficiencies may become critical factors limiting crop growth if fresh plant residues are not returned to the soil. Phytotoxicity of Mn is less common, and it generally occurs on younger acid soils derived from parent materials rich in Mn minerals. Managing such soils for crop production would require the use of lime or other forms of Ca and Mg amendments (Friessen et al., 1981; Sanchez and Salinas, 1981; Sanchez et al., 1983).

Perhaps a key to improving acid tropical soils (i.e. pH <5.0 measured in water) for continuous cropping is the displacement of exchangeable Al from the organic and inorganic exchange complexes with Ca. In some populated areas of the humid tropics, the conventional practice of liming may be economically prohibitive. Alternative soil and crop management systems appropriate to local conditions are yet to be developed. The question remains whether it is ecologically feasible to develop vast areas in the humid tropics for cattle and intensive food crop production (Juo, 1989).

The benefits of soil organic matter as a soil nutrient reservoir has been much emphasized. In particular, the supply of nitrogen through mineralization of the labile organic-matter pool and through the use of green manure in tropical agroecosystems has been reviewed respectively by Sanchez and Miller (1981) and Bouldin et al. (1979).

Soil management options

The management of soils for optimal organic-matter accumulation may be achieved through: (i) return of crop residues to the soil, (ii) optimum use of chemical fertilizers to offset nutrient losses due to leaching and crop removal from the total ecosystem, (iii) integration of useful trees and perennials into annual cropping systems and (iv) minimum disturbance of the soil surface.

Where economically feasible, the role of soil organic matter as a source of plant nutrients may be replaced to a large extent by chemical fertilizers. But the maintenance of adequate soil structure is a long-term process and it can only be attained when the bulk of the annual biomass production is recycled within the agroecosystem. Hence natural bush fallow, green manuring, composting, and the use of farmyard manure and household wastes are all part of the organic matter-based farming systems. As agriculture became more commercialized and cropping practices became more intensive, there was an increased use of manufactured inputs such as chemical fertilizers and large farm machinery to alleviate soil physical and fertility constraints resulting from continuous cultivation.

Few people would dispute the remarkable achievement of industrialized agriculture in the temperate regions in terms of its capacity and efficiency in producing large quantities of food and fiber to meet increasing human needs. The more recent success of the 'Green Revolution' in tropical and subtropical Asia is equally indisputable where high-yielding wheat and rice varieties and fertilizer manufacture and use have been successfully integrated into the ancient but intensive irrigated farming systems.

But the question remains whether such systems can be transferred to a larger part of the tropics, where the predominant soils are those with low-activity clays. Moreover, in terms of global greenhouse-gas emission and the depletion of nonrenewable resources such as fossil fuels and mineral deposits, many national planners and researchers have begun to question whether further spreading of industrialized agriculture is environmentally sound and economically sustainable in the long run.

While the economic and technical feasibility for increasing the use of organic inputs in industrialized agriculture are being reexamined (Chen and Avnimelech, 1986), societies in the tropics may need to take advantage of the high annual biomass production potential to develop low-cost farming systems through maximizing biomass production and recycling. Traditional organic-based soil and crop management practices, such as composting, green manuring, residue mulching, cereal-legume rotation, and intercropping, are well documented. But a challenge facing soil scientists and agronomists is how to modernize these component technologies and integrate them into farming systems appropriate to soil, climate, and socioeconomic environments in the tropics.

To increase biomass production and to discourage biomass burning in the nutrient-depleted tropical agroecosystems, the use of chemical fertilizers is essential. Although the use of inorganic-N fertilizers could be reduced by improving biological-N fixation, it is foreseeable that the demand for P, K, Ca, and Mg from external inorganic sources in the tropics would increase as more productive agroecosystems are developed and adopted in the future.

For upland farming systems, promising organic-matter management systems include the use tree legumes such as leucaena (*Leucaena leucocephala*), gliricidia (*Gliricidia sepium*) and calliandra (*Calliandra calothyrsus*) and annual legumes such as mucuna (*Mucuna utilis*) and kudzu (*Pueraria phaseoloides*) as green manure and as cover crop in rotation with food crops (Kang *et al.*, 1981; Kang *et al.*, 1985; Wilson, 1981; Bowen *et al.*, 1988; Hulugalle *et al.*, 1986; Yamoah *et al.*, 1986; Sanchez and Benites, 1987). Significant findings from these experiments may be summarized as follows: (i) the amount of inorganic N accumulated in the fallow soil under legume cover crop was highly correlated with the aboveground dry-matter yield and N content plus the residual soil N in plots cropped with maize on an Oxisol in Cerrado Brazil; (ii) tree legumes especially leucaena, are capable of fixing a large quantity of nitrogen and of cycling large amounts of mineral nutrients from subsoil horizons in Alfisol ecosystems; and (iii) both alley cropping and cover-crop fallow enhance soil organic matter accumulation through minimizing soil erosion losses and reducing surface soil temperature. Although the experimental evidence produced by these researchers is impressive and convincing, much research and development is needed to integrate these practices into traditional and new farming systems in the tropics.

With regard to green manure and cover crops, perhaps there is a need for research to develop and maintain a greater biodiversity of plant species (NAS, 1979; Bouldin *et al.*, 1979). One benefit of biodiversity would be to maintain a wide range of legume species that can be used as N sources for nonlegume food crops, and to minimize disease and pest infestation. As regards composting, mulching, and intercropping, much research and development is needed in improving farm machinery and implements for land preparation, planting, and harvest that can be adopted by both small- and medium-sized farms. Adoption of mixed tree legume and food-crop systems would depend largely upon the social and economic environments.

Organic matter and land rehabilitation

Perhaps one of the most important tasks in organic-matter management is to increase biomass production and vegetation coverage of vast areas of degraded land in the subhumid and semiarid regions. China has embarked on massive efforts of tree planting and protection in her vast areas of severely degraded subhumid and semiarid lands. India and Pakistan have also taken steps to rehabilitate some of their heavily degraded areas. Deforestation in Latin America and Southeast Asia has become a matter of international as well as national concern. The extent of land degradation in subhumid and semiarid sub-Saharan Africa is equally severe, and collaborative efforts in rehabilitation of agricultural and nonagricultural lands are urgently needed. Moreover, management of organic matter in agroecosystems cannot be restricted to soil management of agricultural land alone. Loss of natural vegetation and biological diversity due to increasing demands for fodder and fuelwood has become a major influence on land degradation in Africa.

Very few research programs in the past dealt specifically with the topic of organic matter and land rehabilitation. Quantitative data on organic-matter dynamics and nutrient cycling within an agricultural watershed are particularly scarce. Results from such studies should provide the much-needed knowledge and information for developing ecologically sound land management systems in the tropics.

References

BLOOM, P.R., McBRIDE, M.B. and WEAVER, R.M. 1979. Aluminum organic matter in acid soils. Buffering and solution. *Soil Science Society of America Journal* 43: 488-493.

BOHN, H.L. 1976. Estimate of organic carbon in world soils. *Soil Science Society of America Journal* 40: 468-470.

BOULDIN, D.R., MUGHOGHO, S., LATHWELL, D.J. and SCOTT, T.W. 1979. *Nitrogen fixation by legumes in the tropics.* Cornell International Agriculture Bulletin no. 75. Ithaca, NY: Cornell University.

BOWEN, W.T., QUINTANA, J.O., PEREIRA, J., BOULDIN, D.R., REID, W.S. and LATHWELL, D.J. 1988. Screening legume green manures as nitrogen sources to succeeding crops: 1. Fallow methods. *Plant and Soil* 111: 75-80.

CHEN, Y. and AVNIMELECH, Y., eds. 1986. *The role of organic matter in modern agriculture.* The Hague: Martinus Nijhoff Publishers. 306p.

FAO (Food and Agriculture Organization of the United Nations). 1975. *Organic materials as fertilizers.* FAO Soils Bulletin no. 27, Rome: FAO. 394p.

FAO. 1977. *Organic materials and soil productivity.* FAO Soils Bulletin no. 35, Rome: FAO.

FAO. 1978. *Organic recycling in Asia.* FAO Soils Bulletin no. 36, Rome: FAO. 417p.

FAO. 1980. *Organic recycling in Africa.* FAO Soils Bulletin no. 43. Rome: FAO.

FRIESSEN, D.K., JUO, A.S.R. and MILLER, M.H. 1982. Residual value of lime and leaching of Ca in a kaolinitic Ultisol in the high rainfall tropics. *Soil Science of America Journal* 46: 1184-89.

GOH, K.M. 1980. Dynamics and stability of organic matter. In: *Soils with variable charge,* ed. B.K.G. Theng, 373-393. New Zealand Soil Science Society. Low Hutt: NZSSS.

GREENLAND, D.J. 1986. Soil organic matter in relation to crop nutrition and management. In: *Proceedings of the International Conference on Management and Fertilization of Upland Soils* (Nanjing, 1986). Nanjing, China: Institute of Soil Science.

HENIN, S. and DUPUIS, M. 1945. Essai de bilan de la matière organique du sol. *Annales Agronomiques* 15: 17-29.

HULUGALLE, N., LAL, R. and TER KUILE, C.H.H. 1986. Amelioration of soil physical properties by Mucuna after mechanized land clearing of a tropical rain forest. *Soil Science* 141: 219-224.

JENKINSON, D.S. 1977. Studies on the decomposition of plant material in soil. *Journal of Soil Science* 28: 424-434.

JENKINSON, D.S. and AYANABA, A. 1977. Decomposition of C-14 labelled plant material under tropical conditions. *Soil Science Society of America Journal* 41: 912-915.

JUO, A.S.R. 1977. Soluble and exchangeable Al in selected Ultisols and Alfisols in West Africa. *Communication in Soil and Plant Analysis* 8: 17-35.

JUO, A.S.R. 1980. Nitrate profile in a kaolinitic Alfisol under fallow and continuous cropping. In: *Nitrogen cycling in West African ecosystems*. Stockholm: SCOPE/UNEP/Royal Swedish Academy of Sciences.

JUO, A.S.R. 1989. New farming systems development in the wetter tropics. *Experimental Agriculture* 25: 145-63.

JUO, A.S.R. and KAMPRATH, E.J. 1979. Copper chloride as an extractant for estimating the reactive Al pool in acid soils. *Soil Science Society of America Journal* 43: 35-38.

JUO, A.S.R. and LAL, R. 1977. The effect of fallow and continuous cultivation on physical and chemical properties of an Alfisol in western Nigeria. *Plant and Soil* 47: 507-584.

KANG, B.T., WILSON, G.F. and SIPKENS, L. 1981. Alley cropping maize and leucaena in southern Nigeria. *Plant and Soil* 63: 165-179.

KANG, B.T., WILSON, G.F. and LAWSON, T.L. 1985. Alley cropping: a stable alternative to shifting cultivation. International Institute of Tropical Agriculture. Ibadan, Nigeria: IITA. 22p.

LAL, R. and KANG, B.T. 1982. Management of organic matter in soils of the tropics and subtropics. In: *Transactions of the XIIth International Congress of Soil Science (New Delhi)*, vol. IV, 152-178. New Delhi, India: Indian Society of Soil Science.

LAL, R., JUO, A.S.R. and KANG, B.T. 1984. Chemical approaches towards increasing water availability to crops including minimum tillage systems. In: *Chemistry and World Food Supplies*, ed. T.W. Shemil, 55-77. Oxford/New York: Pergamon Press.

LATHWELL, D.J. and BOULDIN, D.R. 1981. Soil organic matter and soil nitrogen behavior in cropped soils. *Tropical Agriculture* (Trinidad) 58: 341-348.

MUELLER-HARVEY, I., JUO, A.S.R. and WILD, A. 1985. Soil organic C, N, S and P after forest clearance in Nigeria: mineralization rate and spatial variability. *Journal of Soil Science* 36: 585-591.

NYE, P. 1960. Organic matter and nutrient cycles under moist tropical forest. *Plant and Soil*, 333-45.

NAS (National Academy of Science). 1979. *Tropical legumes: resources for the future*. Washington, DC: NAS.

OFORI, C.S. 1973. Decline in fertility status in a tropical ochrosol under continuous cropping. *Experimental Agriculture* 9: 15-22.

PARTON, W.J., SCHIMEL, D.S., COLE, C.V. and OJIMA, D.S. 1987. Analysis of factors controlling soil organic matter levels in Great Plains grasslands. *Soil Science Society of America Journal* 51: 1173-1179.

PARTON, W.J., STEWART, J.W.B. and COLE, C.V. 1988. Dynamics of C, N, P, and S in grassland soils. *Biogeochemistry* 5: 109-131.

SANCHEZ, P.A. and BENITES, J.R. 1987. Low-input cropping for acid soils in the humid tropics. *Science* 238: 1521-1527.

SANCHEZ, P.A. and MILLER, R.H. 1986. Organic matter and soil fertility management in acid tropical soils. In: *Transactions of the XIIIth ISSS Congress* (Hannover, 1986), vol. 6, 609-625. Wageningen: ISSS.

SANCHEZ, P.A. and SALINAS, J.G. 1983. Low input technology for managing Oxisols and Ultisols in tropical America. *Advances in Agronomy* 34: 279-406.

SANCHEZ, P.A., VILLACHICA, J.A. and BANDY, D.E. 1983. Soil fertility dynamics after clearing a tropical rain forest in Peru. *Soil Science Society of America Journal* 47: 1171-1178.

TATE, R.L. III. 1987. *Soil organic matter: biological and ecological effects.* New York: John Wiley and Sons. 291p.

UEHARA, G. and GILMAN, G. 1981. *The mineralogy, chemistry, and physics of tropical soils with variable charge clays.* Boulder, Colorado: Westview Press. 170p.

VAN VEEN, J.A. and PAUL, E.A. 1981. Organic matter dynamics in grassland soils. I. Background information and computer simulation. *Canadian Journal of Soil Science* 61: 185-201.

VAN VEEN, J.A., LADD, J.N. and FRIESSEL, M.J. 1984. Modeling of C and N turnover through microbial biomass in soil. *Plant and Soil* 76: 257-274.

WETSELAAR, R. and GANZY, F. 1982. Nitrogen balance in tropical agroecosystems. In: *Microbiology of tropical soils and plant productivity*, ed. Y.R. Dommergues and H.G. Diem, 1-33. The Hague: Martinus Nijhoff.

WILSON, G.F. 1981. Mucuna in-situ mulch. *Annual Report.* International Institute of Tropical Agriculture. Ibadan, Nigeria: IITA.

YAMOAH, C.F., AGBOOLA, A.A., WILSON, G.F. and MULONGOY, K. 1986. Soil properties as affected by the use of leguminous shrubs for alley cropping with maize. *Agricultural Ecosystems and the Environment* 18: 167-177.

Manures and organic fertilizers: their potential and use in African agriculture

C.S. OFORI and R. SANT'ANNA[*]

Abstract

The paper considers the potential supply of nutrients from cattle dung, how it can be used efficiently, and the use of other organic materials as sources of nutrients for increasing soil productivity in the sub-Saharan African region. It indicates the need for clearly defined approaches to research to ensure the the best use organic materials is made of.

Introduction

Animal manures, particularly cattle dung, were the main source of plant nutrients for the maintenance of soil fertility in settled agriculture until the advent of mineral fertilizers. The use of mineral fertilizers in many developed countries has increased to the point where they have almost completely superseded animal manures in agricultural production systems. At the same time, specialized livestock production without associated arable farming has resulted in the accumulation of large quantities of manure on such farms, posing serious disposal and environmental problems.

[*]   Respectively: Senior Officer (soil management), FAO, AGLS, Rome, Italy; and Regional Officer (soil resources), FAO, RAFR, Accra, Ghana.

In many developing countries, on the other hand, farmers have limited financial resources and can rarely afford to purchase sufficient mineral fertilizers. Where mixed farming is practiced, although the dung may be used to contribute to the maintenance of soil fertility, the quantities produced are inadequate in most cases. The "compound farms" in West Africa, for example, are kept productive through the application of all the manure and compost available from the kraal and household refuse.

The decline in *per capita* food production as a result of population growth, the shorter recuperation period of the fallows, the loss in soil fertility, and the high cost of mineral fertilizers have made it necessary to maintain attention on the use of manures and other organic materials to sustain soil productivity. An integrated approach to the maintenance of soil productivity, with the complementary use of both mineral and organic fertilizers, offers a good opportunity to the small farmer to maintain yields at reasonable and sustainable levels.

Cattle manure is by far the most important source of nutrients among the various animal dungs. Although the cattle population, and consequently the dung output, have greatly increased in developing countries, there are several constraints on the efficient utilization of the nutrients from the dung. The potential supply of nutrients from cattle dung, its efficient use, and the use of other organic materials as sources of nutrients for increasing soil productivity in the sub-Saharan Africa region are discussed in this paper.

Potential for organic manures

Adequate statistics on animal population in the region are available, but there is a serious lack of reliable data for calculating the potential dung production. However, few calculations have been made using Indian averages. Singh (1975) based his calculations on a dung production of 5.5 t animal^{-1} yr^{-1}, with a nutrient content of 0.29% N, 0.08% P_2O_5 and 0.23% K_2O. Important factors to consider are the average size and the diet of animals in developing countries. Generally, cattle in developing countries are smaller than those in developed countries, and their diet is of low-quality forage, with intakes mostly near maintenance level.

Further, there are wide variations in the nutrient content of the dung reported by various workers, as shown in Table 1. Factors such as husbandry systems, storage methods, and the basis of reporting (whether on a dry or a wet basis) have to date made comparisons difficult.

Table 1. Nutrient content of cattle dung.

N	Percentage P$_2$O$_5$	K$_2$O	Source	Remarks
0.6	0.1	0.5	Berryman (1965)	76% moisture, British
0.6	0.1	0.5	Benne et al. (1961)	79% moisture, USA
1.58	0.70	1.62	Dhanyadee (1984)	Thailand†
1.09	0.53	1.68	Suzuki et al. (1980)	Thailand (exp. stn.)†
0.7-1.3	0.55-0.76	0.96-2.4	Ofori (1962)	Ghana (exp. stn.)†
1.74	0.67	2.40	Kwakye (1980)	Ghana (exp. stn.)†
1.09	0.35	1.58	Hartley (1937)	Nigeria†
0.29	0.08	0.23	Singh (1975)	Indian average (wet-weight basis)
0.53-0.85	0.25-0.42	1.31-2.22	Cooke (1975)	Tanzania†
0.95	0.30	1.26	Cooke (1975)	Malawi†

† Dry-weight basis.

Based on the Indian averages, and using data on animal population in the region (FAO, 1987), estimates for N, P and K availability from cattle dung in Africa are given in Table 2.

Table 2. Potential nutrient supply from cattle manure in Africa.†

Year	N	P$_2$O$_5$	K$_2$O	Total N + P$_2$O$_5$ + K$_2$O	N + P$_2$O$_5$ + K$_2$O (kg ha^{-1} arable land)
		--------------- million t ---------------			
1979/81	2.41	0.66	1.91	4.98	38.8
1986	2.57	0.71	2.04	5.32	40.3

† Africa: sub-Sahara, excluding South Africa.

Although the figures indicate substantial potential amounts of available nutrients, which if effectively utilized would significantly increase soil fertility in the region, the actual amounts available depend on several factors, including the ease of collection, the amount collected, the cost of collection, and the methods used for storage and application. In many parts of Africa, the livestock owners are not necessarily the farmers. The nomadic system of keeping livestock makes collection and conservation of dung difficult.

The dung produced is subjected to the vagaries of the weather, resulting in caking and also nutrient leaching, particularly of nitrogen and potassium, when it rains. In mixed farming systems where dung is produced in kraals, it is used as fertilizers on fields sited in the immediate vicinity of the farmers' homesteads.

Value of N, P, and K in cattle dung

The wide variations found in the literature on the analysis of dung from various sources make precise estimation of the nutrient values difficult. One of the problems encountered in such comparisons is the lack of adequate background of the source, conservation method and, most importantly, the presentation of the analytical data. The variations shown in Table 2 illustrate the difficulties in interpreting the effect of levels of dung application compared with other sources of nutrient supply on crop yields if precise information is lacking.

The renewed interest in all sources of plant nutrients, both organic and mineral, for the maintenance of soil fertility should give a new impetus to a more systematic research approach for evaluating the effect of various organic fertilizers.

Effect of storage on the nutrient content of dung

One of the major factors influencing dung quality is the method of storage. In developed agricultural systems with a strong livestock component, dung, bedding and urine are properly stored to conserve nutrients. In sub-Saharan Africa, most of the dung produced is by animals grazing on free range. The collection and storage of dung under such a system is difficult and labour-intensive. Dung is therefore invariably of poor quality, since it is neither heaped nor covered and is subjected to severe weather conditions. Nitrogen is the main nutrient lost under poor storage conditions.

Kwakye (1980) reported reduced nutrient losses with improvements in storage practices based on a comparison of four methods of storage: loose, compact, loose-compact, and buried. The results presented in Table 3 indicate the highest N, P and K content of the dung, using the buried and compact methods of storage. The greatest nutrient losses occurred when the dung was completely or partially exposed to direct sunlight and rain. The nutrient loss was greatest for nitrogen, followed by potassium, and least for phosphorus. Burying the dung is labour-intensive, and more so when large quantities are involved. Compact storage, with the dung well protected from the direct impact of the weather, would be an improvement over current practices.

Table 3. The effect of different methods of storage on the N, P and K content of cow dung.

Method of storage	Dry matter† (%)		N† (%)		P₂O₅† (%)		K₂O† (%)	
Loose	22‡	28§	0.71‡	59§	0.50‡	28§	1.32‡	45§
Compact	26	16	0.93	47	0.51	27	1.51	37
Loose-compact	24	22	0.79	55	0.55	20	1.45	40
Buried	27	12	1.48	15	0.60	12	2.14	11
LSD (P = 0.05)			0.36		0.07		0.06	
(P = 0.01)			0.51		NS		0.11	

† The original sample contained 31% dry matter, 1.74% N, 0.67% P₂O₅, and 2.40% K₂O.
‡ Actual value in percent for all figures in the column.
§ Loss of the nutrient as a percent of the original sample after 3 months of storage, for all figures in the column.

Other agronomic practices could significantly contribute to the preservation of nutrients, particularly nitrogen in the dung. Kwakye (1980), using different sources of phosphates, showed that nitrogen loss could be reduced significantly by the addition of phosphatic fertilizer to the dung during storage. Single superphosphate was most effective, followed by rock phosphate. The dung was stored by the "compact" method, and the amount of P added was equivalent to 3% of the weight of the dung (Table 4).

Table 4. The effect of phosphate addition on percentage N content of cow dung stored for 3 months.

Treatment	N (%)†	N increase (%)
Dung + no phosphate	0.87	-
Dung + single superphosphate	1.50	73
Dung + triple superphosphate	1.04	21
Dung + rock phosphate	1.29	49
SE	±0.028	
LSD (P = 0.01)	0.09	

† Initial N content 1.74%.

Confinement of the animals will greatly improve the nutrient content of the dung if it is properly stored. If the urine is absorbed by the straw and mixed with the dung, its nitrogen will be conserved. It would appear, however, that the animal husbandry system in most parts of Africa is far from this goal, as its achievement would involve the provision of feed to animals in a feedlot. The full nutrient value of dung cannot therefore be fully exploited at present.

Effect of dung on soil productivity

Whereas the main purpose of applying dung and other organic wastes in subsistence agriculture is to restore soil nutrients, this function is marginal in modern agriculture. However, the application of organic wastes influences soil structure and soil biology, thereby significantly improving productivity.

Effect on soil physical properties

Improvement of soil structure is one of the most commonly cited effects of organic waste application.   Khaleel et al. (1981) found a highly significant correlation between increased organic carbon induced by manure application and a lowering in bulk density of the soil.   Similar results were cited from Rothamsted experimental plots by Jenkinson and Johnston (1977). The bulk density of soils from plots receiving only mineral fertilizers since 1852 was 1.52, compared with a density of 1.29 in plots amended with manure.   Other researchers obtained similar results, significant decreases in soil bulk density and increases in infiltration rates (Sharma et al., 1987). Improvement in soil physical properties is not limited to the plough layer with application of manure, as there is evidence that the subsoil is improved as well. Petterson and von Wistinghausen (1979) reported that subsoil was compacted in plots receiving only mineral fertilizers for a period of 20 years, but in the manured plots the subsoil had a better structure and lower bulk density. Some of the effects of manure application on soil physical properties may be the direct influence of organic matter in soils.   Aina (1979), in Nigeria, reported significant reduction of soil aggregation, aggregate stability, porosity, and hydraulic conductivity, and an increased bulk density after a ten-year continuous cultivation resulting in reduced soil organic-matter content.
The effects of organic-matter content on soil structure appear to be different for soils of different structure and mineralogical compositions. Biswas et al. (1964), working with the alluvial soils of northern India, obtained results showing that with the continuous application of farmyard manure, which

resulted in increased organic-matter content, there was an increase in the percentage of water-stable aggregates and the permeability of soils. However, only negligible improvements in structure resulted from the addition of large quantities of farmyard manure to the sandy soils of Egypt (Abdou and Metwally, 1967), or to the clayey Vertisols in Central India (Venkobarao et al., 1967).

Plant nutrient supply

The main role of animal manure in the maintenance of soil fertility is as a supplier of nutrients, particularly nitrogen and phosphorus. There are in addition numerous micronutrients released from animal manures (Hemingway, 1961; Olsen et al., 1970), and also made available in the soil through the chelating action of the humic acid formed by the decomposition of the manure.

Since nutrients from manures are released relatively slowly compared with those from mineral fertilizers, their residual effects on the subsequent crops are often greater. It is estimated that the efficiency of nitrogen in cattle manure is about 30% of that of nitrogen in mineral fertilizer for the first year of application, but the effectiveness increases in subsequent years due to slow release and therefore a higher residual effect.

Numerous experiments conducted in West Africa using manures showed increases in crop yields (Mokwunye and Stockinger, 1978; Nye, 1952; Kwakye, 1980; Lombin and Abdullahi, 1977; Djokoto and Stephens, 1961; Pieri, 1971; Poulain, 1976). The residual effects of manures on subsequent crops is of particular interest and importance. It has been suggested that it is more a soil physical effect than a question of nutrient release. A combination of both factors is possible. Olsen (1986), in a recent review of the role of organic matter and NH_4^+ on corn yield, pointed out that research results to date indicate that a stable supply of NH_4^+ is essential for the achievement of high corn yields, and one of the advantages of manure is its gradual release of ammonium on decomposition.

Effect of manure on soil biology

In the past, little importance has been accorded to the role played by manures in the biological activities of microorganisms, and their subsequent influence on soil productivity maintenance and crop yield. The addition of manure provides the soil biosystem with a new energy source for organisms that is reflected in changes in microbiological and macrobiological populations.

There is a mineralization of soil organic matter and release of nutrients, as well as improvements in soil structure. The increased activities of soil fauna, such as earthworms and termites, contribute to an increase in soil productivity.

Complementary effects of organic and mineral fertilizers

Numerous studies show an increase in crop yield following the application of manures. The results of work carried out in West Africa show that manures are as effective as mineral fertilizers and, in some cases, better for increasing crop yields (Obi, 1959; Dennison, 1961; Djokoto and Stephens, 1961; Pieri, 1971; Lombin and Abdullahi, 1977). Similar results have been obtained in other parts of the world. However, the low nutrient content of manure means that large quantities are required to produce comparable effects to those of mineral fertilizers. The transport of the manure, and the labour costs of applying it, are still beyond the means of the small farmer.

The residual effects of manure application are important for the maintenance of soil fertility. Hartley (1937) and Djokoto and Stephens (1961) observed that farmyard manure had some "specific effect" that could not be provided by mineral fertilizers. Although improvements in soil physical properties and increased soil fauna activities may contribute, it would appear that the slow nutrient release plays a major role. As noted above, Olsen (1986) showed that NH_4^+ is essential for the achievement of high yields. Manure and compost provide a stable supply of ammonium through slow nutrient release. Musa (1975) obtained results showing that, during the decomposition of cattle dung, the main form of nitrogen release was as NH_4^+. This could partly explain the positive immediate effect of manure on yield, and also its residual effect. The application of manure and mineral fertilizers together will therefore provide the crop with a readily available source of nitrogen from the mineral fertilizer at the time of crop establishment and the early growth stages, supplemented by a slow release of nitrogen from the manure during the later stages of growth.

The data reported on the interaction between manure and mineral fertilizer application in experiments conducted in Africa are scanty. Cooke (1979), reviewing the experience in England, stated that until 1970 most experiments indicated that manure had no effect on yield above that of the nutrients included in the organic matter. However, data collected since 1970 indicate that nutrient utilization is more efficient, and that there are other effects of using manure than those of nutrient supply.

Cooke (1975) reported five years' of results on experiments with potatoes on clay loam at Rothamsted and sandy loam at Woburn, testing farmyard manure and NPK fertilizers. In both sets of experiments, there were large yield responses to nitrogen; however the overall yield levels were raised at both sites when farmyard manure was applied as well as fertilizer. De Haan (1977) observed from long-term experiments that farmyard manure, even at optimum levels of mineral fertilization, substantially increased maximum yields of potatoes and sugarbeet, but not of cereals. It would appear that complex factors such as crops, their rooting system, and improvement in soil physical conditions are important in explaining these results. There is therefore a need for more research on the interaction of mineral and organic fertilizers using various crops and cropping systems.

The combined application of manure and mineral fertilizers, particularly in the tropics, has the additional advantage of buffering the soil against undesirable acidification. Bache and Heathcote (1968) used cattle manure for ameliorating the soil acidity of continuously cropped land in northern Nigeria. In addition to supplying micronutrients directly to the growing plant, the buffering effect of the manure increased the availability of micronutrients.

Use of fertilizer in sub-Saharan Africa

Fertilizer use in Africa is still very low. The main constraint is the farmer's lack of financial resources. The nature of the farming systems, limited infrastructure development, poor pricing policies, and a lack of effective extension services are some of the factors that mitigate against an increased use of fertilizers in the region. Organic sources of nutrients, particularly cattle dung, could contribute to the improvement of soil fertility if their potential is efficiently utilized. The potential nutrients available from cattle manure and the amount of nutrients from mineral fertilizer sources used in selected countries, calculated from FAO data (FAO, 1986; FAO, 1987), are presented in Table 5.

The total nutrients potentially available from cattle manure in some of these countries could have been three to four times the quantity given in the table if the basis of the calculation had been cattle population in relation to the region of the country supporting it. Owing to widespread infestation of tsetse fly, animal husbandry is presently confined to the savanna areas. In Nigeria, for example, 80% of the livestock are confined to the north. A clearer picture would also have emerged if the quantities of mineral fertilizer utilized in the various countries could have been based on cropped area instead of arable land.

Considering the countries listed, the potential total nutrients from dung could contribute significantly to soil fertility maintenance if efficiently utilized. There are, however, severe limitations and, unless the constraints are

Table 5.   Potential nutrients available from cow dung and nutrients utilized as mineral fertilizer (kg ha^{-1} arable land).

	Arable land (x 10^3 ha)	N, P$_2$O$_5$, K$_2$O (from cattle dung)	N, P$_2$O$_5$, K$_2$O (mineral fertilizer)
Angola	2 950	37.8	0.4
Botswana	1 360	58.2	-
Burkina Faso	2 650	38.7	1.3
Cameroon	5 930	24.3	3.4
Chad	3 200	51.7	0.1
Ethiopia	13 200	75.0	1.6
Kenya	1 800	158.0	20.6
Madagascar	2 550	135.7	0.2
Mali	2 073	74.4	1.1
Niger	3 750	29.1	0.2
Nigeria	28 800	13.9	5.6
Sudan	12 420	54.4	1.2†
Tanzania	4 150	113.7	1.0
Zimbabwe	2 680	66.1	20.7

†   Nitrogen only.

seriously addressed, most of the nutrient sources will continue to be wasted. Firstly, the animal husbandry system does not allow for easy collection and preservation of the dung. Secondly, the farming systems still involve separate cropping and livestock enterprises, with most of the livestock kept under nomadic or seminomadic systems. Thirdly, the technology of cultivation on most farms is still poorly developed, and so the application of large quantities of dung poses a major problem. Additionally, we need to recognize the drawbacks of (i) storage, (ii) the difficulty of matching the time of application with maximum crop requirements, and (iii) the difficulty of obtaining a balanced application in terms of nutrient requirements to meet specific crop needs.

Other sources of plant nutrients

Other organic sources of plant nutrients include crop residues, biological nitrogen fixation, green manure, recycled agricultural waste, compost, and sewage sludge. Although these sources contribute little in the way of nutrients at present, if adequately developed they could improve soil productivity in the region.

Crop residues

Besides providing nutrients on decomposition, crop residues play an important role in the maintenance of soil structure and productivity, as well as in moisture and soil conservation practices. Their alternative use as animal feed, with subsequent dung production, contributes to soil fertility maintenance.

Where straw and crop residues provide nutrients on decomposition, it is necessary to incorporate them into the soil as part of the seedbed preparation, and appropriate implements are needed for the purpose. In the traditional agricultural systems in most parts of Africa, the implements used limit the incorporation of crop residues, and this potential source of nutrient supply is therefore lost to the production system. Residues are either burnt as waste or put to other uses such as fuel or fencing materials.

Green manure

Green manuring is an age-old practice in agriculture, particularly in paddy rice production systems and also for increasing the yield of upland crops. As with animal manures, the main role of green manuring is as a provider of plant nutrients to the succeeding crop. The availability of adequate moisture for the growth of the green-manuring crop (without causing water deficiency for the succeeding crop), rapid growth, vigorous root development, and abundant top-growth are prerequisites for its success. Under upland cropping systems in most parts of Africa, good crops of green manure will need to be fertilized, a practice the farmer can rarely afford because of economic constraints. For the farmer, there is also the psychological consideration of ploughing in a crop without getting a direct economic return. This should not be underestimated as it may inhibit acceptance of the technology.

Intercropping

Agboola (1975) suggested intercropping shade-tolerant legumes with cereal crops, especially maize, as a compromise between growing solely green-manuring crops and cash crops. In experiments conducted in Ibadan, he obtained results indicating similar yields for unfertilized maize when intercropped with green gram (*Phaseolus aureus*), cowpea (*Vigna sinensis*) and calopogonium (*Calopogonium mucunoides*) on the one hand, and maize to which mineral fertilizer was applied on the other hand. Such a practice may be more acceptable to farmers than the introduction of green-manuring with upland crops.

Alley cropping

The concept of alley cropping using various leguminous plants, such as *Leucaena leucocephala*, provides a very promising practice for improving soil fertility and controlling soil erosion. The lopped branches are placed on the soil as green manure which, on incorporation into the soil by earthworms and mineralization releases nutrients to crops grown in the alleys. The cover prevents the direct impact of the rain on the soil, thus protecting it from erosion. The multipurpose use of the alley hedges, e.g. as a source of fuelwood, staking material and fodder for animals, makes alley cropping an attractive practice in the farming system.

Rhizobial inoculants

The use of rhizobial inoculants for legume cultivation is not well developed in most of the cropping systems practiced in Africa. The lack of effective nodulation from introduced rhizobial strains because of the presence of indigenous strains, and the poor keeping quality of some of the commercial rhizobial strains sold locally, and other reasons, greatly influence results. Inoculation is however one of the practices that can be potentially exploited profitably to maximize legume production in low-input farming systems. To date, practical applications have only been proven on soybean. The maximization of nitrogen fixation is just as important as the transfer of the fixed nitrogen to the soil. Management systems such as green-manuring, where profitable, and different types of agronomic practices, including mulching, tillage and alley cropping, must be developed in order to exploit fully the benefits of nitrogen fixation.

Legumes provide a major protein source for a large proportion of the population of sub-Saharan Africa. In 1987, the total area under pulse cultivation was 10.6 million ha, with an average yield of 550 kg ha^{-1} (FAO, 1987). Improved management and rhizobial inoculation could increase this yield level substantially.

Sewage sludge and city waste management

The application of city wastes in the form of compost and of sewage sludge could increase the organic-matter content of soils and improve their productivity. However, one of the problems of sludge application is the introduction and subsequent buildup of heavy metals. At present, the availability of sewage sludge in the region is rather limited; but as bigger cities develop, this material will be increasingly available as a source of plant nutrients and soil amendment. Health hazards are, however, important considerations which must be taken into account.

Research needs

Dung produced by the large animal population in sub-Saharan Africa forms a significant source of plant nutrients to supplement the low levels of mineral fertilizers used for soil fertility maintenance. The efficient utilization of this potential nutrient source will depend, however, on the farming and animal husbandry systems practiced, since these determine the collection and conservation practices of the dung.

Developing countries need to adopt strategies for the efficient use of organic materials necessary for increasing soil productivity. Parr and Meyer (1985) emphasized the collection of reliable data as the first step. Most countries lack reliable information on the type, amount, and availability of organic wastes in the agricultural system. The mainly nomadic and seminomadic animal husbandry systems practiced in the region do not allow easy collection or intensive use of the dung produced. Furthermore, precise analytical data on the nutrient content of the dung and the conservation method are essential for the calculation of the amount of nutrients supplied by the dung.

Many of the results of dung experiments reported from West Africa focused mainly on the nutrient effect on individual crops, and only a few dealt with cropping systems and other aspects of soil productivity improvement. The effiency of dung utilization could be greatly increased if assessed on a cropping system basis, since there is often a considerable residual effect on the succeeding crop. Research on the interaction between organic and mineral fertilizers for various crops will need further attention.

The decomposition rates of various organic wastes and the rates at which nutrients are leached under different climatic conditions from the soil-plant-nutrient system remain to be quantified. Composts provide a good source of nutrients and improve soil fertility - particularly under vegetable production systems. Rapid methods of composting need to be further developed to improve its quality.

Nitrogen and phosphorus are the two nutrients most deficient in the soils of sub-Saharan Africa. Increased crop production could be achieved merely with an increased use of fertilizers, particularly those containing nitrogen. The prospects of a substantial increase in fertilizer use in the next decade seem rather doubtful, mainly because of the economic stagnation in most countries of the region. It is therefore important to exploit fully the potential contribution of nitrogen from rhizobia. The following three main areas of research were suggested by Ayanaba (1980).

- studies on the ecology of rhizobia in the major soils: diversity of indigenous rhizobia, their survival, efficiency and competitive abilities;
- assessment of indigenous legumes including trees, plants and forage legumes; and
- more specific studies on the quantification of nitrogen fixed by the various legumes.

The results from these areas of research, if seriously pursued, could contribute significantly to agricultural development in the region, and enhance crop production by making more efficient use of locally available inputs. In addition to these research aspects, there is need for socioeconomic evaluation of the production, use, and environmental impact of organic fertilizers. The economics of the use of appropriate combinations of organic and mineral fertilizers should be assessed.

An intensive manpower training programme has to be initiated. Such programmes must aim at training scientists to cultivate a multidisciplinary approach to problems. This approach assumes greater importance in view of the limited available resources in terms of persons and funding.

Conclusion

The present low level of mineral fertilizer use in sub-Saharan Africa could be partly compensated for by exploiting all the other potential nutrient sources, such as animal manures, biological nitrogen fixation by leguminous plants, and alley cropping practices. Manures could substantially increase soil fertility in the savannas of the region where animal husbandry is currently concentrated because of the incidence of diseases in other areas. However, the present husbandry system does not permit easy collection and storage of the dung. It is therefore necessary to improve the husbandry system in order to preserve dung quality.

Research is needed to assess the decomposition rates of the various organic materials that could be utilized in the region to improve soil productivity. The interaction between these materials, particularly animal dung and mineral fertilizers, should be evaluated on various soils. Exploitation of the potential of biological nitrogen fixation by indigenous and introduced legumes is necessary for the improvement of soil fertility, since the present economic situation in the region limits mineral fertilizer use.

References

ABDOU, F.M. and METWALLY, S.Y. 1967. The effect of organic matter, chemical fertilization and rotation on soil aggregation. *Journal of Soil Science* (United Arab Republic) 7(1): 51-59.

AGBOOLA, A. 1975. Problems of improving soil fertility by the use of green manuring in the tropical farming system. In: *Organic materials as fertilizers*, 147-165. FAO Soils Bulletin no. 27. Rome: FAO.

AINA, P.O. 1979. Soil changes resulting from long-term management practices in Western Nigeria. *Soil Science Society of America Journal* 43: 173-177.

AYANABA, A. 1980. The potential contribution of nitrogen from rhizobia - a review. *FAO Soils Bulletin* 43: 201-210.

BENNE, E.J., HOGLAND, C.R., LONGNECKER, E.D. and COOK, R.L. 1961. Animal manures - what are they worth today? *Michigan Agricultural Experimental Station Circular Bulletin* 231: 1-15.

BERRYMAN, C. 1965. *Composition of organic manures and waste products used in agriculture.* National Agricultural Advisory Service. Advisory Papers no. 2. UK: NAAS.

BISWAS, T.D., DAS, B. and VERMA, H.K.G. 1964. *Bulletin of the National Institute of Science* 26: 142-147.

COOKE, G.W. 1979. Some priorities for British soil science. *Soil Science* 30: 187-213.

COOKE, G.W. 1975. *Fertilizing for maximum yield.* London: Crosby Lockwood Staples. 297p.

DE HAAN, S. 1977. Humus, its formation, its relation with the mineral part of the soil, and its significance for soil productivity. In: *Soil organic matter studies* (Symposium, 1977), vol. I., 21-30. International Atomic Energy Agency. Vienna: IAEA.

DENNISON, E.B. 1961. The value of farmyard manure in maintaining soil fertility in northern Nigeria. *Empire Journal of Experimental Agriculture* 29: 330-336.

DHANYADEE, P. 1984. Organic recycling in Thailand. Paper presented at the Technical Workshop on the Problem of Land with Declining and Stagnating Productivity, 6-9 November 1984, Pattaya, Thailand.

DJOKOTO, R.K. and STEPHENS, D. 1961. Thirty long-term fertilizer experiments under continuous cropping in Ghana: I. Crop yield and responses to fertilizers and manures. *Empire Journal of Experimental Agriculture* 29: 181-195.

FAO (Food and Agriculture Organization of the United Nations). 1987. *Production yearbook 1986.* Rome: FAO.

FAO. 1988. *Fertilizer year book 1987.* Rome: FAO.

HARTLEY, K.T. 1937. An explanation of the effect of farmyard manure in northern Nigeria. *Empire Journal of Experimental Agriculture* 19: 244-263.

HEMINGWAY, R.G. 1961. The mineral composition of farmyard manure. *Emp. Journal of Experimental Agriculture* 29: 14-18.

JENKINSON, D.S. and JOHNSTON, A.E. 1977. Soil organic matter in the Hoosfield continuous barley experiment. *Report of Rothamsted Experiment Station 1976,* pt. 2, 87-101.

KHALEEL, R., REEDY, K.R. and OVERCASH, M.R. 1981. Changes in soil physical properties due to organic waste application. A review. *Journal of Environmental Quality* 110: 133-141.

KWAKYE, P.K. 1980. The effects of method of dung storage and its nutrient (NPK) content and crop yield in the northeast savanna zone of Ghana. *FAO Soils Bulletin* 43: 282-288.

LOMBIN, G. and ABDULLAHI, A. 1977. *Effect of farmyard manure on monocropped cotton, sorghum and groundnuts and a rotation of the three crops under continuous cultivation.* Samaru Miscellaneous Paper Series no. 72. 14p.

MOKWUNYE, U. and STOCKINGER, K.T. 1978. Effect of farmyard manure and NPK fertilizers on crop yields and soil residual phosphorus in 24 years of cropping in Samaru. *Nigerian Journal of Science* 12: 169-179.

MUSA, M.M. 1975. A method for conservation of cattle manure. *FAO Soils Bulletin* 27: 89-96.

NYE, P.H. 1952. Studies on the fertility of Gold Coast soils: IV. The potassium and calcium status of the soils and the effect of mulch and kraal manure. *Emp. Journal of Experimental Agriculture* 20: 227-233.

OBI, J.K. 1959. *The standard DNPK experiments.* Samaru Regional Research Station. Technical Report no. 8. Northern Nigeria: Ministry of Agriculture.

OFORI, C.S. 1962. The role of fertilizers in increasing agricultural productivity in Ghana. In: *Proceedings of the United Nations Conference on the Application of Science and Technology for the Benefit of Less Developed Areas.* Geneva: UN.

OLSEN, S.R. 1986. The role of organic matter and ammonium in producing high corn yields. In: *The role of organic matter in modern agriculture,* ed. Y. Chen and Y. Avnimelech, 29-54. Dordrecht: Martinus Nijhoff.

OLSEN, R.J., HENSLER, R.F. and ATTOE, O.J. 1970. Effects of manure application, aeration and soil pH on soil nitrogen transformations and certain soil test values. *Soil Science Society of America Proceedings* 34: 222-225.

PARR, J.F. and MEYER, R.E. 1985. Strategies for increasing soil productivity in developing countries. In: *Proceedings of a Workshop on Soil, Water and Crop Management Systems: Rainfed Agriculture in Northeast Thailand,* ed. C. Pairintra, K. Wallapopau, J.F. Parr and C.E. Whitman, 252-256. Washington, DC: USAID.

PETTERSON, B.D. and VON WISTINGHAUSEN, E. 1979. *Effects of organic and inorganic fertilizers on soils and crops. Results of long term experiments in Sweden.* Woods End Agricultural Institute. Miscellaneous Publications no. 1. 44p.

PIERI, G. 1971. Survey of the fertilization trials on rainfed cereals in Mali from 1954-1970. IRAT Publication (translated by STRC/OAU, J.P. 26). Dakar: IRAT.

POULAIN, J.F. 1976. Amélioration de la fertilité des sols agricoles du Mali - Bilan de treize années de travaux (1962-1974). *L'Agronomie Tropicale* 31: 403-416.

SINGH, A. 1975. Use of organic materials and green manures as fertilizers in developing countries. *FAO Soils Bulletin* 27: 19-30.

SHARMA, H.L., SINGH, C.M. and MODGAL, S.C. 1987. Use of organics in rice-wheat crop sequence. *Indian Journal of Agricultural Science* 57(3): 163-168.

SUZUKI, M., THEPOOLPON, M., MORAKUL, P., PHETCHAWEE, S. and CHOLIKUL, W. 1980. Soil chemical studies on rotting process of plant remains in relation to fertility of upland soils in Thailand. Joint Research Project between the Tropical Agricultural Research Centre, Japan, and the Department of Agriculture, Thaibud, Thailand. Japan: TARC/Thailand: DOA.

VENKOBARAO, K., NAIR, P.K. and RAO, S.B. P. 1967. Ineffectiveness of farmyard manure in improving soil aggregation in Black Soils of Bellany. *Annales of Arid Zone* 6(1): 138-145.

Organic-matter and soil fertility management in the humid tropics of Africa

AKINOLA A. AGBOOLA[*]

Abstract

The soils of the humid tropics are low in soil organic matter, hence they are inherently infertile. In order to have high crop yield and high soil fertility, soil organic matter must be maintained at a safe level. There is a positive correlation between crop yield and all the major and minor nutrients. As cultivation continues, organic matter declines. Soil organic matter plays such a vital role that most of the soil nutrients are affected by its variability. When the land has just come out of fallow, the soil organic matter and other nutrients are very high. Since most of the nutrients are held by bonds produced by soil organic matter, the elements are detached as soil organic matter declines, and leached down together as cultivation continues.

Reduction in soil organic matter reduces crop yield and affects soils physically, chemically, and biologically. In order to maintain high soil fertility, the rate of soil degradation during cropping must be slowed down. This can be achieved by appropriate farming systems and cropping systems that will maintain an adequate mat layer during cropping.

[*] Department of Agronomy, University of Ibadan, Ibadan, Nigeria.

Introduction

Organic matter can be defined as that active portion of the soil made up of organic materials consisting primarily of carbon, nitrogen, and other elements and which while modifying soil physical properties improve the chemical and biological properties of soil and serve as plant nutrient reserve subject to microbial mineralization.

Organic matter is primarily carbon, about 58% by weight, with lesser amounts of hydrogen, oxygen, and other elements. Small amounts of nitrogen, sulphur, phosphorus, and other elements make up the remaining organic substances, comprising lignins and proteins, amino acids, cellulose, and other carbohydrates, oils, waxes and tannins.

Organic matter makes up 3-5% of the total soil mass, and is found within 15-cm depth from the topsoil surface in the tropical region. Physically, it is responsible for most desirable soil structure, increases soil porosity, improves water retention and aeration, and reduces erosion by facilitating infiltration and reducing surface water runoff. Chemically, it is the soil source of nearly all nitrogen, 5-60% of P, up to 80% of sulphur, and a large part of boron and molybedum.

In Africa, results of research work have indicated that soil fertility and crop productivity are highly dependent on the level of soil organic matter. Soil organic matter is significantly correlated with the CEC, and with all available nutrients, especially N, P, K, Mg, S, Zn, and Cu (Agboola and Corey, 1972). Moreover, the predominant clay mineral, kaolinite, which is low in CEC, is essentially responsible for the high dependence of the soils on their organic-matter content as the major source of CEC and the main nutrient depot. Therefore the need to replenish soil organic matter is very compelling.

The productivity of arable cropland in traditional agriculture is positively correlated with the duration of the bush fallow. Similarly as the fallow period increases, soil organic matter increases. This means that at the onset of cultivation after a fallow period all soil nutrients are at their peak, and as cultivation continues, soil organic matter declines and crop yield declines. Therefore tropical agriculture thrives on organic-matter management. It is the objective of this presentation to discuss the relationship between soil organic matter and soil fertility management in the soils of the African humid tropics.

Some facts about tropical soils

African soils are geologically old, much of the central plateau of Africa is of pre-Cambrian age (over 600 million years). Nearly one-third of the continental surface is an outcrop of rocks of this age. The rest of the surface is covered by sand deposits and alluvium of Pleistocene age (less than 2 million years) (Forth and Schafer, 1980). Also, tropical soils are highly weathered and highly leached, sö they must be given much care and nurturing before they can be productive.

Three major soils are common in the humid tropics, and these three soil types (Alfisols, Ultisols, and Oxisols) occupy 90% of the cultivatable soil. The pecularities of these soils are: they have a low water-holding capacity, a low nutrient-holding capacity, a low-activity clay (or low CEC), they are chemically inert, have a low nutrient reserve, most of their nutrients are in soil organic matter and the biomass and mat layer, and they have a sandy texture and are shallow.

Figure 1 compares the level of organic matter in the humid tropics and the organic matter in temperate soil. In terms of the amount of organic matter of the top 0-15 cm, there is no significant difference between a tropical soil and its temperate counterpart. But once both soils are subjected to cropping, tropical soils rapidly lose organic matter.

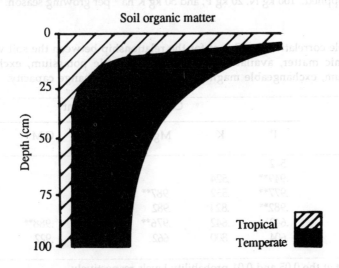

Figure 1.   Comparison of SOM in humid and temperate soils.

High crop yield cannot be maintained with the continuous use of inorganic fertilizer. Table 1 shows the effect of the continuous use of inorganic fertilizer on maize yield. The experiment was conducted for ten growing seasons on three typical soil series cultivated to food crops in southwestern Nigeria. The soils lost over 60% of their organic matter during the period, and the crop yield was reduced to 20% of its initial rate. Furthermore, there is a linear relationship between crop yield and soil organic matter, and there is a high correlation between soil organic matter and soil nutrients (Table 2).

Table 1. Effect·of continuous cultivation of different soil series and fertilizer† use on maize yield and soil OM level.

	Egbeda series		Iwo series		Gambari series	
	OM	Yield	OM	Yield	OM	Yield
1st crop	2.8	3100	3.0	3800	2.1	2500
3rd crop	1.8	2000	2.4	2600	0.9	1000
7th crop	0.8	1000	1.8	1800	0.6	800
10th crop	0.6	700	1.0	1000	0.3	200

† Fertilizer applied: 100 kg N, 20 kg P, and 50 kg K ha^{-1} per growing season.

Table 2. Simple correlation coefficient for the relationship between the soil variables - organic matter, available phosphorus, available potassium, exchangeable calcium, exchangeable magnesium, and cation-exchange capacity.

Soil variables	Correlation coefficient					
	P	K	Mg	Ca	OM	CEC
K	.542					
Mg	.949**	.524				
Ca	.977**	.552	.987**			
O.M.	.982**	.824*	.982	.980**		
CEC	.642	.642	.976**	.982**	.988**	
% clay	.604	.800	.662	.632	.922	.574

*, ** Significant at the 0.05 and 0.01 probability levels respectively.
Source: Agboola and Corey (1972).

Different ways of soil fertility maintenance

In the humid tropics, soil fertility is maintained by building up soil organic matter through natural fallow, organic manure, and the mat-layer mechanism.

The fallow system

In the humid tropics there is a rapid production of large quantities of herbage during fallow as a result of favourable climatic factors. The fallow system has its basis in abandoning exhausted soil uncropped for a period so that natural processes can reestablish the organic-matter content. Herbs, shrubs, and trees supply organic materials in the form of litter to the soil under fallow, thereby hastening the rate at which natural fertility returns to the fallow land.

Fallow makes possible the regeneration of the biomass (the total mass of living matter in the soil, both plants and animals), and crops can be a useful part of this process. The annual increase in humus during fallow will equal the annual amount contributed by decaying litter and root residue minus the losses incurred through mineralization and leaching (Nye, 1958). Table 3 shows the effect of the age of forest fallow on organic-matter accumulation in Ajibode in Ibadan. During the first three years of fallow, there was a slight decline in the level of soil organic matter. Thereafter, there was a steady increase in the organic-matter level as the length of fallow increased up to the ninth year. After the ninth year, the rate of increase was greatly reduced.

Table 3. Effect of age of forest fallow on organic-matter accumulation in Ajibode, Ibadan.

Soil layer	Organic carbon (%) with age of forest					
	1 year	3 years	5 years	7 years	9 years	13 years
0-30 cm	1.02	1.00	2.21	3.41	5.42	5.78
30-60 cm	0.09	0.06	1.06	1.56	1.91	1.90

Source: Agboola (1988).

Organic mulch

One method of reducing soil loss and increasing soil organic matter is the use of organic mulch. Mulching affects the soil in three ways:
-   physically, it conserves soil moisture, improves the infiltration rate, controls runoff and erosion, reduces weed competition, lowers soil temperature and its fluctuation, and improves soil structure;

- biologically, it increase the activity of soil microorganisms and earthworms as well as of nematodes and fungi; and
- chemically, it increases humus and cation-exchange capacity (CEC), as well as the mobilization or immobilization of plant nutrients: prevailing conditions determine whether nutrient deficiencies or toxicity result.

Thus, mulching improves soil fertility, resulting in significantly increased crop yield.

Regular and substantial additions of crop-residue mulch, left on the surface rather than incorporated into the soil, have proved to be a beneficial practice for a wide range of soils and agroecological environments. Djokoto and Stephens (1961) observed in Ghana that the addition of 1 t ha^{-1} of residue over a one-year period increased organic matter by 0.0019 and 0.0011% in the 0-15 and 15-30-cm layers respectively. Any substantial increase in soil organic-matter content would necessitate frequent applications of a large quantity of crop-residue mulch. The high amount of residue required is necessary because 25 to 30% of residue from cereal crops and 40 to 50% of residue from legumes can be decomposed in one month (Lal and Kang, 1982).

It should be noted that the steady state level of soil organic matter can be different for different mulch rates. In other words, the rate of degradation of soil organic-matter content decreases with an increased mulch rate. The high rate of soil organic-matter reduction after land clearing can be slowed down by mulching. Experiments conducted at IITA, in Ibadan, Nigeria, indicated that the rate of decline of organic C during the 18 months after land clearing was respectively 0.103, 0.100, 0.092, 0.083 and 0.078% per month for mulch rates of 0, 2, 3, 6 and 12 t ha^{-1} per season. However, the rate of decomposition 18 months after land clearing is higher in cases of high mulch rates. This implies that a buildup of 0.017% organic C in soil 18 months after land clearing necessitates the application of about 1 t ha^{-1} yr^{-1} of crop residue.

These rates would obviously vary for different regions, depending on the soil moisture regime, the annual rainfall, the length of the dry season, and the temperature regime. On soil previously under grass, no difference in soil organic matter was observed, although the amount of residue returned varied by a factor of about 3. This may be due to the fact that soil under grass had already accumulated the best equilibrium level of organic matter possible for that climate, and it would be difficult to maintain or increase this level. Moreover, many soils have a limited capacity for retaining and maintaining organic-matter reserves (Lal and Kang, 1982).

The mat-layer mechanism

This is a method of generating, maintaining and disposing of organic material by traditional farmers (Figure 2). The litter layer is formed by the decomposing layer of crop residue or litter on the forest floor, or by a layer of organic materials (twigs, leaves and branches) that forms a cover on the soil surface. Under traditional farming practices, a mat layer is always maintained. During the cropping phases a mat layer is also maintained with creeping plants, e.g. melon, cowpea, pumpkin, gourds, and sweet potato. Weed residues, which are turned upside down and neatly left on furrows of yam and cassava plantations, can also serve as a mat layer. The mat layer has two types of effects on the soil:

- a characteristic effect due to reduced surface runoff, leaching and a reduction in soil temperature; and
- a general effect, which it would also have if it were worked into the soil by ploughing, due to the supply and maintenance of the soil organic matter and the release of nutrients as the litter decomposes.

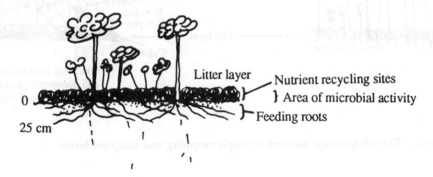

Figure 2. The mat-layer mechanism.

The litter layer mainly occurs in the forest ecosystem, and less so in the savanna. The dense nature of the forest zone with its numerous species, mostly of broadleaves, increases the amount of litter that is found on the soil surface (Figure 3). The number and abundance of species occurring naturally decreases as one moves from the forest towards the savanna. This, coupled with the high rate of mineralization in the savanna and periodic burning, deprives the savanna of much soil cover in the form of litter and mat layer. The naturally occurring forest maintains a large biomass through continuous growth, despite the low fertility of the soil, because the nutrient-conserving mechanism, described as the "mat-layer mechanism", enables the forest to survive on lower amounts of nutrients. Since these mechanisms are part of the living forest, when the trees are cleared for agriculture and the mat layer is removed or destroyed, it gives rise to a situation similar to short-circuiting an electrical current. The nutrients are rapidly released by the increase in soil mineralization, and are leached below the rooting zone, while the exposed topsoil mixed with the mat layer is moved away by soil wash and soil erosion.

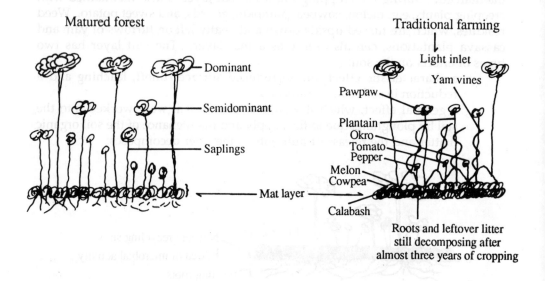

Figure 3.   The relationship between multiple cropping and matured forest.

In undisturbed fallow land, the nutrients move from the soil and the mat layer to the vegetation, and back to the soil and the mat layer through the litter fall. The mat-layer mechanism also serves as a means for nutrient conservation because most of the nutrients are located in the mat of roots and humus that occurs on or near the soil surface. The mat varies from 5 to 30 cm in thickness and contains over half of the small roots. Nutrients are transferred directly from decomposing litter to the roots, and consequently few nutrients move down into the mineral soil. One of the most crucial nutrient-conserving mechanisms in the root mat is the direct absorption of nutrients. Stark and Jordan (1978) found that 99.9% of all ^{45}C and ^{32}P sprinkled on the mat was immediately absorbed, and only 0.1% was leached through the mat.

The rapid growth of small roots in the mat is another mechanism for nutrient conservation. Experiments using ^{32}P have shown that roots take up nutrients efficiently from decomposing litter.

After three years of continuous cropping, the yield declines (Table 1) due to the decline in soil organic matter and the disappearance of the mat layer, which in turn affects the physical, chemical, and biological characteristics of the soil. In Africa, though there is surface and belowground activity, the activity in the surface 0-25 cm is greater. Therefore organic matter management in African humid areas should concentrate on the soil surface activity. Although it is commonly thought that nutrient recycling brings the nutrients from the subsoil, recent work in the University of Ibadan indicates that more than 80% of the recycling takes place on the surface 0-25 cm of the soil. Mat-layer dynamics are, therefore, very important.

Crop-residue management

Crop-residue management to maintain soil organic matter offers the advantage of water conservation. Standing residue conserves water by trapping the rain and by increasing water infiltration. Crop residues must be maintained to prevent bare unprotected soil. In order to ensure groundwater recharge, there must be adequate surface infiltration.

At IITA, it was found that when maize residue was returned, there was an increase in yield of 2 t ha^{-1} over that obtainable when the residue was removed and optimum fertilization applied. Failure to return residues produced a rapid decline of soil organic matter. It was estimated that to maintain organic-matter levels comparable to those under secondary forest, at least two applications totalling 16 t ha^{-1} yr^{-1} of dry maize stover or grass (e.g. *Panicum maximum*) are required when the material is applied as surface mulch (IITA, 1976).

Under normal conditions, the amount of residue mulch returned is only 1 to 4 t ha^{-1} from grains, and this amount can decline to zero with root and tuber crops, and also on fields where termites usually consume all the available residue. Therefore internally generated residues from *in situ* mulch, live mulch, and alley cropping are better alternatives.

In long-term fertility trials carried out since 1972 on an Entisol, it was observed that in a maize--cowpea annual rotation, yields, particularly of cowpea, were greatly reduced with the annual removal of the maize residue (Table 4). The effect of maize crop residue removal is more pronounced than not applying fertilizers. Continuous removal of maize-crop residue with fertilizer use greatly reduces soil pH and results in pronounced Mn toxicity of the cowpea crop and depletion of the soil K status (Lal and Kang, 1982).

Table 4. Effect or removal of maize-crop residue and continuous fertilizer application on grain yield of maize and cowpea grown on a Psammentic Ustorthen in southern Nigeria in 1980.

Treatment	Maize (kg ha^{-1})	Cowpea (kg ha^{-1})
No fertilizer, maize crop residue removed	2067	257
No fertilizer, maize crop residue retained	1200	325
N, P and K, maize crop residue removed	3743	83
N, P and K, maize crop residue retained	5460	178
LSD (0.05)	1040	158

Source: Lal and Kang (1982).

Organic manure

Agboola and Odeyemi (1972) have indicated that the best fertilizer mixture for humid tropical soils is a mixture of organic and inorganic fertilizers. A study on the effect of continuous application of ammonium sulphate alone and in combination with organic manures in long-term experiments on a lateritic sandy loam soil has indicated that the application of organic manure at 45 kg N ha^{-1} equivalence gave yields equal to those obtained with 67.5 kg N in ammonium sulphate (Table 5). A combination of 45 kg N in ammonium sulphate and the same amount of N in farmyard manure gave the highest yield, indicating their synergistic effect. The study clearly indicates the importance of organic manure on lateritic sandy loam soils.

Agboola (1988) studied the effect of organic manures and N, P, and K fertilizers on the grain yield of maize. The results showed that different forms of organic manures did not increase the yield significantly. The combined application of N, P_2O_5 and K_2O at 120, 60 and 60 kg ha^{-1} respectively, over a basal dressing of farmyard manure at 25 t ha^{-1} gave the highest yield (4078 kg ha^{-1})

Besides N, P and K, organic manures also supply the soil with micronutrients such as B, Zn, Cu, Mo, Ca and Si, thus preventing micronutrient deficiencies in spite of a long history of cultivation. A micronutrient deficiency

Table 5.   Effect of organic manure with inorganic fertilizers on the yield of rice.

Ammonium sulphate	Yield of rice (kg ha⁻¹)				
	No organic manure	Farmyard manure†	Green manure†	Groundnut cake†	Mean
No fertilizer	2147	2443	2511	2402	2376
22.5 kg N ha⁻¹	2345	2468	2400	2364	2394
45.0 kg N ha⁻¹	2361	2667	2496	2390	2479
67.5 kg N ha⁻¹	2411	2396	2267	2179	2313
90.0 kg N ha⁻¹	2231	2280	2021	2200	2183
Mean	2299	2451	2339	2307	

† Rates of organic matter adjusted to supply 45 kg N ha⁻¹.

that occurs in an area may be due to (i) the parent material of the soil being poor in micronutrients; (ii) a new variety of crops with a greater demand for certain micronutrients having replaced a traditional variety; (iii) antagonism between nutrients (e.g. Zn deficiency resulting from excessive application of phosphatic fertilizer); or (iv) changes in the soil water regime.

Other sources of organic materials

Sewage sludes and other municipal and agricultural wastes are promising future sources soil organic matter. Large quantities of these materials have traditionally been wasted. With the recent interest in decreasing air and groundwater pollution, agricultural scientists should advocate the utilization of these wastes on cropped land, rather than just allowing them to be disposed of. However, the composition of waste materials must be determined before land application to avoid potentially hazardous high levels of trace metals and toxic organic compounds.

Techniques for increasing soil organic matter

The soil organic-matter level can be maintained by: (i) increasing the amount of organic material added to the soil, and (ii) preventing the decline of natural organic matter. The amount of organic material added to the soil can be increased by:

- ensuring a permanent ground cover;
- choosing plants which will give maximum biomass production for the climatic zone concerned;
- maximizing the return of crop residue, i.e. threshing grain on site if possible, returning the chaff, and allowing cattle to graze on the stubble and returning their manure; and
- using externally produced mulch whenever this is technically and economically feasible.

Similarly, the decline of soil organic matter can be slowed down by:

- keeping the ground cool by maintaining a permanent ground cover (i.e. growing a live mulch such as intercropped melon, calabash, cowpea etc.), practicing relay cropping and multistorey layer cropping, and planting a crop or a cover during the dry season;
- keeping the ground as moist as possible for as long a time as possible (drying kills the soil biomass as fumigation and organic matter is rapidly mineralized);
- not using pesticides which will harm the earthworm population, which is very important in pulling surface-applied residues deep into the soil where they are kept cool and moist, and which creates channels for roots to grow deep in the subsoil, thus maximizing infiltration rates;
- minimizing soil erosion, one of the most rapid ways of losing organic matter from cultivated soil;
- maximizing the amount of woody and lignaceous residue added to the soil, as these have the slowest rate of decay and leave the highest percentage of C in the soil; and
- tilling the soil as little as possible.

References

AGBOOLA, A.A. 1988. Soil fertility maintenance in Ajibode, Ibadan. Department of Agronomy, University of Ibadan, Nigeria. Mimeo.

AGBOOLA, A.A. and COREY, R.B. 1972. Nutrient deficiency survey and maize in Western Nigeria. *Nigerian Journal of Science* 10(1): 1-18.

AGBOOLA, A.A. and ODEYEMI, O. 1972. The effect of different land use on the soil organic matter, exchangeable phosphorus, exchangeable potassium, calcium, magnesium, and mineral elements in the maize tissue. *Soil Science* 115: 367-376.

DJOKOTO, R.K. and STEPHENS, D. 1961. Long-term fertilizer experiments under continuous cropping in Ghana: II. Soil studies in relation to the effects of fertilizer and manures on crop yield. *Empire Journal of Experimental Agriculture* 29: 245-256.

FORTH, H.D. and SCHAFER, J.W. 1980. *Soil geography and land use.* New York: John Wiley and Sons.

IITA (International Institute of Tropical Agriculture). 1976. *Annual Report, 1976.* Ibadan, Nigeria: IITA.

LAL, R. and KANG, B.T. 1982. Management of organic matter in soil of the tropics and undertropics: Non-symbiotical N fixation and organic matter in the tropics. In: *Proceedings of the XIIth International Congress of Soil Science* (New Delhi).

NYE, P.H. 1958. The relative importance of fallows and soils in storing plant nutrients in Ghana. *Journal of the West African Science Association* 4: 31-41.

STARK, N. and JORDAN, C.F. 1978. Nutrient retention by the root mat of an Amazonian rainforest. *Ecology* 59: 434-437.

The use of cover crops, mulches, and tillage for soil water conservation and weed control

J.F. PARR, R.I. PAPENDICK, S.B. HORNICK and R.E. MEYER*

Abstract

Conservation tillage systems provide the most practical and effective means for controlling soil erosion by both wind and water, and for conserving soil water. This is achieved by reducing the number of tillage operations and maintaining cropping residues as mulches on the soil surface which, in turn, reduce runoff, evaporation, energy use, and mechanical disturbance of the soil. Conservation tillage systems also utilize crop rotations and cover crops to ensure effective soil and water conservation practices, which help to offset soil degradative processes and maintain soil productivity at an acceptable level. Small amounts of crop residues in these systems can effectively control both wind and water erosion. However, considerably greater amounts are needed to significantly reduce the evaporative loss of soil water. Cover crops as living mulches can effectively control certain weeds by blocking out sunlight and suppressing their growth. Although surface mulches of dead organic materials such as crop residues and composts can control weeds, they must be applied at very high rates to suppress weed growth. In most cases, such control measures would be impractical, especially in developing countries where are important competitive uses of these residues as fuel, fodder, and fiber. A number of

* Respectively: U.S. Department of Agriculture, Beltsville, Maryland; U.S. Department of Agriculture, Pullman, Washington; and U.S. Agency for International Development, Washington, DC, USA.

research needs and priorities are discussed that could lead to the development of conservation tillage systems that are more productive, stable, environmentally-sound, and sustainable over the long term.

Introduction

Effective soil and water conservation practices are essential to a sustainable agriculture on most of the world's farmlands. Traditional methods to conserve soil and water have generally been developed with the sole objective of maximizing the available water for crop transpiration and minimizing soil erosion.

Increasingly, there are strong indications that many current farming practices are not sustaining the productivity of soils and are causing serious environmental degradation. This is especially the case in the dryland regions, where increasing human populations now require greater productivity per capita and unit of land. Often the basic principles of soil and water conservation have not been adopted and applied, which has resulted in excessive soil erosion and nutrient runoff losses and a decline in soil productivity. The ultimate goal of a sustainable agriculture is to develop farming systems that are productive and profitable, conserve the natural resource base, protect the environment, and enhance health and safety over the long term. How we achieve this goal will depend on creative and innovative conservation and production practices that provide farmers with economically viable and environmentally sound alternatives or options in their farming systems.

Dynamics of soil productivity

The "key" to a sustainable agriculture is to maintain or improve soil productivity. An important relationship (Figure 1), often overlooked, is that for most agricultural soils degradative processes such as soil erosion, nutrient runoff and leaching losses, and loss of soil organic matter are occurring simultaneously with the beneficial effects of soil conservation practices such as conservation tillage, crop rotations, and crop residue management (Hornick and Parr, 1987; Stewart et al., in press). As soil degradative processes intensify, there is a concomitant decrease in soil productivity. Conversely, soil conservation practices tend to offset the effects of the degradative processes and increase soil productivity. The potential productivity of a particular soil at any point in time is the result of ongoing degradative processes and applied

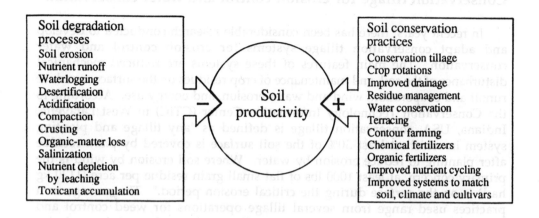

Figure 1. Relationship of soil degradative processes and soil conservation practices (Hornick and Parr, 1987).

conservation practices. Thus a sustainable farming system is one in which the beneficial effects of various conservation practices at least equal, or more than offset, the adverse effects of degradative processes.

In most rainfed or dryland farming systems the most serious degradative processes are soil erosion by wind and water, loss of soil organic matter, and loss of nutrients through runoff. Collectively these processes contribute significantly to the loss of precipitation through excessive runoff and evaporation. The vital component in this dynamic equilibrium (Figure 1) is soil organic matter which must be maintained and crop residues (Parr and Colacicco, 1987; Parr et al., 1989; Parr et al., 1990) and composted municipal wastes (Hornick and Parr, 1987). The proper use of organic amendments is of vital importance in maintaining the tilth, fertility, and productivity of agricultural soils, minimizing wind and water erosion, and preventing nutrient losses through runoff and leaching.

Intensive tillage accelerates soil erosion, the loss of soil water through increased evaporation, and the loss of soil organic matter through increased biological activity and oxidative processes. Thus, conservation tillage provides the best opportunity for halting these degradative processes and for restoring and improving soil productivity.

Conservation tillage for erosion control and water conservation

In recent years, there has been considerable research conducted to develop and adapt conservation tillage systems for erosion control and water conservation. The main features of these systems are reduced mechanical disturbance of the soil, and maintenance of crop residues on the surface to reduce runoff and evaporation, wind and water erosion, and energy use. According to the Conservation Technology Information Center (CTIC) in West Lafayette, Indiana, USA, conservation tillage is defined as "any tillage and planting system in which at least 30% of the soil surface is covered by plant residue after planting to reduce erosion by water. Where soil erosion by wind is the primary concern, at least 1000 lbs of flat small grain residue per acre (1120 kg ha^{-1}) is on the surface during the critical erosion period." The management practices used range from several tillage operations for weed control and seedbed preparation, to one-pass no-till planting. It mainly excludes conventional clean tillage operations that invert the soil and bury crop residues.

The CTIC identifies five types of conservation tillage systems: no-till, reduced tillage, mulch tillage, strip tillage, and ridge till. The first three types, and to a lesser extent strip tillage, have application to dryland farming systems. In the USA, ridge till is used mainly in the more humid northern regions with row crops such as maize (*Zea mays* L.) and soybeans (*Glycine max* L.)

Erosion control

According to estimates by the CTIC, the average reduction in soil erosion for conservation tillage systems is about 50% of that for most conventional tillage practices (Brosten, 1988). However, the range may vary from 30 to 90% depending on the particular type of conservation practice, and factors such as the soil type and the previous tillage system. For example, no-till with a substantial residue cover may reduce erosion by 90% or more, whereas stubble mulching may be considerably less effective (e.g. only 30 to 40%) compared with clean cultivation.

One of the basic principles in controlling wind and water erosion is to keep the soil surface in a rough condition and/or protected with surface residues. The amount of residue required to control erosion has been estimated in a number of studies. As shown in Figure 2, the relationship between the soil loss ratios (i.e. the ratio of soil loss from residue-covered soil compared with bare soil) and percent surface cover for wind and water erosion are quite similar. Small

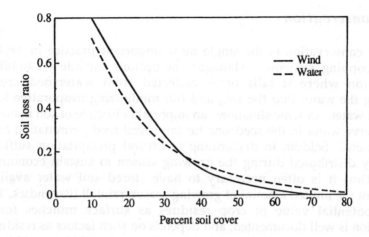

Figure 2.   Relationship between the soil loss ratio (SLR = soil loss from residue-covered soil divided by soil loss from bare soil) and percent surface cover (adapted from Laflen *et al.* [1981] for water erosion and Fryrear [1985] for wind erosion).

quantities of residues are highly effective for controlling both wind and water erosion. For example, approximately one t ha^{-1} of wheat (*Triticum aestivum* L.) straw uniformly distributed on the surface would provide about 50% surface cover. According to Figure 2, this amount of cover would reduce both wind and water erosion to less than 10% of that for bare soil. The data indicate that the lower quantities of surface cover (<30%) are more effective for controlling water erosion than wind erosion. The reverse is true when the residue cover exceeds 40%.

A soil surface with either random or oriented roughness is also highly effective for erosion control. The greater the random roughness, the more depression storage for precipitation, and hence the less water available for runoff. However, because of raindrop impact, the effect of surface roughness may be short-lived for weak-structured soils, unless there is a vegetative cover or surface mulch. The maximum benefit of surface roughness for control of wind erosion is due to the combined influence of oriented roughness and cloddiness. Both research and actual field practices have shown that ridges formed perpendicular to the wind direction will control soil erosion by trapping erodible soil particles in the furrow between the ridges. Ridge height is critical because unless all loose erodible soil is trapped during an erosion event, wind erosion will not be controlled. Relationships between the ridge soil loss to flat soil loss ratio and ridge height have been established (Fryrear, 1984).

Water conservation

Water conservation is the single most important practice in dryland or rainfed cropping systems. Management options include (i) holding the precipitation where it falls or is collected from watershed areas; (ii) infiltrating the water into the soil; and (iii) minimizing evaporative losses of stored soil water. In some situations, an important objective of soil management is to conserve water in the seedzone for improved seed germination and crop establishment. Seldom in dryfarming is natural precipitation sufficient or adequately distributed during the growing season to sustain economic crop yields. Thus it is often necessary to have stored soil water available to supplement the limited amount of growing-season rainfall (Papendick, 1989).

The potential value of crop residues as surface mulches for water conservation is well documented, and depends on such factors as residue color, surface configuration (standing stubble vs. flattened stubble), stubble height, and rate of decomposition. It was cited earlier that relatively small amounts of crop residues - e.g. one t ha^{-1} of wheat straw which would provide 50% surface cover - are effective in enhancing water infiltration by reducing runoff, erosion, and surface crusting. However, considerably larger amounts are required to significantly reduce the evaporative loss of soil water, as much as 4 to 5 t ha^{-1} or more. Moreover, surface residues are most effective in slowing evaporation during the wet season, and have much less effect on evaporative loss in the dry season, or after soil drying has occurred. Thus surface residues have their greatest value for water conservation during the rainy season.

Greb (1983) reported the results of some studies conducted at several locations in the Great Plains, USA, on soil water storage during the fallow period for different mulch rates using residues in an alternate wheat-fallow cropping system. These long-term field experiments, summarized in Figure 3, showed that as the surface mulch rate increased there was a significant increase in soil water storage over the range of 2.2 to 6.6 t ha^{-1}, and that the relationship tended to be linear except at the very high mulch rates.

Conservation tillage can also increase the available water during the growing season through increased infiltration and reduced evaporation, and thus enhance crop yields where water is in limited supply. Table 1 shows water storage efficiencies for different fallow systems at Akron, Colorado, USA (Central Great Plains). Storage efficiency is defined as the percent of precipitation that is stored during the fallow period (in this case 12 months) and available to the subsequent crop. Storage efficiencies increase as the number of tillage operations decrease and more crop residues are retained on the soil surface. Water storage with the minimum-tillage and no-till systems was also improved through better weed control with herbicides.

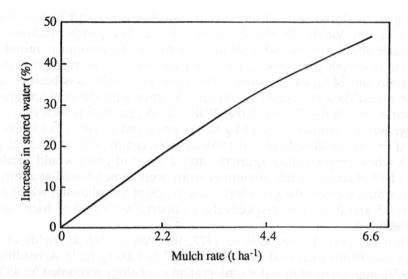

Figure 3.   Increase in soil water storage during the fallow period as affected by the rate of surface mulch in the Great Plains, USA (adapted from Greb [1983]).

Table 1.   Effect of different tillage systems on fallow efficiency at Akron, Colorado, USA.

Tillage system	Years in use	Number of tillage operations	Fallow efficiency†
Maximum tillage, plow and harrow	1915-30	7 to 10	16 to 22
Conventional bare shallow disk, rodweed, or harrow	1931-45	5 to 7	20 to 24
Modified conventional disk (once), chisel, rodweed	1946-56	4 to 6	24 to 27
Stubble mulch, sweep, rodweed	1957-70	4 to 6	27 to 33
Minimum tillage, herbicide to replace one or more tillages	1968-77	2 to 3	33 to 38
No-till	1975-77	0	45 to 55

†   Defined as the percentage of precipitation stored in the soil.
Source: Greb (1979).

The gain in stored water from surface mulches can result in significant increases in crop yields. In the Pacific Northwest, USA, each millimeter of stored water above a threshold level of 100 mm (i.e. the amount required for the crop to reach the grain production stage) will produce an average of 18 kg ha^{-1} of grain and 24 kg ha^{-1} of straw. The grain yield value has been revised from the earlier data of Leggett (1959) for relevance with the new semidwarf wheat varieties. In the Central Great Plains, USA, the threshold level is 150 mm, after which approximately 13 kg ha^{-1} of grain and 20 kg ha^{-1} of straw are produced for each millimeter of water stored (pers. comm., D.E. Smika). In this region, a wheat crop yielding approximately 2 t ha^{-1} of grain would produce about 3 t ha^{-1} of straw. If this amount of straw were to be utilized as a surface mulch, it would increase the grain and straw yields of the following wheat crop by about 0.5 and 0.75 t ha^{-1} respectively, compared with that of bare fallow (Greb, 1983).

During the period from 1936 to 1977, the average wheat yields in the Central Great Plains increased from 750 kg ha^{-1} to 1800 kg ha^{-1}. According to Greb (1979), improvement in water conservation technology accounted for 45% of the increase. Other technologies and their estimated percentage contribution to this yield increase were wheat varieties (30%), harvesting equipment (12%), planting equipment (8%), and fertilizer practices (5%).

The importance of improved tillage and residue management on water conservation and wheat yields was demonstrated by changes in fallow practices in western Nebraska. As shown in Figure 4, wheat yields doubled (increased from 0.7 to 1.5 t ha) during the late 1930s when animals were removed from the land and stubble grazing was eliminated, and mulch tillage

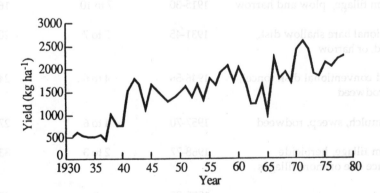

Figure 4. Average winter wheat yields in Nebraska over time.

and improved weed control were introduced in the wheat-fallow cropping systems (Fenster, 1990). Fertilizer usage was an important but secondary factor contributing to this yield increase. Without animal grazing there was more surface cover and, combined with good weed control practices, there was a substantial increase in available water. This, in turn, produced more crop residues for additional water conservation benefits and higher yields for the subsequent crop. The increase in yield was sufficient to justify separating the production of livestock and grain crops on the same land. This is largely the practice which prevails today in most dryland wheat areas in the USA. Moreover, retention of surface stubble reduces wind and water erosion which helps to sustain long-term soil productivity.

No-tillage system

The no-till system is a specialized type of conservation tillage consisting of a one-pass planting/fertilizing operation in which the soil and the surface residues are minimally disturbed during seed and fertilizer placement. Because there is no tillage, the surface residues are of critical importance for soil and water conservation and, without them, conservation objectives would be difficult if not impossible to achieve. In most cases weed control is achieved with herbicides or in some cases with crop rotations.

Though the use of no-till is increasing, farmer adaptation of this practice has been slow. In the USA, no-till is practiced on less than 10% of the farmland that is in some form of conservation tillage. Research results indicate that the success of no-till is soil-dependent, and that the practice works better for some crops than others. Poorly structured and poorly drained soils appear to benefit from some form of tillage, and therefore may not be well adapted to a no-till system. The same is true where hard pans or compacted subsoils restrict root growth or water infiltration, as is the case with some sandy loam soils in Botswana (Willcocks, 1981). Other limitations of no-till include poorly understood biological factors that limit crop growth in surface residue systems; increased weed infestations and costs of controlling weeds; increased plant diseases associated with surface residues; and problems associated with equipment and stand establishment in planting and fertilizing in trashy seedbeds and compacted soil.

Long-term no-till promotes a build up of organic matter on the soil surface. This helps to reduce surface crusting and increase the infiltration rate of many soils. Roose and Piot (1984) reported increased termite activity in uncultivated areas with small quantities of surface residues. The utilization of residues as a food source by these burrowing insects leads to the formation of stable vertical channels. This results in increased water infiltration and reduced runoff losses from intense rains on soils that otherwise are relatively impermeable.

Adaptability of conservation tillage systems in developing countries

The farming systems in most developing countries are considerably different from those in the developed countries such as the United States or Australia with similar agroecological zones. In the USA, crops and livestock are generally not raised on the same land, and consequently crop-residue management is principally directed toward the conservation of soil and water. In many developing countries low yields and competitive uses of crop residues often limit their availability for soil and water conservation. For example, crop residues are the main source of animal feed in many dryland areas of the Near East, Africa and South Central Asia. Often, the land owner owns neither the herds nor controls the grazing rights. As a result, overgrazing is common and there is little residue left during the periods when the need for water conservation and erosion control are critical.

Another factor limiting the adoption of conservation tillage in many developing countries is the lack of proper equipment and adequate power where some tillage is needed. There is evidence, however, that mechanization is increasing, although at a slow rate in some countries. More needs to be done in adapting animal power to conservation tillage systems. It is essential that the tillage implements be kept basic and simple in design because often a farmer with small holdings cannot afford to own more than one or two pieces of equipment. There is also a need to establish criteria for determining the appropriate tillage equipment for different soils and cropping systems, and to provide training in the proper use of these machines.

Crop rotations can significantly affect the success of conservation tillage systems. The development of rotations that produce adequate quantities of residues during the cropping period can help to minimize the constraints caused by competitive uses of residues for fodder, fuel, fiber, and building materials. Crop rotations can also be an effective means for long-term control of weeds, insects, and diseases in conservation cropping systems.

The production of forage in the rotation could provide an alternative feed source for livestock, and reduce their dependence on crop residues. Perennial grasses such as buffelgrass (*Cenchrus ciliaris* L.) have the potential to produce several tons per hectare of high quality forage annually under very dry conditions, but often cannot withstand heavy grazing pressure. Therefore, to maximize the production potential of this particular grass, as well as others, alternatives to uncontrolled or communal grazing must be explored. A promising approach that appears to have considerable merit is the cut and carry practice that is being used by some farmers in India, both in dryland and irrigated areas.

Various crops produce different amounts of residues with varying characteristics. For example, the amount of residue produced by a grain legume crop such as lentil (*Lens culinaris* L.) is considerably less, and decomposes faster, than the residues from most cereal grain crops. As a result, soil erosion is often greater following a grain legume crop irrespective of the tillage system.

Management options for conservation tillage systems

No-till fallow

There has been considerable research in the USA on technology development for complete no-till fallow in the dryland areas. The success of the practice depends on the availability of adequate amounts of surface residue during the fallow period. The potential benefits of no-till fallow, compared with other tillage systems, are more effective control of soil erosion, increased water storage, lower energy costs per unit of production and higher grain yields. No-till fallow has been most successful in summer rainfall areas such as the Great Plains. In the winter precipitation zones such as the Pacific Northwest, there is a substantial loss of water moisture during the summer without tillage (Lindstrom et al., 1974). This reduces the chances for early-fall establishment of wheat, and the farmer must then rely on late planting and a lower yield potential. Under these conditions, no-till fallow for water conservation would be a recommended practice on sandy soils that do not have restrictive hardpans (Hammel et al., 1981). These soils tend to self-mulch upon drying, which makes the effect of tillage less important in reducing evaporation than for finer-textured soils.

A major disadvantage of no-till fallow (sometimes referred to as chemical fallow) is tis heavy dependence on herbicides for weed control. Currently, there is a lack of cost-effective, broad-spectrum herbicides that will provide season-long weed control without residual carry-over and potential phytotoxic effects on subsequent crops. Moreover, there is a growing concern as to the environmental effects and health risks of these chemicals on surface and groundwater quality, wildlife, natural predator insects, human and animal health, food safety and quality, beneficial soil microorganisms and soil microbiological diversity.

Cover crops

The value of proper and regular additions of organic matter to soils was elaborated upon earlier. However, it is well to point out that organic matter can be added to soil in only three ways: (i) by incorporation of crop residues, (ii) by application of animal manures and composted organic wastes, and (iii) by growing grass and legume cover crops. For many farms in both developed and developing countries, cover crops may offer the best practical means of providing the organic matter needed to maintain and improve the tilth, fertility, and productivity of soils. Animal manures and crop residues often cannot meet the requirements of extensive areas of cropland (Miller *et al.*, 1989).

Cover crops can be highly effective for controlling erosion, improving water conservation, and controlling certain weeds. However, it is important to follow proven practices and techniques so as to maximize the positive effects of cover crops and minimize their negative consequences. This will depend on site-specific considerations, including the soil type, climatic factors, and other crops in the farming system. The advantages and disadvantages of cover crops are listed in Table 2.

Table 2. Advantages and disadvantages of cover crops.

Advantages	Disadvantages
Provides a source of organic matter that improves and maintains soil tilth, fertility, and productivity	May deplete soils of available water and nutrients which limit yields of succeeding crops
Provides available nutrients for succeeding crops	Living mulches may pose problems of planting or seeding of cash grain and vegetable crops, and competition for available water and nutrients
Provides effective control of soil erosion, evaporation, and weeds	Weed control may be achieved at lower cost with herbicides
Reduces the need for fertilizer nitrogen	May increase populations of rodents and insect pests
Provides a source of forage for integrated crop/livestock systems	May require special equipment and increased draft/power to incorporate the cover-crop biomass into the soil

In some areas, grass or legume cover crops are grown during the winter months, or non crop period, to reduce runoff, erosion, and nutrient loss. These crops are especially beneficial in preventing erosion on steeplands, which would otherwise lack cover during the noncrop period. Cover crops can also be grown in combination with row crops for controlling soil and water loss. A study conducted in the northeastern USA showed that soil loss was 86% and 99% less in no-till maize (*Zea mays* L.) with living legume cover crops of birdsfoot trefoil (*Lotus corniculatus* L.) and crown vetch (*Coronilla varia* L.) respectively, compared with conventionally tilled maize (Hall *et al.*, 1984). The cover crop was especially effective for intercepting raindrops and preventing splash erosion.

In some cases the cover crop vegetation is allowed to mature naturally or is killed with herbicides in advance of the growing season to conserve soil water. Crops can then be directly seeded into the mulch, which helps to prevent runoff and evaporation. Under some conditions, cover crops are a disadvantage because they may reduce the amount of available water for the succeeding crop, especially in water-deficient areas or in dry years. A practice that has been adopted by farmers in some areas is the direct seeding of grain crops into chemically killed sod following take out of a grass or legume crop grown for seed or forage (Elliott and Papendick, 1983).

Cover crops as living mulches are able to suppress or control weeds mainly by blocking out sunlight and preventing them from growing. Certain legumes can be very effective for weed control, especially those that have an aggressive growth pattern and are more competitive than weeds for available water and nutrients in unfertilized environments. The application of fertilizers, especially when broadcast, will often negate the situation by making weeds more competitive than the legume, in which case the weeds soon take over.

Surface mulches of various organic materials such as leaves, sawdust, cereal straws, composts, and dead (shredded) weeds can also be used to control weeds. However, most of these materials must be applied at very high rates (possibly 10 t ha^{-1} or more) to provide the necessary biomass thickness to block sunlight effectively and suppress weed growth. In most developing countries, there is a continuing shortage of available organic materials due to competitive uses. Thus, it is unlikely that the use of nonliving mulches at such high rates would be a very practical weed-control method.

Ridge till with surface residues

Ridge till (i.e. planting row crops on a preformed ridge) has been gaining popularity as a conservation practice for maize and soybean production in the

USA. Advantages are decreased erosion, increased water conservation, reduced fuel and equipment costs, controlled traffic patterns that reduce soil compaction, faster seedbed warmup in the spring, and effective nonchemical weed control. The row crops are sown in the spring on ridges formed by cultivation of the crop during the previous growing season. The residues from the previous crop are usually shredded in the fall to provide an even distribution of surface cover during the nongrowing season.

The planting operation places seed and starter fertilizer in the ridge top and at the same time moves 2 to 3 cm of soil, weeds, and organic materials, such as animal manures and crop residues, into the interrow or wheel-track area. As the maize or soybeans grow, a later cultivation is used to control weeds and preform the ridges for the following crop. The mulch in the interrow area helps to prevent runoff and erosion.

Alley cropping

Alley cropping is essentially an agroforestry system in which food crops are grown in alleys formed by hedgerows of nitrogen-fixing leguminous trees or shrubs planted on contour or across slope (Kang et al., 1981). The hedgerows are cut back when the food crop is planted and kept pruned during the cropping season to reduce competition with the crop for available water and plant nutrients. During the noncrop period, the hedgerows are allowed to grow to protect the land from erosion and to produce biomass.

Alley cropping has been referred to as "controlled bush fallow" and retains the basic features of the bush fallow or shifting cultivation systems. Trees and shrubs in the alley cropping system:

- provide green manure or mulch for companion food crops, and facilitate nutrient uptake and recycling from deeper soil layers;
- provide prunings or loppings as mulch and shade during the fallow period to suppress weeds;
- provide contour barriers on sloping land to reduce soil erosion and nutrient runoff losses;
- provide biologically fixed nitrogen to the companion crop;
- provide favorable conditions for macro- and microorganisms; and
- provide forage for ruminant animals, construction material, and firewood.

The major advantage of alley cropping over the traditional shifting cultivation or bush fallow systems is that the crop and fallow phases can take place concurrently on the same land. This allows the farmer to crop for an extended period without returning the land to bush fallow.

Goals and research needs for development of conservation tillage systems and a more sustainable agriculture

Conservation tillage must be considered as a system that is site-specific. Its successful application and use over a wide range of soil conditions and geographic locations will depend on the farmer's management expertise, and his innovativeness in matching a particular tillage system with soil type, crop cultivar, climatic factors, and the environment. Therefore the principal goals of conservation tillage research are:

- to develop conservation tillage systems that effectively control soil erosion and nutrient runoff, conserve water and energy, minimize surface and groundwater pollution by agricultural chemicals, and enhance the economics of production;
- to improve and systematize research techniques for incorporating conservation tillage technology into conservation production systems; and
- to develop conservation tillage systems that (i) are stable, sustainable and productive; (ii) use plant nutrients more efficiently; and (iii) require minimal and infrequent applications of herbicides and insecticides for control of weeds and insects.

How successfully we meet these goals will depend on the means that are utilized in achieving them. The following list of research needs should be given high priority for the development of conservation tillage systems for a more productive, stable and sustainable agriculture.

- Develop and/or refine computer models, expert systems, or other methodology for assessing the combined effects of climate, soils, and conservation management practices on soil erosion and productivity, crop water use efficiency, and surface and groundwater quality.
- Develop crop cultivars that are adaptable to conservation tillage systems and have characteristics that aid in reducing wind and water erosion, improving water quality, and maintaining food (feed) production.
- Develop fertilizer management practices for conservation cropping systems that that maximize nutrient and water-use efficiency, and minimize the loss of chemicals through leaching and runoff.
- Design and adapt crop rotations with cover or green manure crops to control erosion and weeds, and reduce the need for pesticides and chemical fertilizers in the production system.
- Develop acceptable agronomic and economic practices for pest management that maximize the use of the natural, physical, and/or biological controls, while conserving soil and water and improving water quality.

- Determine the effect of weed growth (density, species, and time of growth) on water loss during fallow, and evaluate the effects of tillage-herbicide combinations on water conservation and weed control.
- Determine the effects of fertilizer practices, crop-residue management, organic amendments, stress factors (i.e. water, nutrient and temperature) on the nutritional quality of crops and the bioavailability of food nutrients.
- Determine the relative agronomic and economic value of crop residues for soil and water conservation, maintenance of soil fertility and productivity, and feed for livestock. Determine what compromises and trade-offs are possible to achieve multipurpose benefits from residues.

Literature cited

BROSTEN, D. 1988. How much can we lose? *Agrichemical Age*, 32(10): 6-8. San Francisco, CA: Farm Publications, Inc.

ELLIOTT, L.F. and PAPENDICK, R.I. 1983. Direct seeding into bluegrass sod for erosion control. *Journal of Soil and Water Conservation* 38: 436-439.

FENSTER, C.R. 1990. Fifty years of tillage practices for winter wheat. In: *Proceedings of the International Conference on Dryland Farming*, ed. P.W. Unger, W.R. Jordan, T.V. Sneed, and R.W. Jensen. Texas: Texas A&M University Press.

FRYREAR, D.W. 1984. Soil ridges, clods and wind erosion. *Transactions of the ASAE* 27 (2): 445-448.

FRYREAR, D.W. 1985. Wind erosion on arid croplands. In: *Science reviews on arid zone research*, 31-48. Jodhpur, India: Scientific Publishers.

GREB, B.W. 1979. Technology and wheat yields in the central Great Plains: Commercial Advances. *Journal of Soil Water Conservation* 34: 269-273.

GREB, B.W. 1983. Water conservation in the central Great Plains. In: *Dryland agriculture*, ed. H.E. Dregne and W.O. Willis, 57-72. American Society of Agronomy. Agronomy no. 23. Madison, WI: ASA.

HALL, J.K., HARTWIG, N.L. and HOFFMAN, L.D. 1984. Cyanazine losses in runoff from no-tillage corn in "living" and dead mulches vs. unmulched conventional tillage. *Journal of Environmental Quality* 13; 105-110.

HAMMEL, J.E., PAPENDICK, R.I. and CAMPBELL, G.S. 1981. Fallow tillage effects on evaporation and seedzone water content in a dry summer climate. *Soil Science Society of American Journal* 45: 1016-1022.

HORNICK, S.B. and PARR, J.F. 1987. Restoring the productivity of marginal soils with organic amendments. *American Journal of Alternative Agriculture* 2: 64-68.

KANG, B.T., WILSON, G.F. and SIPKENS, L. 1981. Alley cropping maize (*Zea mays* L.) and Leucaena (*Leucaena leucocephala* Lam) in southern Nigeria. Plant and Soil 63: 165-179.

LAFLEN, J.M., MOLDENHAUER, W.C. and COLVIN, T.S. 1981. Conservation tillage and soil erosion on continuously row-cropped land. In: *Crop production with conservation in the 1980's*, 121-133. American Society of Agricultural Engineers. ASAE Publication, 7-81. St. Joseph, Michigan: ASAE.

LEGGETT, G.E. 1959. Relationship between wheat yield, available moisture, and available nitrogen in eastern Washington dryland areas. Washington Agricultural Experiment Station. Bulletin no. 609. Pullman, Washington: Washington State University.

LINDSTROM, M.J., KOEHLER, F.E. and PAPENDICK, R.I. 1974. Tillage effects on fallow water storage in the eastern Washington dryland region. *Agronomy Journal* 66: 312-316.

MILLER, P.R., GRAVES, W.L., WILLIAMS, W.A. and MADSON, B.A. 1989. Cover crops for California agriculture. Division of Agriculture and Natural Resources, University of California. Publication no. 21471. California: University of California. 24p.

PAPENDICK, R.I. 1989. Storage and retention of water during fallow. In: *Proceedings of a Workshop on Soil, Water and Crop/Livestock Management Systems for Rainfed Agriculture in the Near East Region*, 260-269. Washington, DC: Government Printing office.

PARR, J.F. and COLACICCO, D. 1987. Organic materials as alternative nutrient sources. In: *Energy in plant nutrition and pest control*, ed. Z.R. Helsel, 81-89. Amsterdam: Elsevier Science Publishers.

PARR, J.F., PAPENDICK, R.I., HORNICK, S.B. and COLACICCO, D. 1989. Use of organic amendments for increasing the productivity of arid lands. *Arid Soil Research and Rehabilitation* 3: 149-170.

PARR, J.F., PAPENDICK, R.I., YOUNGBERG, I.G. and MEYER, R.E. 1990. Sustainable agriculture in the United States. In: *Proceedings of the International Symposium on Sustainable Agricultural Systems*, 50-67. Soil and Water Conservation Society. Ankeny, Iowa: SWCS.

ROOSE, E. and PIOT, J. 1984. Runoff, erosion, and soil fertility restoration on the Mossi Plateau (Central Upper Volta). In: *Proceedings of the Symposium on Challenges in African Hydrology and Water Resources* (Harare, 1984). International Association of Hydrological Services. IAHS Publication no. 144. Wallingford, England: IAHS.

STEWART, B.A., LAL, R. and EL-SWAIFY, S.A. In press. Sustaining the resource base of an expanding world agriculture. In: *Proceedings of a Workshop on Mechanisms for a Productive and Sustainable Resource Base* (Edmonton, 1989). Soil and Water Conservation Society. Ankeny, Iowa: SWCS.

WILLCOCKS, T.J. 1981. Tillage of clod-forming sandy loam soils in the semi-arid climate of Botswana. *Soil and Tillage Research* 1: 323-350.

Biological activities and soil physical properties

H.W. SCHARPENSEEL, H.U. NEUE and B. HINTZE[*]

Abstract

The soil organic matter-biological activity interaction is undoubtedly one of the most important soil parameters, but is not included in the soil 'fertility capability classification' (FCC) system as a system modifier. Its relation to soil physical properties, such as structure, water retention, pore space, and resistance to erosion, is indisputable, and does not therefore enjoy priority status in experimental planning. This report presents the wider aspects of the topic, and considers central issues like carbon source-sink relations, biological activity, and organic-matter decomposition in different climate belts. There is also some discussion of the impact of organic-matter contents and biological activity on the greenhouse effect and active trace-gas emissions. Finally, in the major part of the paper, the soil physical facts relating to biological activity in tropical agroecosystems are reviewed, and it is suggested that harmonizing and even optimizing soil physical properties with soil biological activities should be one of the major aims of improved agricultural management.

[*] Respectively: Institute of Soil Science, University of Hamburg, Germany; International Rice Research Institute, Los Baños, Philippines; and International Board for Soil Research and Management, Bangkok, Thailand.

Introduction

It can be reasonably assumed that the release of CO_2 is indicative of a soil's biological activity, and it can be observed that this activity usually reaches an optimum at about 80 to 100% of field capacity moisture level, which is higher in more sandy soils, and lower in more clayey soils. It is well known that higher organic-matter contents and biological activity improve the soil structure and give stability against peptization. Furthermore, the soil's water-holding capacity and resistance to erosion are also enhanced. In sedimentology, it is considered axiomatic that the physical and mechanical properties of a sediment, such as its water-retention capacity, porosity, and compressibility, are to some extent controlled by the amount of organic material within the sediment (Weller, 1959).

While soil organic matter, as a combined factor comprising soil organic matter and the related biological activity, is not included in the list of modifiers of the soil 'fertility capability classification' (FCC) system (Sanchez et al., 1982), and is not considered as being among the most critical parameters modifying soil fertility capability, Lal (1981a) points out that milieu degradation and aridization occur even in tropical rainforests, due to the increase in insolation and the air and soil temperature, and the decrease in soil moisture and relative humidity in consequence of forest removal and organic-matter degradation. In the tropics, it is considered normal for organic-matter decay to exceed a factor of 2.5 to 4, if compared with temperate-climate organic turnover (Jenkinson and Ayanaba, 1977). Subsequently, there is further depreciation of soil fertility due to loss of colloidal fractions, and the consequent erosion leads to acidification, dramatic losses of N, and a decline of yield potential. With the exception of riziculture, attempts in West Africa to establish intensive agriculture have not measured up to expectations (Ruthenberg, 1976). However, mulch with *Pennisetum* or legumes like *Pueraria Javanica*, *Stylosanthes* and *Calapogonium* (Jurion and Henry, 1969) have proved quite successful in Zaire. Sanchez (1981) proposes "simply to overcome soil fertility depletion by judicious liming and fertilizer practices, coupled with stress-tolerant "cultivars", and suggests a new strategy of complete fertilization "coupled with mulching to control soil compaction", which was leading to considerable yield increases of corn and soybean in cleared land of the Amazon Basin. In Yurimaguas, he observed remarkably beneficial effects from mulching with *Panicum maximum* and kudzu (*Pueraria phaseolides*).

The beneficial effect of organic matter and biological activity in making the epipedon friable, in improving structure, in providing most of the desirable physical properties, and in replenishing the nutrient status of soils, is tacitly implied in the observations of most agricultural scientists. It is, though, considered to be a rather obvious and old-fashioned topic, compared with problems of strategic mineral fertilization, acidity control by liming, or maintaining soil fertility by the choice of optimal cropping or farming systems and high-yielding cultivars. Also, due to the growing environmental consciousness, source-sink relations in connection with soil organic matter focus more on nutrient recycling, groundwater protection, and curtailing greenhouse active trace-gas emissions than on the improvement of soil physical properties, except when these relate to erosion control. The subject assigned to me is not therefore particularly topical, however important it may be, and it should not be confined to very narrow areas of discussion.

Carbon source-sink relations

Every year, 115 billion t of C (as CO_2) is turned into organic matter by terrestrial photosynthesis and photolyzed O and H (from water), after about 50% of the initial gross primary production has been released again by respiration. The areal (per hectare) gross net production depends on the climate and the type of ecology. The same holds true for the percentage of respiration by heterotrophic organisms in the food web and for the remainder added to the soil (>80% in fully grown forests, around 50% in good grassland, and about nil in arid lands). Soil is part of an ecosystem with producers (autotrophs), reducers (especially microflora), and consumers (higher organisms). All of them, producers (autotrophs by root respiration), reducers, and consumers (as predators by respirative digestion of the ingested organic body substances), release CO_2, which may be partly reduced by metanomonas bacteria under strongly reduced conditions into methane, and the production of both CO_2 and methane is characteristic of a soil's biological activity. Van Breemen and Feijtel (1989) have given an overview of source and sink relations for CO_2 and other soilborne trace gases against the background of the FAO/ Unesco soil classification.

Biological activity

Measurement of the biological activity, based on $^{14}CO_2$ release of soil from uniformly labelled plant substance, has been excellently reviewed with a historic introduction by Paul and van Veen (1978). Jenkinson and Ayanaba (1977) were the first authors to compare the ^{14}C-labelled ryegrass decomposition in temperate Rothamstedt and tropical Nigeria, and confirmed the well-known 1:4 relationship between decomposition speed and intensity. When comparing several soils in Germany and Costa Rica, Saverbeck and Gonzales (1977) found a less pronounced domination of the reaction speed by the climate/temperature, but also a strong dependence on soil acidity.

Decomposition of labelled rice straw in flooded rice soils as well as in upland rice soils was measured by Krishnappa and Shinde (1978). In flooded soils with rice-rice rotation, these authors used ^{15}N-labelled rice straw. In the Philippines, Neue and Scharpenseel (1978) compared rice straw decomposition in upland and lowland soils over a period of four years, and found slightly but not dramatically accelerated decomposition under upland conditions, and about 15 to 20% remaining organic carbon in the soil after one year (Figure 1). Tsutsuki and Kuwatsuka (1989) tested organic-matter decomposition in humid Andosols by chemical methods and radiocarbon dating.

Haider and Martin (1981) and Ellwardt et al. (1981) tested the microbial lignine decomposition in different environments by radiometric and ^{13}C-NMR methods. Labelled lignine decomposition studies in a steady state model paddy are being undertaken by Neue at the International Rice Research Institute in the Philippines, and decomposition studies with ^{14}C-labelled sorghum straw and ^{14}C-groundnut straw in a semiarid environment are being carried out at the International Crops Research Institute for the Semi-Arid Tropics in India. Singer (1990) describes the decomposition of ^{14}C-groundnut straw in semiarid Rhodustalf and Chromustert soils in India. In addition to biotic decomposition, the total organic-matter turnover comprises abiotic-protolytic (Laura, 1975), and photochemical decay components (Scharpenseel et al., 1984), which can both be considerable - especially under conditions of acidity and high-sunlight intensity on soils without vegetation cover (Figure 2).

Radiocarbon dating of soil samples reveals that biotic, abiotic, and photochemical degradation of soil organic matter spares the chemically and physically protected fractions, most of which survive as oxide or clay organic complexes, eventually over thousands of years, whereas the young continuously/annually replenished organic residues or humic substances, which do not find an appropriate mineral adsorption matrix (e.g. clay domains) are mostly decomposed and recycled. Figure 3 shows the fixation of humic substances via polyvalent metal ions, which act as bridges to the clay minerals (Theng and Scharpenseel, 1975). Layer by layer soil profile ^{14}C dating makes it possible to plot a ^{14}C date of individual layers divided by the maximum age ^{14}C date of the profile vs. depth curve. The steepest curve, with the coefficient of the x member of the function closest to unity (Scharpenseel et al., 1986a, b), indicates the most intensive and fastest decomposition, which of course implies high biological activity (Figure 4).

UPLAND SOIL

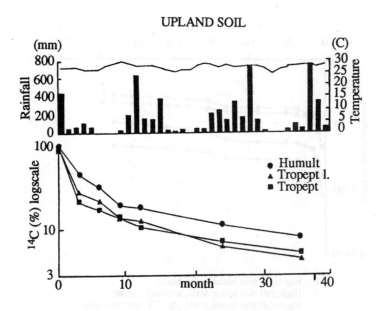

FLOODED LOWLAND SOIL

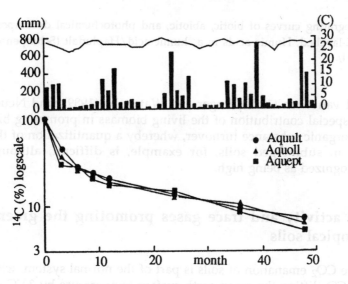

Figure 1. Decomposition pattern of ^{14}C-labelled rice straw in tropical soils (Neue and Scharpenseel, 1987).

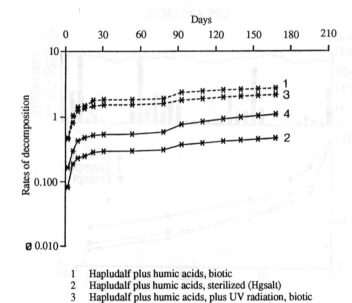

1	Hapludalf plus humic acids, biotic
2	Hapludalf plus humic acids, sterilized (Hgsalt)
3	Hapludalf plus humic acids, plus UV radiation, biotic
4	Hapludalf plus humic acids plus UV radiation, sterilized

Figure 2. Integrating curves of biotic, abiotic, and photochemical decomposition of ^{14}C-labelled, Humic acids in a humic acid/Hapludalt (Scharpenseel *et al.*, 1984).

Paul and van Veen (1978), Roger and Watanabe (1989), and Neue (1989) point to the special contribution of the living biomass in promoting biological activity and organic substance turnover, whereby a quantitization of the effect of tubefices in submerged soils, for example, is difficult, although it is generally recognized as being high.

Biological activity and trace gases promoting the greenhouse effect in tropical soils

While the CO_2 emanation of soils is part of the normal system, with water vapour and CO_2 lifting the mean earth surface temperature by 33°C to +15°C instead of -18°C, methane from reduced soils is about 70% biotic in origin, and is caused by the irrigation or bonding of rainfed fields. Methane increases annually by 1.1%, adding about 18 ppb to the present 1.7 ppm of methane in the

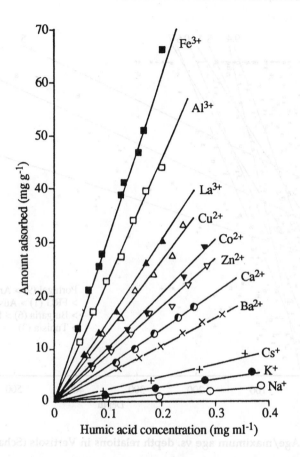

Figure 3. Isotherms at 20°C for the absorption of ^{14}C-humic acid by montmorillonite saturated with different cations (Theng and Scharpenseel, 1975).

lower atmosphere. It must be properly assessed due to its high rate of efficiency in increasing the greenhouse effect - an efficiency of 32 compared with CO_2 on a molar basis (Neue *et al.*, 1989). N_2O can promote the greenhouse effect with a factor of efficiency of 150 as compared to CO_2, and as a stratospheric ozone killer has to be controlled, since the annual increase of the present 0.3 ppm gross concentration by about 3% (1 ppb per year) is mainly caused by nitrification ($NH_4 \rightarrow NH_2OH \rightarrow N_2O + NO_2$), denitrification [$C_{org.} + NO_3 \rightarrow (CH_2O)_n + CO_2 + N_2O$], and nitrogenase-activated diazotrophic N-collecting systems (Umarov, 1989). N_2O is distributed and analyzed in the atmosphere over tropical rainforests and savanna as well as

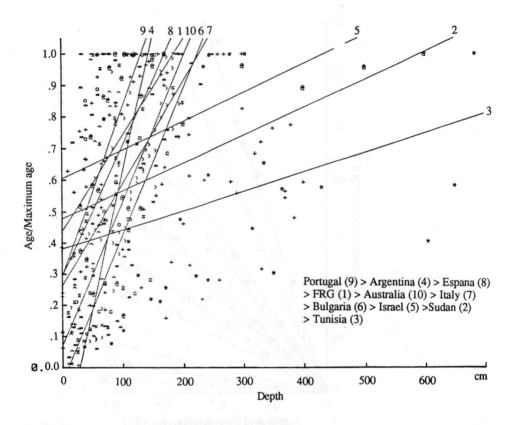

Figure 4.   Age/maximum age vs. depth relations in Vertisols (Scharpenseel et al., 1986).

over extended rice fields, and often in even higher concentrations. The aggravating factor is that the biological activities leading to CH_4 and N_2O emissions have, in the case of rice plants, sedges, etc., a very efficient bypass to avoid reoxidation by virtue of the thin oxidized horizon of gleyey or submerged land caused by migration, which makes it possible for these emissions to go straight through the plant's nodes and internodes by means of the chimney-pipe effect.

Behind the anthropogenic trace-gas emissions, mankind is creating an exponential population increase, which means that man is becoming a major geological agent in the tropical/subtropical region. Increasing the temperature increases the mineralization as well as the form of the organic matter, as well as the water balance, the salt balance, and the soil temperature. This then will increase the susceptibility of the soil to physical degradation, including erosion.

Soil physical factors relating to biological activity - soil organic matter

The annual organic substance turnover in global soils amounts to about 50 to 70×10^9 t of C in a total C pool of 1050×10^{12} kg (Keeling, 1973) to 3000×10^{12} kg (Bolin, 1976) as soil humic substances.

Jenny (1949, 1950) demonstrated the relationship between mean annual temperature and soil organic matter for the temperate climate zone, and Laudelot *et al.* (1960) confirmed that this was an exponential relationship for tropical soils as well - though with a stronger variation in organic matter under the influence of the temperature, and a discontinuity in the transition zone (temperate-tropical). The temperature optimum for the humus synthesis is below the temperature prevailing in the tropics, but organic-matter synthesis as well as decomposition are both highest in the tropics.

In its Agro-Ecological Zones Project, FAO (1978) operates on the basis of the "growing period", i.e. the period in days during a year when precipitation exceeds half of the potential evapotranspiration, plus the time required to evaporate an assumed 100 mm of soil moisture reserve (see Figure 5). The aridic zone (<75 G.P.) is the zone with the lowest biological activity, and the highest photochemical decomposition, and is associated with the formation of extremely low humic saline, sal-sodic, calcic, and gypsic crust. The arid-ustic zone (75-149 G.P) is a zone with rather low humic but high-activity clays, which have a high cation-exchange capacity and are in very stable clay organic complexes. The udi-ustic zone (150-269 G.P.) is intermediate to the udic zone (>270 G.P.), and has low-activity clays, oxidic or deep-argillic B horizons, and biotic plus protolytic abiotic respiration caused by high temperature. The more recent literature (TropSoils, 1989; van der Heide, 1989) reveals many ongoing activities related to the effects of the soil organic matter-biological activity interaction in connection with cropping system studies. Further, the advantage of nutrient management by mixed systems of fertilizer plus crop residue management, the use of legume green manure, mulch, or alley cropping is universally accepted. But specific studies focused on the effect of soil physical properties on soil biological activity - other than those mentioned in the chapter on "biological activity" in the TropSoils report - are rare. The mutually beneficial effect of high biological activity on soil chemical and physical properties - and conversely of good soil physical conditions, such as structure, water-holding capacity, pore space, and aeration, on biological activity - seems to be taken for granted as a kind of generally known, self-evident axiom.

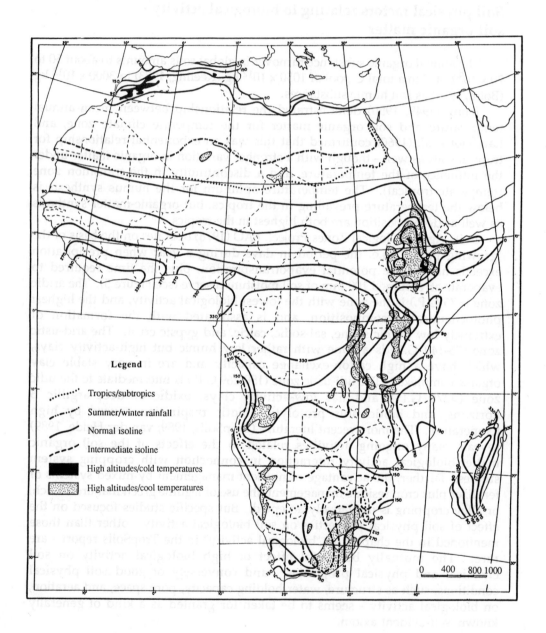

Legend

.... Tropics/subtropics

———— Summer/winter rainfall

———— Normal isoline

– – – – Intermediate isoline

▆▆▆▆ High altitudes/cold temperatures

░░░░ High altitudes/cool temperatures

0   400   800 1000

Figure 5.   Chart of growth periods (in days) in Africa (FAO, 1978).

The most dramatic loss of organic matter, soil humic substances, biological activity (ecological as well as soil biological activity), and soil physical quality is undoubtedly associated with tropical deforestation and subsequent soil management techniques. Lal (1981) shows in Table 1 the great changes brought about by these events, and their effects, in physical terms, on soil degradation. In contrast to the list of modifiers given in the FCC (Sanchez et al., 1982), Eswaran (1981) suggests that organic C and N should be determined and included in the analysis of class Ia, which is required for all classification work. In class Ib, a limited set of physical properties for some soil orders and suborders should be included, notably bulk density for Andepts and humic suborders, 15-bar water for Inceptisols, Alfisols, Ultisols, and Oxisols; COLE value for Vertisols and vertic subgroups; and conductivity for some families of Aridisols. Moreover, in class IIa, the following soil physical measurements should be applied for the purpose of soil surveys: infiltration, permeability, and available water-bearing capacity. Van Wambeke (1981) has proposed that for "planing objectives and soil resource inventories", a set of "limiting soil properties" for tropical crops should be produced. He quotes, as an example, a study by Sys (1976) for the oil palm, where soil organic matter and some physical properties are listed as being essential (Table 2).

Table 1. Effects of methods of deforestation and tillage techniques on soil erosion.

Method of vegetation removal	Sediment density $(g\ L^{-1})$	Water runoff $(mm\ yr^{-1})$	Soil erosion $(t\ ha^{-1}\ yr^{-1})$
Traditional farming - incomplete clearing, no-till	0.0	3	0.01
Manual clearing - no-till	3.4	16	0.4
Manual clearing - conventional tillage	8.6	54	4.6
Shear blade - no-till	5.7	86	3.8
Tree pusher root rake - no-till	5.6	153	15.4
Tree pusher root rake - conventional tillage	13.0	250	19.6

Source: Lal (1981b).

Table 2. Limiting soil properties for tropical tree crops.

Properties to be evaluated	Critical limits
1. Topography (slope)	not more than 30% slope
2. Drainage	not more poorly drained than imperfectly drained
3. Flooding	not more than slight flooding
4. Texture	not lighter than sandy loam
5. Stoniness	not more than 75%
6. Effective soil depth	not less than 50 cm
7. Cation-exchange capacity	not less than 16 cmol kg^{-1} clay; no net positive charge
8. Base saturation	not less than 15% in the surface horizons
9. Organic matter	not less than 0.8%

Source: Sys (1976).

Regarding the kinds of organic material which obtain in the "development countries" and the amount of recycling nutrients which they represent (both for the organic matrix and in terms of the released nutrients which promote the soil's biological activity), an estimate and precast for 1980 by Singh (1975) is indicated in Table 3. A greater use of the ability of organic matter to recycle nutrients is urgently required to avoid growing eutrophication by a sustained import of available nutrients into the pedosphere from the atmosphere (N) and lithosphere (P,K) as a result of the unbridled use of mineral fertilizer.

Table 3. Total annual production of nutrients through organic wastes in the "development world" (precast for 1980).

Source	Mil t of nutrients		
	N	P	K
Human	15.26	3.57	3.25
Cattle	22.25	6.14	17.65
Farm compost	11.93	4.18	11.93
Urban compost	0.60	0.48	0.71
Urban sewage	1.79	0.36	1.08
Other (bones, bagasse, oil cake, press mud)	8.29	5.55	14.19
Total	60.12	20.28	48.81

Source: Singh (1975).

Musa (1975) points to the advantage of storing cattle manure until it is time to use it, so as to prevent it from drying out in the tropical sun or to avoid losses by rain leaching. Agboola (1975) has made a critical review of the supporting effect of fallow vegetation in improving soil physical properties and enhancing soil biological activity, particularly with regard to stabilizing the nutrient balance due to the nutrient import from the subsoil layers.

Ayanaba and Okigbo (1975) point to the advantages of mulching for water conservation, the humus content-biological activity interaction, the CEC, and the nutrient supply, and note the possibility of deficiencies or toxicities occurring in relation to the source of the mulch. Agboola et al. (1975) opt for a combination of organic manure (farmyard plus compost) and fertilizer, with a view to producing legumes in mixed-cropping systems and thereby assuring sustainable fertility in tropical land use.

There is some disagreement with regard to basic management approaches. In tropical humid Africa, which has an isothermic or isohyperthermic soil temperature regime, and also for the most part has low nutrient levels, capital-intensive arable foodcrop farming in the lowland regions has rarely been successful, and in fact has led to considerable land deterioration (Moormann and Greenland, 1979; Greenland, 1986). Thus, in the humid tropics zero tillage with crop and weed residues as mulch, and eventually with intercropping, would seem to be preferable. Charreau (1975), on the other hand, demonstrated that in the dry tropical zone of West Africa, deep ploughing with the incorporation of vegetative material is better for yields, soil structure, and porosity than minimum or zero tillage.

Classic studies on the perennial problem of shifting cultivation and land rotation in the humid tropics of Africa are reflected in assessments by Nye and Greenland (1960) and Ripley (1975). These authors estimate that the annual addition of fallow organic matter from forest litter, dead roots, root slough, and exudate is 15 000 lbs per acre. Ripley (1975) even considers that with grass legume mixtures + roots, as much as 40 000 lbs per acre of organic matter is produced every year. Annual grass burning and forest burning at the end of the fallow period reduces the level of the remaining nutrients dramatically, especially N, S, and C. Due to the long-standing tradition of shifting cultivation - since times before recorded history - these authors warn of the danger of replacing traditional methods with new immature management approaches, which do not control the many facets of the highly differenciated system of soil biological, physical, and chemical stability. Bartholomew et al. (1953) observed, that during fallow-regeneration periods, nutrient immobilization and root development were very rapid at the first onset of the fallow period, and were not uniformly spread. The nutrient gain in grass plots compared favourably with that in bush/forest fallow.

Table 4 (Sanchez, 1987) shows the annual erosion losses in four locations in Africa, comparing forestland, cultivated land, and bare soil (i.e. the same land transformed by management into different biotopes). Lungu (1987) indicates the need for research to assess the true value and possible advantages of legumes compared with shifting cultivation, chitimene (cut branches, twigs, and leaves), and grass composting systems. The specificity of soil organic matter, i.e. the form of the humus in Ferralsols and Oxisols ("sols ferrallitiques") was tested by Riquier (1966) with a view to establishing criteria for distinguishing soil classes. The attempt failed because of the influence of many detailed pedogenetic processes in their synthesis, which results in a loss of marked individuality.

Table 4. Magnitudes of annual soil erosion.

	Soil loss (t ha⁻¹)		
Site	Forest land	Cultivated land	Bare soil
Oagadougou, Burkina Faso	0.1	0.6-8.0	10-20
Sofa, Senegal	0.2	7.3	21
Bouaké, Senegal	0.1	0.1-26	18-30
Abidjan, Ivory Coast	0.03	0.3-90	108-170

Source: P.A. Sanchez, in Dover and Talbot (1987).

According to Oygard (1981) and Hargrove and Thomas (1981), an important aspect of increasing soil organic matter by manure, mulch, and green manure, is that they have a depressing effect on exchangeable aluminium and decrease the soil pH, the threshold where Al toxicity begins. There is a consequent improvement in the soil structure and in aluminium fixation by the soil organic matter.

A study carried out near Lamto in Côte d'Ivoire (Martin, 1989) provides an example of ecological change from savanna to secondary woodland in Ferralsols. The study involved scanning the $\delta^{13}C$ change when C-4 grasses ($\delta^{13}C$ -10‰) were replaced by C-3 bush and trees ($\delta^{13}C$ -25‰), and calculating the organic-matter turnover from these data. Organic-matter use in Chinese agriculture has a long tradition, based on the empiric recognition of the improved structure, water retention, nutrient supply, and improved P availability in southern Ultisols in competition either with organic ligands, or with fixation by protonization at Al edges of oxides and LAC clays. Tanaka *et al.* (1987) give an overview of the amounts of organic substrates used in some areas, e.g. 80 to 200 t ha^{-1} yr^{-1} of pig manure, pit manure, night soil, farmyard manure, and green manure. Neue (1984), after evaluating all the data on N inputs, organic-matter content and yield at his disposal, demonstrated that for rice soils soil organic-matter contents followed a yield minimum-optimum-maximum relationship. The optimum of the yield curves was about 2 to 2.5% C, depending on the supply of N, and the optimum C level was reached as the N supply increased (Figure 6).

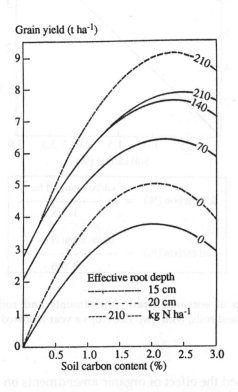

Figure 6.   Relationship between soil carbon content, N-fertilizer rate, and grain yield of rice in puddled and submerged soil (Neue, 1984).

Figure 7 (Neue, 1985) shows the regression curves between rice straw + roots + stubble, the related organic-C input, and the resulting soil carbon percentage. Studies regarding the organic matter-soil biological activity interactions and soil physical properties in rice soils have been formalized since 1984 in the

Physical Aspects of Soil Management in Rice-Based Cropping Systems (PASMIRKS) Programme, coordinated by IRRI (Los Baños, Philippines).

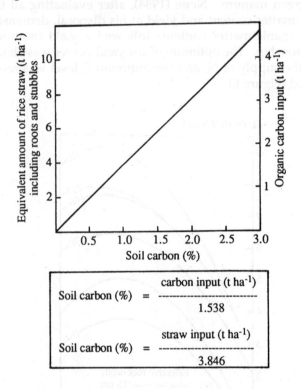

Figure 7. Relationship of seasonal straw amendments and soil carbon content in tropical lowland soils, with (two rice crops a year followed by dry fallow (Neue, 1985).

Voss (1988) studied the effect of organic amendments on soil properties and rice production, and demonstrated the effect of organic matter and legumes like *Sesbania* or *Aeschynomene*, added to rice soils, on moisture retention, linear shrinkage, aggregate stability, pore volume, and bulk density. Table 5 shows the increase in the available water capacity (AWC) and total pore volume (TPV) with the addition of straw. The rather low bulk density in rice fields - which do not have the lowest organic-matter contents - was the reason for inducing soil compaction by using a heavy roller (Ghildyal, 1978), which led to considerable gains in bulk density (up to 1.88 g cm^{-3}) and almost doubled grain yields (up to 2.2 t, as opposed to 1.2 t without previous compaction).

Table 5. Effect of straw addition to rice fields on bulk density, available water capacity and total pore volume.

Treatment	BD (g cm⁻³)	AWC (vol%)	TPV (vol%)
Straw composted	0.80	30.88	71.46
No straw	0.88	21.30	66.48
Straw mixed	0.84	33.03	67.99
Straw burned	0.87	29.38	66.93

Source: Voss (1988).

The organic matter-soil biological activity interaction and soil physical properties are interrelated in most basic reactions within a diversified and dynamic soil-water-plant system, and this soil fertility aggregate is basic to terrestrial life. But most of these relationships are so much part of the empirical basis of soil management that they are not repeatedly questioned, reconfirmed, or debated. The complex interactions of soil biological activity and physical properties are nonetheless of prime importance for soil conservation and productivity.

References

AGBOOLA, A. 1975. Problems of improving soil fertility by the use of green manuring in the tropical farming system. In: *Organic materials as fertilizers*, 147-163. FAO/SIDA Soils Bulletin no. 27. Rome: FAO.

AGBOOLA, A., ABIGBESAN, G.O. and FAYEMI, A.A.A. 1975. Interrelations between organic and mineral fertilizer in the tropical rainforest of Western Nigeria. *Organic materials as fertilizers*, 337-351. FAO/SIDA Soils Bulletin no. 27. Rome: FAO.

AYANABA, A. and OKIGBO, B.N. 1975. Mulching for improved soil fertility and crop production. *Organic materials as fertilizers*, 97-119. FAO/SIDA Soils Bulletin no. 27. Rome: FAO.

BARTHOLOMEW, W.V., MEYER, J. and LAUDELOT, H. 1953. *Mineral nutrient immobilization under forest and grass fallow in the Yangambi region (Congo)*. Institut National d'Etudes Agronomiques du Congo. Série Scientifique no. 57. Brazzaville, Congo: INEAC.

BOLIN, B., DEGENS, E.T., DUVIGNEAUD, P. and KEMPE, S. 1979. The global biogeochemical carbon cycle. In: *The global carbon cycle*, 1-56. Scientific Committee on Problems of the Environment (SCOPE). Publication no. 13. Chichester: John Wiley and Sons.

VAN BREEMEN, N. and FEIJTEL, T.C.J. 1989. Soil processes and properties involved in the production of greenhouse gases with special relevance to soil taxonomic systems. Preprint for the International Conference on Soil and the Greenhouse Effect, International Soil Reference and Information Centre, Wageningen, August 1989.

CHARREAU, C. 1975. Organic matter and biochemical properties of soil in the dry tropical zone of West Africa. *Organic materials as fertilizers*, 131-335. FAO/SIDA Soils Bulletin no. 27. Rome: FAO.

DOVER, M. and TALBOT, L.M. 1987. To feed the earth: Agro-ecology for sustainable development. In: *Properties and management of soils in the tropics*, ed. P.A. Sanchez. World Resources Institute. New York: WRI.

ELLWARDT, P.C., HAIDER, K. and ERNST, L. 1981. Untersuchungen des mikrobiellen Liginnabbaus durch ^{13}C-NMR-Spektroskopie an spezifisch ^{13}C angereichertem DHP-Lignin aus Coniferylalkohol. *Holzforschung* 35: 103-109.

ESWARAN, H. 1981. Soil analysis for soil surveys. In: *Soil Resource Inventories and Development Planning on Proceedings of the Workshop* (Cornell University, 1977-1978). 75ff. Soil Conservation Service, U.S. Department of Agriculture. Technical Monograph no. 1, Soil Management Support Services. Washington, DC: Government Printing Office.

FAO (Food and Agriculture Organization of the United Nations). 1978. *Report on the agro-ecological zones project. Methodology and results for Africa.* World Soil Resources Report no. 48, vol. 1. Rome: FAO.

GHILDYAL, B.P. 1978. In: *Soils and rice*, 317-336. International Rice Research Institute. Los Baños, Philippines: IRRI.

GREENLAND, D.J. 1986. Soil organic matter in relation to crop nutrition and management. *Proceedings of the International Conference on Management and Fertilization of Upland Soils* (Nanjing, 1986). International Potash Institute. Canada: IPI.

HAIDER, K. and MARTIN, J.P. 1981. Decomposition in soil of specifically ^{14}C-labelled model and corn stalk lignins and coniferyl alcohol over two years as influenced by drying, rewettig and additions of an available substrate. *Soil Biology and Biochemistry* 13: 447-450.

HARGROVE, W.L. and THOMAS, G.W. 1981. Effect of organic matter on exchangeable aluminium and plant growth in acid soils. American Society of Agronomy. Special Publication no. 40. Madison, WI: ASA.

VAN DER HEIDE, J. ed. 1989. *Nutrient management for food crop production in tropical farming systems* (Indonesia, 1987), The Netherlands: Haren.

JENNY, H., GESSEL, S.P. and BINGHAM, F.T. 1949. Comparative study of decomposition rates of organic matter in temperate and tropical regimes. *Soil Science* 68: 419-432.

JENKINSON, D.S. and AYANABA, A. 1977. Decomposition of carbon-14 labelled plant material under tropical conditions. *Soil Science Society of America Journal* 41: 912-915.

JENNY, H. 1950. Causes of the high nitrogen and organic matter contents of certain tropical soils. *Soil Science* 69: 63-75.

JURION, F. and HENRY, J. 1969. *Can primitive farming be modernized?* Institut National d'Etudes Agronomiques du Congo. Brazzeville, Congo: INEAC.

KEELING, C.D. 1973. The carbon dioxide cycle. In: *Chemistry of the lower atmosphere*, ed. S.I. Rasool, 251-329. New York: Plenum Press.

KRISHNAPPA, A.M. and SHINDE, J.E. 1978. The turnover of ^{15}N-labelled straw and FYM in a rice-rice rotation. In: *Proceedings of the FAO/IAEA International Symposium on the Use of Isotopes and Radiation in Research on Soil-Plant Relationships*. International Agricultural Economics Association. Colombo: IAEA.

LAL, R. 1981a. Deforestation of tropical rainforest and hydrological problems. In: *Tropical agricultural hydrology*, ed. R. Lal and E.W. Russel, 131-140. Chichester, UK: John Wiley and Sons.

LAL, R. 1981b. Management of the soils for continuous production, controlling erosion and maintaining physical condition. In: *Characterization of soils*, ed. D.J. Greenland, 188-201. Oxford: Clarendon Press.

LAUDELOT, H., MEYER, J. and PEETERS, A. 1960. Les rélations quantitatives entre la teneur en matière organique du sol et le climat. *Extrait de Agricultura* 8(1): 103-140.

LAURA, R.D. 1975. The role of protolytic action of water in the chemical decomposition of organic matter in the soil. *Pédologie* 25: 157-170.

LUNGU, O.J. 1987. A review of soil productivity research in high-rainfall areas of Zambia, 42. Department of Soil Science, University of Zambia. Skriftserie A no. 8. Oslo: NORAGRIC.

MARTIN, A. 1989. Effet des vers de terre tropicaux géophages sur la dynamique de la matière organique du sol dans les savanes humides. Thèse, Centre d'Orsay, Université Paris-Sud.

MOORMANN, F.R. and GREENLAND, D.J. 1979. Major production systems and soil-related qualities in humid tropical Africa. In: *Proceedings of a Conference on Alleviating Soil-Related Constraints to Food Production in the Tropics*. Los Baños, Philippines: IRRI.

MUSA, M.M. 1975. A method for conservation of cattle manure. Organic materials and fertilizers, 89-95. FAO/SIDA Soils Bulletin no. 27. Rome: FAO.

NEUE, H.U. 1984. Annual report (1984) to IRRI on the management of organic manures. Los Baños, Philippines: IRRI. Mimeo.

NEUE, H.U. 1985. Annual report (1984) to IRRI on the management of organic manures. Los Baños, Philippines: IRRI. Mimeo.

NEUE, H.U. In press. Holistic view of chemistry of flooded soils. Preprint in: Proceedings of the First International Symposium on the Fertility of Paddy Soils (Chiang Mai, 1988). Bangkok: IBSRAM.

NEUE, H.U. and SCHARPENSEEL, H.W. 1987. Decomposition pattern of ^{14}C labelled rice straw in aerobic and submerged rice soils. *The Science of the Total Environment* 62: 431-434.

NEUE, H.U., BECKER-HEIDMANN, P. and SCHARPENSEEL, H.W. In press. Organic matter dynamics, soil properties, and cultural practices in rice lands and their relationship to methane production. In: *Proceedings of the International Conference on Soil and the Greenhouse Effect* (Wageningen, 1989), ed. A.F. Bouwman, 457-467. Chichester, UK: John Wiley and Sons.

NYE, P.H. and GREENLAND, D.J. 1960. *The soil under shifting cultivation.* Commonwealth Agricultural Bureau International. Technical Communication no. 51. Wallingford, UK: CABI.

OYGARD, R. 1987. In: *Economic aspects of agricultural liming in Zambia,* 98-99. Department of Agricultural Economy, Agricultural University of Norway. Skriftserie A no. 7. Oslo: Agricultural University of Norway.

PAUL, E.A. and VAN VEEN, J.A. 1978. The use of tracers to determine the dynamic nature of organic matter. In: *Transactions of the XIth ISSS Congress* (Edmonton, 1978), vol. 3, 61-102. Symposium papers. International Society of Soil Science. Wageningen:: ISSS.

RIPLEY, O. 1975. Shifting cultivation and burning. In: *Proceedings of the Joint Commissions (I, IV, V, VI) of the ISSS Conference on Savannah Soils of the Sub-Humid and Semi-Arid regions of Africa and their Management,* ed. H.B. Obeng, 196-201. Kumaṣi: Soil Research Institute.

RIQUIER, J. 1966. La matière organique dans les sols ferrallitiques. *Cahiers ORSTOM, série Pédologie* 4(4): 33-37.

ROGER, P.O. and KURIHARA, Y. In press. Floodwater biology of tropical wetland fields. Preprint in: Proceedings of the International Symposium on Paddy Soil Fertility (Chiang Mai, 1988). Bangkok: IBSRAM.

RUTHENBERG, H. 1976. *Farming systems in the tropics,* 2d ed. Oxford, UK: Oxford University Press.

SANCHEZ, P.A. 1981. Soil management in the Oxisol savannahs and Utisol jungles of tropical South America. In: *Characterization of soils,* ed. D.J. Greenland, 214-253. Oxford: Clarendon Press.

SANCHEZ, P.A., COUTO, W. and BUOL, S.W. 1982. The Fertility Capability Soil Classification system: interpretation, applicability and modification. *Geoderma* 27: 283-309.

SAUERBECK, D. and GONZALES, M.A. 1977. Decomposition of carbon-14 labelled plant residues in various soils of the FRG and Costa Rica. In: *Soil organic matter studies* (Brunswig, FRG, 1976), 159-170. Rome: FAO/Vienna: AEA.

SCHARPENSEEL, H.W., FREYTAG. J. and BECKER-HEIDMANN, P. 1986. C-14-Altersbestimmungen und $\delta^{13}C$-Messungen an Vertisolen unter besonderer Berücksichtigung der Geziraböden des Sudan. *Zeitschrift für Pflansinernährung und Bodenkunde* 149: 277-289.

SCHARPENSEEL, H.W., TSUTSUKI, K., BECKER-HEIDMANN, P. and FREYTAG, J. 1986. Untersuchungen zur Kohlenstoffdynamik und Bioturbation von Mollisolen. *Zeitschrift für Pflansinernährung und Bodenkunde* 149: 582-597.

SCHARPENSEEL, H.W., WURZER, M., FREYTAG, J. and NEUE, H.U. 1984. Biotisch und abiotisch gesteuerter Abbau von organischer Substanz in Boden. *Zeitschrift für Pflansinernährung und Bodenkunde* 147: 502-16.

SINGER, S. 1990. *Report on the decomposition of uniformly labelled groundnut straw and organic matter turnover in the SAT soils, alfisols and Vertisols.* Mid-term report to ICRISAT and GTZ. Patancheru, India: ICRISAT/Bonn, FRG: GTZ.

SINGH, A. 1975. Use of organic materials and green manures as fertilizers in developing countries. *Organic materials as fertilizer*, 21. FAO/SIDA Soils Bulletin no. 27. Rome: FAO.

SYS, C. 1976. *Land evaluation*, pts. I and III. International Training Centre for Post-Graduate Soil Scientists. Ghent: State University of Ghent.

TANAKA, A., YUANCHANG, X. and L. and HASEGAWA, M. 1987. In: *Studies on plant nutrients in rice-oriented farming systems in Jiangsu Province*, China, 15ff. Kali Kenhyukai.

TROPSOILS 1989. *TropSoils technical report for 1986-87*. TropSoils Management Entity. Raleigh, NC: North Carolina University.

THENG, B.K.G. and SCHARPENSEEL, H.W. 1975. Clay organic complexes between montmorillonite and C-14-labelled humic acid. In: *Proceedings of the International Clay Conference* (Mexico City, 1974), 643-653. Wilmette, Illinois: Applied Publications Ltd.

TSUTSUKI, K. and KUWATSUKA, S. 1989. Degradation and stabilization of the humus in buried humic Andosols. In: *Proceedings of the IVth ISSS Meeting on the Science of the Total Environment* (Sevilla, 1988). Madison, WI: ISSS.

UMAROV, M.M. 1989. Biotic sources of nitrous oxide (N_2O) in the context of the global budget of nitrous oxide. In: *Proceedings of the International Conference on Soil and the Greenhouse Effect* (Wageningen, 1989), ed. A.F. Bouwman, 263-268. Chichester, UK: John Wiley and Sons.

VAN WAMBEKE, A. 1981. *Planning objectives and the adequacy of soil resource inventories*. Soil Conservation Service, U.S. Department of Agriculture. Technical Monograph no. 1. Washington, DC: Government Printing Office.

VOSS, G. 1988. *Effect of organic amendments on soil properties and rice production*. International Rice Research Institute. Terminal report to IRRI, 12ff. Los Baños, Philippines: IRRI. 12p.

WELLER, J.M. 1959. Compaction of sediments. *Bulletin of the American Association of Petroleum Geologists* 43: 273-310.

Agroforestry for the management of soil organic matter

ANTHONY YOUNG*

Abstract

An objective of sustainable land-use systems should be to maintain soil organic matter at between 40 and 60% of its level under natural vegetation. It is the labile humus fraction that is primarily affected by soil management. As compared with the usual decline in soil organic matter under agriculture (annual cropping), agroforestry offers four ways to increase inputs, or reduce losses, of soil organic matter: higher biomass production, removal of fewer parts of the plants (tree and crop), a lower humus decomposition constant, and reduction of loss of organic matter through erosion control. The computer model SCUAF shows the superiority of agroforestry over agriculture for organic matter maintenance and sustained production, with best performance in systems that have a substantial area under trees. The limited available experimental evidence supports this conclusion. Three agroforestry technologies appear to have a high potential for soil organic-matter maintenance: multistorey tree gardens, plantation crop combinations, and hedgerow intercropping. Other agroforestry technologies have a qualified or limited, but positive, potential. Soil monitoring and specialized soil research are needed to confirm this potential.

*   ICRAF, Box 30677, Nairobi, Kenya.

How much soil organic matter should be maintained?

The importance of maintaining soil organic matter, as a major factor contributing to sustained land use, is now so widely accepted that it need not be re-stated at length (Young, 1989, p. 105). It is the main management-dependent factor affecting a wide range of soil physical properties, and thereby erosion resistance and water-holding capacity. There are biological effects, in the form of enhanced faunal activity (Swift and Sanchez, 1984). Finally, there is a wide range of effects on soil chemical properties and nutrient supply, applicable to both smallholders and high-technology farming: in low-input systems, a steady release of nutrients from decomposing plant residues, and in systems with high fertilizer inputs, a greater efficiency of use, as has been demonstrated by experiments combining with/without organic manure and fertilizer.

Although the optimum organic-matter content for cultivated soils is no doubt close to that under natural vegetation, it is unrealistic to attempt to achieve this. Such levels do, however, provide a basis from which to estimate what is desirable, as a percentage of that under vegetation. Levels under natural conditions vary with climate (higher with higher rainfall and lower temperature) and texture (higher for clays than sandy soils). Andosols are high in organic matter, and levels in soils derived from basic rocks (FAO Nitosols) are relatively high. Taking soils of medium to heavy texture, assuming soil organic matter to equal organic carbon x 2 (rather than the formerly used figure of x 1.72), and using round figures, representative values for climatic zones are given in Table 1.

What fraction of these levels should be the target to maintain in an agroecosystem? Few statements on this have been made, and these are mostly based on surmise rather than evidence. Most people would agree that to maintain 60% of the level under natural conditions is sufficient; whilst some would suppose that degradation of physical properties and lowering of fertility was likely if levels fell below 40%. Let us therefore take 50% of the organic-matter levels found under natural vegetation as a target for maintenance under productive use, i.e. half the levels shown in Table 1.

In passing, one may note the potential and limitations of one method of maintenance - farmyard manure. In the temperate zone, with its lower decomposition rates, 'muck-spreading' was an effective method until labour costs restricted its application. Research in the tropics some years ago demonstrated that manure could maintain organic matter under continuous cultivation, but only at rates of application of some 10-25 t ha^{-1} yr^{-1}, possible on intensively cultivated fields but impracticable for the main arable area of farms.

Table 1. Typical values of soil organic matter for humid (rainforest), subhumid (savanna), and semiarid zones.†

	Topsoil (approximately 0-20 cm)			
	Organic carbon		Organic matter	
	(%)	(t ha⁻¹)	(%)	(t ha⁻¹)
Humid tropics	2.0	50 000	4.0	100 000
Subhumid tropics	1.0	25 000	2.0	50 000
Semiarid tropics	0.5	12 500	1.0	25 000

† Values for sandy soils are about a half of these levels.
Source: Young (1976).

Gains and losses of soil organic matter

The terms 'soil organic matter' and 'humus' will be treated as synonymous. There is some measure of agreement on the fractions of organic matter on or in the soil, although nomenclature varies, as follows:

Fresh litter   Plant residues, aboveground and roots; the L horizon of soils.

Decomposing litter   Comminuted, partly humified, plant residues: the F horizon of soils.

Active humus   Organic matter contained, for a period, in the bodies of soil microorganisms.

Labile humus   Fully humified material, decomposing at a rate of the order of 3-5% per year.

Stable humus   Fully humified, relatively inert material, decomposing slowly.

Litter is sometimes divided into a decomposable or high-quality fraction, and a resistant or low-quality fraction, with high and low nitrogen:lignin ratios respectively. This may be a transition, rather than clearly separated fractions.

Leaving aside biochemcal complexities, there is some measure of consensus between the various models that have been proposed for the plant-soil carbon cycle (Figure 1) (e.g. Jenkinson and Rayner, 1977; Parton *et al.*, 1987; Young, 1989, 113; Greenland *et al.*, in press). Fresh plant residues are progressively comminuted and decomposed. In the final stage of decomposition, these pass into the bodies of soil microorganisms, becoming converted to fully humified organic matter. During this process of humification, more than half the plant

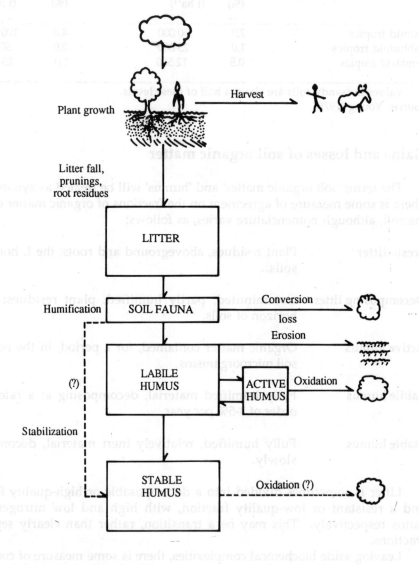

Figure 1. The plant-soil carbon cycle.

carbon is lost to the atmosphere as carbon dioxide - the litter-to-humus conversion loss. Humus forms the substrate of other soil microorganisms, such that some 3-5% (of its original amount) is lost per year by oxidation. Little is known about how stable soil organic matter originates; it may be by physical, chemical or biological transformation processes, and it is possible (although this is an unorthodox view) that some lignin-rich plant residues may be converted directly to the stable fraction. We have almost no information on the rate at which this fraction is lost, other than that it is at a very much slower rate than for the labile fraction, part of the stable humus remaining resident in the soil for over 100 years.

In the earlier model of soil organic matter, the oxidation of carbon during humification of litter was treated as a percentage loss, estimated as 80-90% for aboveground plant residues and 50-80% for roots (Nye and Greenland, 1960). More recent studies based on isotopic labelling show that this initial loss is in fact a negative exponential decay curve with a rate such that, in the tropics, about half the original litter carbon is lost in 4-6 months; when combined with the subsequent, slower, decay of humus, a two-stage curve is clearly distinguishable.

Labile humus decays at a rate dependent on the amount of initial humus present, according to the relation:

$$C_1 = C_0 - KC_0 \quad \text{or} \quad C_1 = C_0 (1 - K)$$

where C_0 is initial humus carbon, C_1 is humus carbon after one year, and K is the humus decomposition constant, with a value of about 0.03-0.05. This leads to a negative exponential decay curve (Figure 2). Kc, the decomposition constant under cultivation, is generally believed to be higher then Kf, that under forest or other natural vegetation.

It may be supposed that the loss of stable humus follows a similar pattern, but with a very much lower rate of decomposition. Since over a period of 20 or even 50 years, stable humus is little affected by management, we need not concern ourselves too much over this, other than by recognizing that it forms possibly 50-75% of the total soil carbon as determined by standard analysis. When talking about stable humus, G.K. Chesterton's poem on the microbe is applicable:

> *But scientists, who ought to know,*
> *Assure us that it must be so.*
> *Oh, let us never, never doubt*
> *What nobody is sure about.*

For modelling soil organic matter, however, some assumptions are needed. As recent plant residues (aboveground and shallow roots) enter the topsoil, most of the topsoil humus consists of the labile fraction. Two alternative working hypotheses are suggested:

either: - model changes in topsoil humus only and assume that all of this is labile, on the basis that whatever topsoil humus belongs to the stable fraction is approximately balanced by labile humus in the lower horizons.

or - model the whole soil profile, and assume that between 50 and 75% of the humus belongs to the stable fraction.

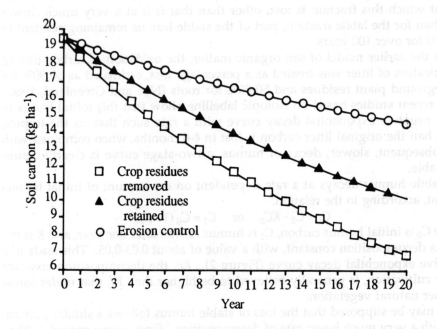

Figure 2. Modelling of soil carbon losses under different agricultural systems.

Thus, the balance of soil organic matter in an agroecosystem is affected by the following factors:

1. The biomass production (above- and belowground) of plants in the system.
2. Which plant parts are removed from the system.
3. The litter-to-humus conversion losses, for above- and belowground residues.
4. The humus decomposition constant, for labile humus.
5. The loss of humus through soil erosion.
6. The proportion of soil humus that is in stable form.
7. To a small degree only, the rate of loss of stable humus.

Factors 1-5 can be modified by management, but factors 6 and 7 can only be modified to a slight degree.

Soil organic matter under agriculture

Biomass production under continuous annual cropping may, with fertilizer inputs, approach that under natural vegetation, but it is often lower (factor 1 above). The fruit or other harvested part of the plant is removed as harvest, roots retained in the soil, and the crop residue removed or retained (factor 2). The humus decomposition constant is probably higher than under natural conditions, owing to high temperatures under bare soil and increased aeration through tillage (factor 4). These effects of management lead to a negative exponential decline in soil organic matter. The lowered fertility leads to poorer crop growth and hence a reduced supply of plant residues. Even with this soil-plant feedback effect, a new low-level equilibrium will eventually be reached if erosion is prevented, but only at unacceptably low levels of soil organic matter and crop production.

Under some perennial (plantation) cropping systems, soil organic matter may be indefinitely maintained at acceptable levels, particularly if there is a cover crop or pasture between the trees.

Soil organic matter under agroforestry

In considering whether agroforestry systems have the potential to maintain soil organic matter at acceptable levels, we are looking at the capacity of such systems to modify factors 1-5 above. The standards for comparison are natural vegetation and annual cropping. Thus, the questions to ask for each of these five factors are first, how much poorer is the agroforestry system compared with natural vegetation, and secondly, how much better is it compared with annual cropping? What we are hoping for is that in agroforestry systems, biomass production will be higher than in agriculture, fewer plant parts will be removed, the litter-to-humus conversion loss and/or the humus decomposition constant will be lower than under agriculture, and loss of humus through soil erosion will be lower. Opportunities for management of organic matter in agroforestry arise through the addition of tree residues (above- and belowground) to the soil; shading of the ground surface by tree litter or prunings; the manner of addition of such prunings (as mulch, buried mulch, or compost); and control of erosion.

Biomass production

Many of the fast-growing trees commonly used in agroforestry are able, in pure stands, to equal or exceed the biomass production rates of natural vegetation in their ecozones. Thus in the humid tropics, *Leucaena leucocephala* commonly reaches 20 000 kg DM ha^{-1} yr^{-1} (DM = dry matter), a typical figure for rainforest, whilst *Acacia mangium* can exceed 15 000 kg DM ha^{-1} yr^{-1} (data sources for this section are given in Young, 1989, pp. 121-25). In the subhumid tropics, *Leucaena* and most of the trees commonly used in agroforestry, such as *Gliricidia sepium*, *Cassia siamea* and *Flemingia congesta*, can equal or exceed the biomass production of natural savanna, with some *Sesbania* species greatly exceeding it during the first few years of growth. The same applies to *Prosopis* spp. in the semiarid zone. The reason is evident from ecological principles: the climax trees in natural vegetation have slower rates of growth than the pioneer and early successional species, and it is from the latter that many multipurpose trees originate.

The relevant aspect, however, is not the growth rate of multipurpose trees in pure stands, but the biomass production of practical agroforestry systems as a whole - where 'practical' means productive and acceptable to the farmer. This depends on the respective areas occupied by the tree and crop components and the rates of growth of each. Here there is an effect that weighs in favour of agroforestry, namely that isolated trees or trees in rows grow faster, per unit area, than uniform plantations; they obtain more light, and have less competition from their own kind. Growth of the associated crop may not be correspondingly reduced if the tree canopy is either low or open, and if tree/crop competition for moisture and nutrients can be kept low by good agroforestry design.

The most substantial evidence to date comes from two agroforestry technologies: plantation crop combinations of coffee or cocoa with (so-called) shade trees, and hedgerow intercropping. For the dense, mixed intercropping systems of *Erythrina*, *Inga* and/or *Cordia* with coffee or cocoa, found under humid tropical environments in Latin America, aboveground biomass production alone has been recorded as 9000-15 000 kg DM ha^{-1} yr^{-1}. Hedgerow intercropping trials, initially at Ibadan, Nigeria, under a moist subhumid bimodal rainfall climate, but now finding support from other sites, show biomass production *from the tree component alone* of 3000-8000 kg DM ha^{-1} yr^{-1}, even though this component may only occupy 20% or less of the ground surface area. If crop biomass production is added, such values exceed that of natural vegetation under the same rainfall.

For the general case, we can expect the biomass production of agroforestry systems to exceed that of equivalent agriculture in proportion to the percentage of tree cover. On this criterion, different agrosylvicultural systems can be rated as follows:

Potential for high biomass production:
Multistorey tree gardens
Plantation crop combinations
Hedgerow intercropping

Qualified potential for biomass production:
Shifting cultivation (only with high fallow ratio)
Improved tree fallow (only with moderate fallow ratio or very fast-growing species, e.g. *Sesbania*)
Trees in cropland (only if moderately dense)

Limited potential for biomass production:
Trees in cropland (if widely spaced)
Trees on erosion-control structures
Windbreaks and shelterbelts

Plant residues returned to the soil

Agroforestry systems contain two plant components, trees and crops, each of which may have four parts: leaf (herbaceous matter), fruit (reproductive matter), wood, and roots. When the crop is an annual, crop wood is absent.

In both agriculture and agroforestry, crop fruit (e.g. maize cobs) is normally removed from the system as harvest, crop leaf (i.e. residues) may or may not be, whilst crop roots remain in the soil unless, as in root crops, parts of them are harvested. For soil organic matter maintenance, it is far better to retain crop residues, but farmers may have many reasons for removing them (e.g. for fodder, fencing or thatching, fuel, or to prevent carry-over of pests and diseases), and there is no reason to suppose that their decision on whether to remove or retain these residues will differ as between agriculture and agroforestry.

In agroforestry systems, tree roots remain in the soil, whilst tree wood is nearly always harvested, as fuelwood or timber. Tree fruit may be harvested for food or fodder, but its biomass is relatively small. The critical question is whether tree leaf, consisting of natural litter fall plus prunings, is harvested (usually as fodder) or remains on the soil. Farmers with livestock will usually harvest this excellent fodder, often green and nutritious, when grass is unpalatable in the dry season. For hedgerow intercropping systems in particular, it is crucial whether prunings are removed as fodder; this is nearly always done in the Indian subcontinent, a fact which may help to account for the almost invariably negative effects of these hedgerow intercroppings on crop yields (S. Chinnamani and Deb Roy, pers. comm.). In mixed systems, such as plantation crop combinations and multistorey tree gardens, much tree leaf reaches the soil as natural litter fall.

In summary, all agroforestry systems provide to the soil an additional plant part in the form of tree roots, whilst some also supply tree leaf residues.

Litter-to-humus conversion loss

The loss of carbon by oxidation during the process of conversion from litter to humus is substantial, probably over 50% for root residues and perhaps over 75% for aboveground plant litter. Management options which could affect this loss are to leave plant residues on the soil surface, incorporate them into the soil, or compost them. It is thought that incorporation may reduce the conversion loss, although this practice is less desirable from the point of view of erosion control. However, these options are available in both agriculture and agroforestry, so do not constitute a reason for potential superiority of the latter.

Humus decomposition constant

It is quite widely accepted that the humus decomposition constant is higher under agriculture than under natural forest. Reasons are the disturbance and aeration of the soil through cultivation, causing enhanced microbiological activity, and the high topsoil temperatures caused by exposure of bare soil during early crop growth. We know also that mulching and minimum tillage can dramatically reduce soil surface temperatures.

There is an obvious potential in agroforestry systems for soil temperature reduction, either by shading from a tree canopy or by a surface mulch of tree litter or prunings. Canopy shading and litter cover are both found where the tree component is allowed to grow, as for example in trees amid cropland and plantation crops with shade trees, whilst the tree litter effect occurs where the trees are pruned, as in hedgerow intercropping, provided that the prunings remain on the soil.

Soil erosion

A substantial amount of soil organic matter is lost in eroded sediment. For example, with erosion at a rate of 20 t ha^{-1} yr^{-1}, a topsoil carbon content of 1% and a carbon enrichment factor in eroded sediment of 2%, then the loss is 400 kg C ha^{-1} yr^{-1}. To replace this requires some 3200 kg ha^{-1} of plant residues which, if some plant parts are harvested, calls for a biomass production of several tonnes per hectare.

Under agroforestry, erosion can be reduced to the level commonly regarded as acceptable, about 10 t ha^{-1} yr^{-1}. There is strong indirect evidence, coupled with limited but invariably positive direct experimental evidence, that a number of agroforestry systems can effectively control erosion (Young, 1989). These include dense mixed systems, such as multistorey tree gardens, together with contour-aligned hedgerow intercropping. Advantages of agroforestry over conventional methods are the taking up of less land, lower labour requirements or construction costs, and production from the conservation works themselves.

Modelling soil organic matter changes

Constants and variables

The computer model, 'soil changes under agroforestry' (SCUAF) provides a means of comparing the effects on the soil of agriculture and agroforestry. Version 2 of SCUAF models changes in erosion, soil organic matter (represented by carbon), and soil nitrogen. The structure and functioning of the model have been described elsewhere (Young et al., 1987; Young, 1989, p. 197; Young and Muraya, 1989). The organic-matter model is based on that shown in Figure 1.

For the present purpose, to compare agriculture with agroforestry, we shall standardize the environmental conditions and the agricultural component of the systems. The environment chosen is a tropical lowland subhumid climate (savanna zone), a soil of medium texture and free drainage, and a gentle slope. For the agricultural system, and also the crop component of the agroforestry system, a cereal crop (e.g. maize or upland rice) with a grain yield of 2000 kg ha^{-1} yr^{-1} is assumed. Initial carbon in the topsoil (0-20 cm) is set to 0.75%, or 19 500 kg C ha^{-1}.

Within this standardized framework, certain features of system design and management critical to soil organic matter are varied (Table 2, columns 1-7). In the agricultural system, crop residues are either removed or retained, and the system is modelled without and with control of soil erosion; it is assumed that erosion-control structures take up 20% of the land surface. The agroforestry systems assume that crop residues are retained on the soil. Two systems are modelled: one with 15% trees and 85% crops, typical of hedgerow inter-cropping; and one with 40% tree cover and 80% crop cover, i.e. 20% of the surface area is crops growing beneath trees, the situation found in dense mixed multistorey systems. Each of these systems is modelled with tree leaf residues removed or retained, but in all cases (as is likely in practice) with the wood fraction of the trees harvested. It is assumed that the agroforestry systems intrinsically possess erosion control.

Table 2. Results of modelling three agricultural and five agroforestry systems on SCUAF, under a lowland subhumid climate, on a gentle slope, with initial topsoil carbon = 19 500 kg C ha^{-1}.

Parameters	Model no.							
	1	2	3	4	5	6	7	8
Area under								
- Trees (%)	0	0	0	15	15	40	40	40†
- Crops (%)	100	100	80	85	85	80	80	80
- Erosion structures (%)	0	0	20	0	0	0	0	0
Removal of								
- Crop leaf	Yes	No	No	No	No	No	No	No
- Tree leaf	-	-	-	Yes	No	Yes	No	No
Erosion control	No	No	Yes	Integral	Integral	Integral	Integral	Integral
Soil carbon (0-20 cm) Year 20 (kg C ha^{-1})	6 899	10 128	14 579	15 517	16 068	17 427	18 975	19 009
Crop harvest								
- Year 1 (kg ha^{-1})	2 000	2 000	1 600	1 700	1 700	1 600	1 600	1 600
- Year 20 (kg ha^{-1})	1 337	1 499	1 396	1 527	1 550	1 513	1 574	1 579
Other harvest (Year 1)								
- Crop (kg ha^{-1})	4 000	0	0	0	0	0	0	0
- Tree (kg ha^{-1})	-	-	-	480	0	1 280	0	0
- Tree (kg ha^{-1})	-	-	-	990	990	2 640	2 640	18 930‡

† Trees allowed to grow, cut every 8 years.
‡ Years 8 and 16.

Results

The results of modelling are shown in Figures 2-5 and Table 2. Column 8 of the table shows topsoil carbon after 20 years, whilst the last five columns give harvest from the system.

Models 1-3 are agricultural systems. There is a substantial decline in soil carbon in all three, with the greatest decline occurring if crop residues are removed (Figure 2). The addition of soil conservation structures (e.g. bunds) reduces the rates of humus decline, but a consequence is a reduction in harvest (Table 2), caused by the area taken up by these structures. Indeed, the crop harvest with soil conservation does not surpass that without erosion control until Year 17, a powerful disincentive to the farmer against adopting (conventional) conservation.

Models 4 and 5 are agroforestry systems with 15% trees and 85% crops, respectively with and without harvest of tree leaf prunings (Figure 3). Erosion control is intrinsic to the system, without the need to allot land specifically to conservation structures. Both systems are more favourable for soil organic matter than the agricultural systems. Under these standardized assumptions,

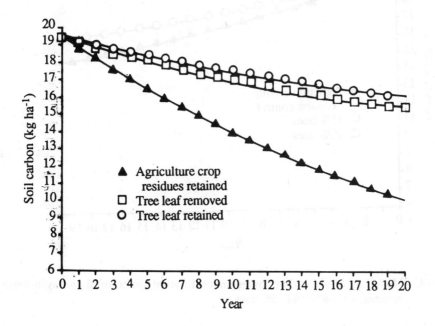

Figure 3. Modelling of soil carbon losses under different agricultural and agroforestry systems, with 15% trees.

however, they do not entirely prevent decline, the basic reason being insufficient biomass production. Crop production is initially lower than under agriculture without erosion, but higher then under agriculture with erosion control, since hedgerows take up less land than bank-and-ditch structures. Crop yield rises above that of agriculture without conservation in Year 11, and there is an additional harvest of 1 t ha⁻¹ yr⁻¹ of fuelwood.

The agroforestry system with 40% trees, Models 6 and 7, comes very close to sustaining soil organic matter - indeed, within the margins of error inherent in modelling it is fully sustainable in this respect (Figure 4). This is brought about by the higher biomass reaching the soil, particularly tree root biomass, together with a lower humus decomposition constant caused by canopy and litter shading. Crop yield is sustained at 1600 kg ha⁻¹ yr⁻¹, and there is an additional harvest of 2.6 t ha⁻¹ yr⁻¹ of fuelwood and, if tree leaf prunings are removed, 1.3 t ha⁻¹ yr⁻¹ of fodder.

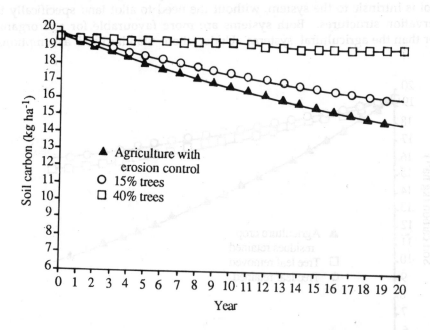

Figure 4.   Modelling of soil carbon losses under different agricultural and agroforestry systems, with 40% and 15% trees.

A more realistic assumption for dense mixed agroforestry systems is that the trees are allowed to grow for some years and then felled. Model 8 (Figure 5) shows the changes in soil organic matter with this assumption. There is a decline whilst organic matter is accumulating in the growing trees, but this is restored by the return to the soil of accumulated tree root biomass when the trees are felled, giving an oscillating but steady state system. This system produces a wood harvest of 19 t ha^{-1} every eight years - a sizeable contribution to the farmers' requirements, for on-farm use or sale.

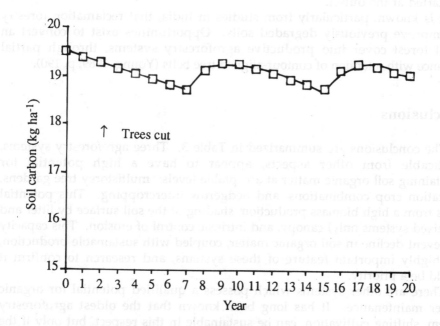

Figure 5. Modelling of soil carbon losses under an agroforestry system of 40% trees, cutting trees after eight years of growth.

Experimental evidence

The very limited experimental evidence available supports the conclusion that dense mixed agroforestry systems are fully sustainable with respect to soil organic matter. Studies of plantation crop combinations of cacao with *Cordia alliodora* and *Erythrina poeppigiana* in a humid tropical environment in Costa Rica show gains equal to losses, and no significant change in soil humus levels over 5 years (Alpizar *et al.*, 1986, 1988; Young, 1989, p. 176). For the system of home gardens, qualitative observation, together with a few measurements, support this finding (Soemwarto, 1987).

The data for hedgerow intercropping, however, do not support the modelled decline. In moist subhumid environments at Ibadan, Nigeria and Lilongwe, Malawi, soil monitoring over 6 and 2 years respectively shows an apparent increase in soil carbon (Kang *et al.*, 1985; Chiyenda and Materechera, 1989). It is doubtful whether these increases are statistically significant. It is important that in the many trials of hedgerow intercropping currently being initiated, statistically rigorous programmes of soil sampling and monitoring are started at the outset.

It is known, particularly from studies in India, that reclamation forestry can improve previously degraded soils. Opportunities exist to convert an initial forest cover into productive agroforestry systems, through partial clearance with retention of contour-aligned tree belts (Young, 1989, p. 190).

Conclusions

The conclusions are summarized in Table 3. Three agroforestry systems, practicable from other aspects, appear to have a high potential for maintaining soil organic matter at acceptable levels: multistorey tree gardens, plantation crop combinations, and hedgerow intercropping. This potential arises from a high biomass production, shading of the soil surface by litter and (in mixed systems only) canopy, and intrinsic control of erosion. This capacity to prevent decline in soil organic matter, coupled with sustainable production, is a highly important feature of these systems, and research to confirm it should be a priority.

There are other systems which possess a qualified potential for organic matter maintenance. It has long been known that the oldest agroforestry system, shifting cultivation, can be sustainable in this respect, but only if the fallow-to-cultivation ratio is sufficiently high (Young and Wright, 1980; Young, 1989, p. 86). The same necessarily applies to systems of improved tree fallow, although there may be some potential for short-term (1-3 year) fallows of very fast-growing trees such as *Sesbania* spp.. This has been demonstrated for fallows between crops of swamp rice (Weerakoon and Gunasekera, 1985) but is unproven for rainfed cropping. Among problems are the nutrients removed by the trees, and the absence, in rotational agroforestry systems, of erosion control during the cropping period.

In most systems of trees in cropland the tree cover is too sparse to substantially affect soil organic matter, but there may be a potential if the trees are planted more densely, say over 20% cover, as is being tried with *Acacia albida*.

Table 3. Agroforestry systems with potential for maintenance of soil organic matter.

High potential

Dense spatial mixed systems:	Multistorey tree gardens
	Plantation crop combinations
Spatial zoned system:	Hedgerow intercropping

Qualified potential

Rotational systems:	Shifting cultivation: only with high fallow-to-cultivation ratio
	Improved tree fallow: only with moderate fallow-to-cultivation ratio and/or very fast-growing trees (e.g. *Sesbania* spp.)
Open spatial mixed system:	Trees in cropland: only if tree cover is moderately dense, over 20% (possible with, e.g. *Acacia albida, Sesbania sesban*)

Limited, although positive, potential

Spatial zoned systems:	Boundary planting
	Trees on erosion-control structures
	Windbreaks and shelterbelts

The 'taungya' system, if well managed, is satisfactory for soil organic-matter maintenance, although slightly less favourable than pure forestry. Biomass transfer can have limited favourable effects, although is environmentally adverse to the forest from which biomass is removed. Sylvopastoral systems are satisfactory if well managed.

Other systems can be expected to have a positive tendency for conservation of soil organic matter, but to be insufficient on their own to maintain it in the intervening areas of cropping. These are boundary planting, trees planted on conventional erosion-control works (bank-and-ditch structures, grass strips), windbreaks and shelterbelts, together with widely spaced systems of trees in cropland. The reason is that tree biomass production is insufficient to balance humus decomposition.

To complete the coverage of agroforestry technologies, it may be noted that the 'taungya' system can be satisfactory with respect to soil organic matter, provided the main period for forest cover does not suffer erosion. In sylvopastoral systems, soil conservation cannot be achieved solely through the presence of the trees, but requires good pasture management.

There are many reasons for the adoption of agroforestry, including both production and service functions. Among the latter, soil conservation is certainly the most important. Potential benefits to the soil of agroforestry systems include erosion control, efficient recycling of nutrients and, as outlined here, maintenance or improvement of soil organic matter and physical properties. Whilst the apparent potential is high, experimental evidence is very sparse. Research of two kinds is needed. First, monitoring of soil organic matter changes, conveniently at 3-year intervals, should be undertaken in all substantial agroforestry trials, whatever their primary objectives. Secondly, soil specialists should undertake comparative studies of the effects on soils of agricultural and agroforestry systems within different climates and soil types. In the soil studies, it is important to apply to the design of soil sampling and analysis the standards of statistical control normally expected in agronomic experimentation.

References

ALPIZAR, L., FASSBENDER, H.W., HEUVELDOP, J., FOLSTER, H. and ENRIQUEZ, G. 1986. Modelling agroforestry systems of cacao (*Theobroma cacao*) with laurel (*Cordia alliodora*) and poro (*Erythrina poeppigiana*) in Costa Rica, pt. 1. *Agroforestry Systems* 4: 175-190.

CHINYENDA, S. and MATERECHERA, S.A. 1989. Some results from alley cropping *Leucaena leucocephala, Cassia siamea*, and *Cajanus cajan* with maize at Bunda College of Agriculture. In: *Trees for development in subsaharan Africa*, ed. J.N. Wolf, 135-142. Stockholm: International Foundation for Science.

GREENLAND, D.J., WILD, A. and PHILLIPS, D. In press. Organic matter in soils of the tropics from myths to complex reality. In: *Myths and science of soils of the tropics*, ed. R. Lal and P.A. Sanchez. American Society of Agronomy. Madison, WI: ASA.

JENKINSON, D.S. and RAYNER, J.H. 1977. The turnover of soil organic matter in some of the Rothamsted classical experiments. *Soil Science* 123: 298-305.

KANG, B.T. and VAN DER HEIDE, J., eds. 1985. Nitrogen management. In: *Farming systems in humid and subhumid tropics*. Haren, Netherlands: Institute of Soil Fertility. 361p.

NYE, P.H. and GREENLAND, D.J. 1960. *The soil under shifting cultivation*. Technical Communication no. 51. Harpenden, UK: Commonwealth Bureau of Soils. 144p.

PARTON, W.J., SCHIMEL, D.S., COLE, C.V. and OJIMA, D.S. 1987. Analysis of factors controlling soil organic matter levels in Great Plains grasslands. *Soil Science Society of America Journal* 51: 1173-79.

SOEMWARTO, O. 1987. Homegardens: a traditional agroforestry system with a promising future. In: *Agroforestry: a decade of development*, ed. H.A. Steppler and P.K.R. Nair, 157-170. Nairobi: ICRAF.

SWIFT, M.J. and SANCHEZ, P.A. 1984. Biological management of tropical soil fertility for sustained productivity. *Nature and Resources* 20: 2-10.

WEERAKOON, W.L. and GUNASEKERA, C.L.G. 1985. *In situ* application of *Leucaena leucocephala* (Lam) de Wit as a source of green manure in rice. *Sri Lankan Journal of Agricultural Sciences* 22: 20-27.

YOUNG, A. 1976. *Tropical soils and soil survey*. UK: Cambridge University Press. 468p.

YOUNG, A. 1989. *Agroforestry for soil conservation*. Wallingford, UK: CAB International/Nairobi: ICRAF. 276p.

YOUNG, A. and MURAYA, P. 1989. Soil changes under agroforestry (SCUAF): a predictive model. In: *Land conservation for future generations* (ISCO Conference, Bangkok, 1989), ed. Sanarn Limwanich, 655-667. Department of Land Development, Ministry of Agriculture and Cooperatives. Bangkok: DLD.

YOUNG, A. and MURAYA, P. 1990. *SCUAF: soil changes under agroforestry. A predictive model. Version 2. Computer program with user's handbook*. Nairobi: ICRAF. 124p.

YOUNG, A. and WRIGHT, A.C.S. 1980. Rest period requirements of tropical and subtropical soils under annual crops. In: *Report of the second FAO/UNFPA expert consultation on land resources for the future*, 197-268. Rome: FAO.

Section 4: Tillage and soil conservation

Tillage and residue management in rainfed agriculture: present and future trends

PAUL W. UNGER[*]

Abstract

Soil degradation and sustaining crop productivity are of widespread concern and are receiving considerable attention in the agricultural community. To avoid degradation and at the same time to sustain production, the balance between negative and positive factors affecting soil degradation must be tilted in favor of the positive factors. One factor on the positive side of the balance is crop-residue management.

Satisfactory soil erosion control and water conservation can be achieved in many cases with clean tillage in conjunction with one or more appropriate support practices. However, even better erosion control and greater water conservation are possible where crop residue management is incorporated into the crop production system. Retaining crop residues on the soil surface with conservation tillage systems is effective for controlling erosion, both by water and wind. It is effective also for increasing soil water storage, which enhances the potential for improved crop yields, especially in arid, semiarid, subhumid regions. Crop yields in these regions usually are greatly influenced by water stored in soil at planting.

Major limitations to the widespread use of crop residues in crop production systems are low residue production by nonirrigated crops, especially in the drier

---

[*] U.S. Department of Agriculture, Agricultural Research Service, Conservation and production Research Laboratory, P.O. Drawer 10, Bushland, Texas 79012, USA.

regions, and residue removal for feed, fuel, building materials, and manu-facturing. In addition, some producers choose to use clean tillage and eliminate surface residues by burning or inversion types of tillage. To minimize future land degradation, and therefore sustain the productive potential of croplands, management strategies must be developed and/or adopted that reduce competition for residues and permit more residues to be retained on land for resource conservation.

Introduction

Sustaining the productive capacity of land is currently receiving much attention in the agricultural and environmental communities. To sustain production and avoid land degradation, negative factors that cause soil degradation must be overshadowed by positive factors that conserve the soil (Figure 1). One positive factor that has a major influence is crop-residue management. With good management, crop residues can greatly reduce soil erosion. They also improve water conservation, which can lead to higher crop yields.

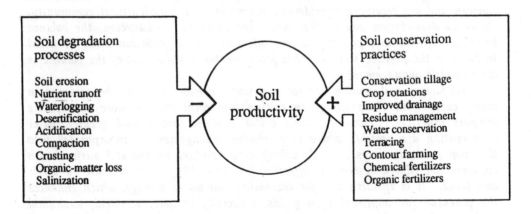

Figure 1. Relationship of soil productivity to soil degradation processes and soil conservation practices (Hornick and Parr, 1987).

The tillage method or system employed for producing a crop has a major impact on residues remaining from previous crops. Tillage methods are available that incorporate most or all residues with soil, retain some residues on the surface, or retain virtually all residues on the surface, as with no tillage. Besides the effects of tillage, the amount of residues retained is influenced also by such management decisions as whether to burn or remove the residues for feed, fuel, building materials, or manufacturing. Whatever the choice, residue management can have a major impact on soil and water conservation. This report assesses the value of crop residues for various purposes, and reviews the effects of tillage and residue management practices on soil and water conservation under rainfed conditions. It suggests some alternatives to residue management for soil and water conservation, and finally mentions some management techniques to help meet the demand for residues.

Value of crop residues

For grain, fiber, or nut crops, about one-half or more of the production effort results in residues. If these residues are not utilized to their fullest extent, this production of residues, although unavoidable, results in a tremendous waste of effort and expense in the form of labor, fuel, water, and plant nutrients. By nonutilization, I am referring to residue destruction by burning or other poor management practices rather than utilization for feed, fuel, building materials, manufacturing, and resource conservation. Economic gains may be derived from utilization of residues for the latter purposes. Indirect economic gains may be derived also from residue destruction (e.g. burning) due to reduced tillage requirements and weed, insect, and disease control. However, residue burning has immediate economic disadvantages because it results in nitrogen losses and may have long-term economic disadvantages because it contributes to placing particulates in the atmosphere that may contribute to the "greenhouse effect" and the associated global warming.

Assigning economic values to crop residues is difficult. When used for feed, residues of many crops have limited value because of their low nutrient content and digestibility. While survival may be possible, animals may lose weight if their diet consists solely of residues of crops such as wheat (*Triticum aestivum* L.), rice (*Oryza sativa* L.), and barley (*Hordeum vulgare* L.) because they cannot ingest sufficient residues to derive the nutrients and energy needed for weight maintenance (N.A. Cole, pers. comm., 1987). Sorghum (*Sorghum* sp.), maize (*Zea mays* L.), and fababean (*Vicia faba* L.) residues may be more nutritious (Kernan *et al.*, 1981; Klopfenstein and Owen, 1981; Thorlacius *et al.*, 1979). In addition, chaff, leaves, and seedheads are more nutritious than plant stems (Coxworth *et al.*, 1981; Kernan *et al.*, 1984). Thus, selectively harvesting residues could result in improved animal performance when their diet consists primarily of crop residues. Performance is improved also when animals have access to growing vegetation or when fed residues that have been treated with alkali, ammonia, acid, or steam (Coxworth *et al.*, 1981; Garrett *et al.*, 1981; Kernan *et al.*, 1981; Sankat and Bilanski, 1982; Winugroho *et al.*, 1984). These latter treatments, however, may not be practical where grazing or foraging is practiced, as is the case for many small-scale animal production operations. Improved animal performance is also possible by supplementing the residues with proteins and minerals (Bhaskar *et al.*, 1980; Hofmeyr *et al.*, 1981; Klopfenstein and Owen, 1981).

As with residue use for feed, the economic value of residue use for fuel as practiced by small-scale landholders in many countries is difficult to assess. Certainly, burning residues for fuel can replace burning other fuels such as gas, oil, and charcoal, that must be purchased. However, such fuels may not be available or too costly. Thus, using residues may be the only practical alternative, and hence may be of considerable economic value. Residue use for large-scale power generation, gas and alcohol production, and direct burning for grain drying and heating is generally practical and economically feasible when sufficient amounts are available near the site of use so that harvesting and transportation costs are low, or when the residues are by-products of processing operations (e.g. cotton [*Gossypium hirsutum* L.] gin trash, and sugarcane [*Saccharum* sp.] bagasse) (Bhagat *et al.*, 1979; Flaim and Urban, 1980; Heid, 1984; Hitzhusen and Abdallah, 1980). However, large concentrations of residues usually are not available under small-farm conditions, and thus residue use for large-scale operations has limited potential.

Crop residues contain chemicals, and methods for extracting some of them for commercial use are available (Antonopoulos, 1980; Chen, 1984; Miller, 1980; Taylor, 1981). Residues are also used to make paper (Guha and Dhawan, 1978) or building materials (Gerjejansen, 1976). The economic feasibility of crop residue use for manufacturing and building materials depends on the amount and type available, the proximity to the site of use, and the value of the final product.

While it is difficult to establish the economic value of residues used for feed, fuel, building materials, and manufacturing, it is relatively easy to do so for water conservation. However, it is extremely difficult to place an economic value on residues for soil conservation, because usually the benefits are not immediately apparent. For water conservation, Wilhelm *et al.* (1986) developed regression equations that related surface residue amounts to soil water storage at planting. The equations were:

$$Y = 173 + 5X \; (r^2 = 0.84) \text{ (for maize) and}$$

$$Y = 175 + 8X \; (r^2 = 0.71) \text{ (for soybean, } Glycine \; max \text{ L.)}$$

where Y is the available soil water content at planting (mm) and X is the amount of surface residues remaining or applied after harvest of the previous crop (t ha^{-1}). Thus, water storage was increased 5 and 8 mm by each t ha^{-1} of maize and soybean residue respectively. The additional water furthermore increased subsequent crop yields, with the overall effect of surface residues on crop yields being shown by the following regression equations:

$$Y = 2.90 + 0.12X \ (r^2 = 0.18) \text{ (for maize) and}$$
$$Y = 1.53 + 0.09X \ (r^2 = 0.84) \text{ (for soybean)}$$

where Y is the grain yield (t ha^{-1}) and X is the crop residue (t ha^{-1}) remaining on the surface or applied after harvest of the previous crop. These relationships indicate that, on the average, each t ha^{-1} of residues increased maize and soybean grain yields by 0.12 and 0.09 t ha^{-1} respectively. Therefore if the monetary value of the grain is known, the value of residues for water conservation and grain production can be calculated.

The value of residues for soil conservation purposes is widely recognized, but placing an economic value on residues is virtually impossible. Certainly, if soils become degraded through continued erosion, decreases in soil organic-matter and nutrient concentration, and soil structure deterioration, all of which are influenced by crop-residue management, then the productive capacity of the soil will decline. When this occurs, crop production will become more difficult and expensive because of the need for greater inputs of fertilizers, erosion-control measures, and possibly tillage to alleviate undesirable soil conditions. These changes, however, usually occur slowly and often are not recognized until major soil degradation has occurred. As a result, short-term economic advantages of residue use for other purposes frequently overshadow the long-term advantages of soil conservation. Such practices must be discouraged if the productive capacity of soils is to be maintained for future generations.

Tillage and residue management

Tillage has a direct effect on crop residues that are not removed for other purposes. In this section, the effects of residue management through the selection of tillage methods on soil and water conservation and on crop yields will be presented. Although important in any crop production region, the emphasis in this section is on soil and water conservation in the drier regions (subhumid, semiarid, and arid). However, appropriate examples from more humid regions will be used to illustrate certain points. Tillage systems that result in 30% or more of the soil surface being covered with crop residues at planting of the next crop are frequently referred to as "conservation tillage" systems.

Soil erosion by water is a complex process that involves soil particle detachment and transport. Detachment results from raindrop impact and from flowing water. Some transport results from raindrop splash, but most results from water flow across the soil surface. Therefore to minimize erosion, both soil detachment and transport must be decreased. If sufficient amounts are present, surface residues control erosion because they dissipate the energy of falling raindrops, thereby reducing particle detachment and transport due to raindrop splash. Residues on or near the surface also reduce the rates and amounts of water flow across the surface, thus further reducing particle detachment and transport. Reduced water flow across the surface also increases water storage in soil because more time is provided for water infiltration. Approximate amounts of wheat residues needed on the surface to maintain erosion by water at or below 11 t ha^{-1} on various soils are given in Table 1. This amount is considered "tolerable" because it is the amount of soil regeneration that occurs annually on many soils. The approximate amounts of wheat and sorghum residues needed to maintain erosion by wind at or below 11 t ha^{-1} on various soils are included in Table 1.

Table 1. Approximate amounts of residue needed to maintain erosion below a tolerable level of 11 t ha^{-1} on various types of soil.

Soil texture†	Water erosion	Wind erosion					
	Flat wheat residue	Wheat residue		Growing wheat		Sorghum residue	
		Standing	Flat	Standing	Flat	Standing	Flat
				t ha^{-1}			
Silts	1.6	0.5	1.0	0.6	0.5	2.0	2.9
Clays and silty loams	2.1	0.9	1.8	1.1	0.9	3.7	5.3
Loamy fine sands	1.1	1.2	2.4	1.3	1.0	4.7	6.9

† Silts with 50%, clay and silty clay with 25%, and loamy sand with 10% nonerodible fractions (greater than 0.84-mm diameter).
Sources: Anderson (1968); Fenster (1973).

Water conservation due to surface residues results from decreased particle detachment, which decreases surface sealing, and hence helps to maintain favorable water infiltration rates. Water conservation also results from decreased water flow across the surface which, as mentioned previously, provides more time for water infiltration. Surface residues reduce both soil and water losses, but soil loss is generally reduced more than water loss, as was shown by Harrold and Edwards (1972) (Table 2) and Rockwood and Lal (1974) (Table 3).

Table 2.   Runoff and sediment yield from maize watersheds at Cochocton, Ohio, during a severe rainstorm on 5 July 1969.

Tillage	Slope (%)	Rainfall (mm)	Runoff (mm)	Sediment yield (t ha^{-1})
Plowed, clean-tilled sloping rows	6.6	140	112	50.7
Plowed, clean-tilled contour rows	5.8	140	58	7.2
No tillage, contour rows	20.7	129	64	0.07

Source: Harrold and Edwards (1972).

Table 3.   Effect of tillage on runoff and soil losses from land cropped to maize in Nigeria.†

Slope (%)	Bare fallow		Plowed		No tillage	
	Runoff (%)	Soil loss (t ha^{-1})	Runoff (%)	Soil loss (t ha^{-1})	Runoff (%)	Soil loss (t ha^{-1})
1	18.8	0.2	8.3	0.04	1.2	0.0007
5	20.2	3.6	8.8	2.16	1.8	0.0007
10	17.5	12.5	9.2	0.39 (sic)‡	2.1	0.0047
15	21.5	16.0	13.3	3.92	2.2	0.0015

†   Rainfall was 44.2 mm.
‡   Probably an error.
Source: Rockwood and Lal (1974).

Besides reducing water losses (runoff), another important aspect of water conservation is evaporation control. As is the case for runoff, evaporation decreases with increasing amounts of crop residues on the soil surface (Bond and Willis, 1969; Russel, 1939; Smika, 1976a; Unger and Parker, 1968, 1976). The value of surface residues for reducing soil water evaporation under field conditions was shown by Smika and Unger (1986, adapted from Smika, 1976a). In this study, water loss as affected by conventional-, minimum-, and no-tillage practices were compared after a 165-mm rainstorm. Soil water contents were similar on all areas on the day after the rainstorm (Figure 2a), but after 34 rainless days, it was lowest on the conventional-tillage area and highest on the no-tillage area (Figure 2b). Surface residue amounts when the rain occurred were 1.2, 2.2, and 2.7 t ha on the conventional-, minimum-, and no-tillage treatment areas respectively. This study, as well as others, have shown that water conservation generally increases with increasing amounts of surface residues (Tables 4 and 5) (Greb *et al.*, 1967, 1970; Unger, 1978). The immediate effect of surface residues is less runoff, as illustrated in Tables 2, 3, and 6. Runoff was usually lower with reduced tillage methods than with plowing, but plowing decreased runoff when residues were absent (Tables 3 and 7) or when surface residue amounts were low (O.R. Jones, pers. comm.).

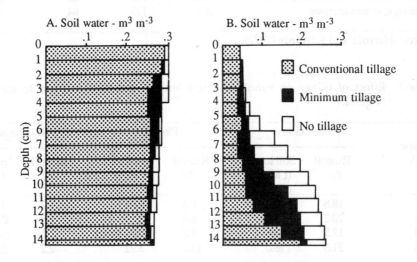

Figure 2. Soil water contents to a 15-cm depth 1 day (A) and 34 days (B) after 165 mm of rainfall as influenced by tillage treatments (Smika and Unger, 1986).[1]

Tillage method has a major effect on the percentage of crop residues retained on the soil surface. Approximate amounts remaining on the surface after various types of tillage were given by Anderson (1968) (Table 8). Even

---

1   Reproduced from *Cropping strategies for efficient use of water and nitrogen*, ASA Special Publication no. 51 (1988), 69-100, by permission of the American Society of Agronomy, the Crop Science Society of America, and the Soil Science Society of America.

Table 4. Straw-mulch effects on soil water storage efficiency at Sidney, Montana; Akron, Colorado; and North Platte, Nebraska, USA (1962 to 1965).

Mulch rate (t ha⁻¹)	Fallow period precipitation (mm)	Water storage efficiency (%)
0	355	16
1.7	355-549	19-26
3.4	355-648	22-30
6.7	355-648	28-33
10.1	648	34

Source: Greb *et al*. (1967, 1970).

Table 5. Straw-mulch effects on soil water storage during fallow[†], water-storage efficiency, and dryland grain sorghum yield at Bushland, Texas, USA (1973-1976).

Mulch rate (t ha⁻¹)	Water storage[‡] (mm)	Storage efficiency[‡] (%)	Grain yield (t ha⁻¹)	Total water use (mm)	WUE[§] kg m⁻³
0	72 c[#]	22.6 c	1.78 c	320	0.56
1	99 b	31.1 b	2.41 b	330	0.73
2	100 b	31.4 b	2.60 b	353	0.74
4	116 b	36.5 b	2.98 b	357	0.84
8	139 a	43.7 a	3.68 a	365	1.01
12	147 a	46.2 a	3.99 a	347	1.15

[†] Fallow duration of 10 to 11 months.
[‡] Water storage determined to 1.8-m depth. Precipitation averaged 318 mm.
[§] Water use efficiency based on grain produced, growing season precipitation, and soil water changes.
[#] Column values followed by the same letter are not significantly different at P = 0.05 level (Duncan's multiple range test).
Source: Unger (1978).

greater retention than that shown for subsurface cultivators occurs where a no-tillage cropping system is used. In such systems, herbicides are used for weed control, and crop planting is accomplished with equipment that minimizes soil and residue disturbance. With no tillage, usually less than 10% of the surface is disturbed by the planting operation (Lessiter, 1982). Because water conservation generally increases with increasing amounts of residues retained

Table 6. Runoff and soil loss as influenced by water applied,† time of application, type of fall tillage, and previous crop, Illinois, USA.

Time‡ (mins.)	Water applied (cm)	Tillage method					
		Fall moldboard plow		Disk-chisel		No-fall tillage	
		Corn	Soybean	Corn	Soybean	Corn	Soybean
		------------- Runoff (mm) -------------					
60	64	30	39	1	21	23	32
90	95	58	69	8	51	46	60
120	127	86	96	29	83	71	88
		------------- Soil loss (t ha^{-1}) -------------					
60	64	42	109	0.6	28	4	14
90	95	86	180	4	52	8	26
120	127	127	256	14	75	11	39

†   Simulated rainfall was applied at a rate of 65 mm h^{-1} after overwinter soil weathering, but before any spring tillage.
‡   Cumulative time from start of water application.
Source: Griffith et al. (1977).

Table 7. Effect of tillage-induced plow layer porosity and surface roughness on cumulative infiltration of simulated rainfall.

Tillage treatment‡	Potential water storage volume due to		Total water storage volume	Cumulative infiltration† to		
	Pore space§	Surface roughness		Initial runoff	25-mm runoff	50-mm runoff
	--- mm ---					
Untilled	81	8	89	9	21	24
Plow	137	50	187	171	217	230
Plow-disk-harrow	124	25	149	53	73	84
Cultivated	97	29	126	57	83	91
Rotavated	117	15	132	24	38	41

†   Water applied at a 127 mm h^{-1} rate.
‡   Plowing and rotavating performed to a 15-cm depth; cultivating to a 7.5-cm depth on previously untilled soil.
§   Measured to tillage depth.
Source: Burwell et al. (1966).

Table 8. Effect of tillage machines on small-grain residues remaining on the soil surface after each operation.

Tillage machine	Approximate residue maintained (%)
Subsurface cultivators	90
Wide-blade cultivator and rodweeder	
Mixing-type cultivators	75
Heavy-duty cultivator, chisel, and other type of machines	
Mixing and inverting disk machines	50
One-way flexible disk harrow, one-way disk, tandem disk, offset disk	
Inverting machines	10
Moldboard plow, disk plow	

Source: Anderson (1968).

on the surface, it should be higher with the less intensive tillage methods such as no, subsurface, and chisel tillage than with the more intensive tillage methods of disking and moldboard plowing.

The combined effect of greater infiltration and lower evaporation resulting from retention of surface residues is greater soil water storage - provided, of course, that weeds are adequately controlled and that the soil has the capacity to store the water. The following are examples of tillage effects on soil water storage and, if available, associated crop yields.

Smika (1976b) summarized the effects of clean (little or no residue retention) or stubble-mulch (subsurface) tillage practices on soil water gain during fallow periods at several Great Plains (USA) locations (Table 9). At wheat planting time at the end of the fallow period, the average water content was 27 mm greater with stubble mulch than with clean tillage. Soil water storage and wheat grain yields resulting from the progressive improvements in tillage practices since 1916 in Colorado (USA) are shown in Table 10.

In Nebraska (USA), reduced-tillage treatments involving both tillage and herbicides for weed control significantly increased water storage and grain sorghum [Sorghum bicolor (L.) Moench] yields, but not wheat yields, in a three-year wheat-fallow-sorghum rotation (Table 11). However, water storage and wheat yields were higher for the tillage-herbicide combination and a herbicide-only (no-tillage) treatment than for plowing in a two-year wheat-fallow rotation (Tables 12 and 13). The yield increases were attributed to increased water storage during the preceding fallow period (Smika and Wicks,

Table 9. Net gain in soil water during fallow with clean and stubble mulch tillage at seven central great plains locations (USA).

		Net gain in soil water (mm)	
		Tillage method	
	Years of data	Clean	Stubble mulch
Akron, Colorado	6	142	173
Colby, Kansas	4	115	141
Garden City, Kansas	6	86	90
Oakley, Kansas	4	82	131
North Platte, Nebraska	8	146	203
Alliance, Nebraska	8	29	32
Archer, Wyoming	2	28	42
Weighted average		95	122

Source: Smika (1976b).

Table 10. Progress in fallow systems with respect to water storage and wheat yields, Akron, Colorado, USA.

		Fallow water storage		Wheat yield (t ha^{-1})
Years	Tillage during fallow†	mm	Percent of precipitation	
1916-1930	Maximum tillage; plow, harrow (dust mulch)	102	19	1.07
1931-1945	Conventional tillage; shallow disk, rod weeder	118	24	1.16
1946-1960	Improved conventional tillage; began stubble mulch in 1957	137	27	1.73
1961-1975	Stubble mulch; began minimum tillage with herbicides in 1969	157	33	2.16
1976-1990	Projected estimate; minimum tillage; began no tillage in 1983	183	40	2.69

† Based on a 14-month fallow, from mid-July to mid-September.
Source: Adapted from Greb (1979).

Table 11. Effect of tillage and herbicide treatments on soil water contents at the end of the fallow period† and on wheat and sorghum yields in a three-year wheat-fallow-sorghum rotation, Nebraska, USA.

Treatment from wheat harvest to sorghum planting		Treatment from sorghum harvest to wheat planting	Soil water gain‡ (mm)	Grain yields (t ha⁻¹)	
Fall	Spring			Wheat	Sorghum
Subtillage	Disk	Subtillage (5)§	186 b#	3.49 a	4.08 b
Subtillage	Atrazine	Subtillage (4)	213 ab	3.76 a	4.20 b
Atrazine	Atrazine	Subtillage (4)	211 ab	3.63 a	4.58 ab
Atrazine	Atrazine	Contact herbicide (4-6)	223 a	3.49 a	4.89 a
Subtillage	Atrazine	Contact herbicide (4-6)	216 ab	3.63 a	5.02 a

† Fallow duration of about 11 months.
‡ Determined to a 3-m depth.
§ Values in parenthesis denote number of operations.
# Average values in a column followed by the same letter or letters are not significantly different at P = 0.05 level.
Source: Smika and Wicks (1968).

Table 12. Effect of tillage and herbicide treatments on soil water contents at the end of the fallow period† and on wheat yields in a two-year wheat–fallow rotation, Nebraska, USA.

Operations during fallow		Soil water gain‡ (mm)	Grain yield (t ha⁻¹)
Initial operation following wheat harvest	Subsequent operation		
Plow	Subtillage (5)§	186 c#	3.09 b
Subtillage	Subtillage (5)	238 b	3.36 ab
Atrazine followed by subtillage	Subtillage (5)	272 b	3.29 ab
Atrazine	Subtillage (4)	275 b	3.36 ab
Atrazine	Contact herbicides (4-6)	325 a	3.56 a

† Fallow duration of about 14 months.
‡ Determined to a 3-m depth.
§ Values in parenthesis denote number of operations.
# Average values in column followed by the same letter or letters are not significantly different at P = 0.05 level.
Source: Smika and Wicks (1968).

Table 13. Effect of tillage and herbicide treatments on operations to control weeds, surface residues, and soil water storage during fallow, and wheat yields in a two-year wheat-fallow rotation, Nebraska, USA.

Treatment	Operations during fallow†			Soil water gain# (mm)	Grain yield (t ha⁻¹)
	Tillage‡ no.	Herbicide application no.	Residues maintained§ (%)		
Plow	8.5	0	0	146	2.69
Stubble mulch	8.7	0	21	203	2.88
Atrazine + stubble mulch	7.6	1.4	21	215	2.91
Atrazine + contact herbicide + stubble mulch	5.1	2.8	25	237	3.04
Atrazine + contact herbicide	0.0	6.0	46	274	3.17

†   Fallow duration of about 14 months.
‡   The plow treatment included one moldboard plowing in the spring; other tillage was with a sweep implement.
§   Average amount of residues at the start of fallow was 6.6 t ha⁻¹.
#   Determined to a 3-m depth.
Source: Wicks and Smika 91973).

1968; Wicks and Smika, 1973). Similar results were reported by Morishita *et al.* (in press) and Norwood *et al.* (in press).

Cotton residues (stalks) have limited value for controlling erosion or enhancing soil water storage. However, Keeling *et al.* (1988) and Lyle and Bordovsky (1987) showed that irrigated cotton yields were not affected by the tillage method (conventional, minimum, or no tillage), but that nonirrigated cotton usually yielded more with minimum- and no-tillage methods, especially when it was grown in rotation with sorghum, wheat, or terminated wheat (wheat used as a cover crop, then killed before planting the cotton). These results indicate a major potential for controlling erosion where cotton is grown, especially when it is grown in rotation with wheat or sorghum. Favorable net returns indicate that these cotton production systems are economically sound, and hence should be acceptable to producers. The studies were conducted in a region of west Texas (USA) on high sand-content soils that are highly subject to wind erosion, and where mean annual precipitation is between 400 and 500 mm.

Crop-residue (stubble) management is an important aspect of soil and water conservation in the northern Great Plains (USA) and the Canadian Prairies, with standing stubble being highly effective for controlling wind erosion and for trapping snow for water conservation. Water storage in the upper 1.2 m of soil in northeastern Montana (USA) averaged 76 mm, with 30 cm-high standing wheat stubble and 48 mm-high stubble where the fields were disked in the fall (Black and Power, 1965). In North Dakota (USA), fall moldboard plowing resulted in 5 to 54 mm, and no tillage resulted in 32 to 90 mm of overwinter water storage; disking and V-blade tillage resulted in intermediate amounts of storage (Black and Bauer, 1985).

Besides stubble *per se*, water storage is also affected by stubble height (Table 14) (Black and Siddoway, 1977) and alternating strips of tall and short stubble. In the latter case, alternating strips of tall and short stubble, usually 5 m wide, increased snow depth and density, which resulted in 30% more water storage than uniform medium-height stubble (Willis and Frank, 1975). Similar results were reported also by Kachanoski *et al.* (1985) and Zentner *et al.* (1988) in Canada, who showed that soil water storage averaged 12 mm more for the trap strip area than for an area of uniform-height short stubble. A limitation to water storage from trapped snow is that frozen soil reduces infiltration and enhance runoff. There are, however, indications that no-tillage enhances water infiltration from snow melt (Zentner *et al.*, 1988).

Table 14. Snow depth, water equivalent of snow, and total soil water content as affected by stubble management, Montana, USA.

Date	Stubble height (cm)			
	0†	15	28	38
	Snow depth (mm)			
2 Apr. 1975	61	119	198	259
	Water equivalent of snow (mm)			
2 Apr. 1975	13	25	43	56
	Total soil water content (mm)			
12 May 1975	267	277	287	307

† This treatment was conventional disk tillage.
Source: Black and Siddoway (1977).

The Pacific Northwest (USA) is a region that has a Mediterranean climate with wet winters and dry summers, and some croplands with slopes greater than 50%. The soils may freeze to a depth of 40 cm or more, especially when there is no snow cover. As a result, heavy runoff and soil losses can occur, even with low-intensity precipitation or melting snow when the soil is frozen (Papendick and Miller, 1977). Although runoff losses are large, even greater water losses result from evaporation (Ramig et al., 1983), with close to 75% of the water being lost from bare, uncropped soils (Bristow et al., 1986). In wheat-fallow systems, water storage efficiencies range from 50 to 75% for the first winter, but only 10 to 50% for the second winter. Less water storage during the second winter results from higher soil water contents and lower amounts of surface residues to suppress evaporation (Leggett et al., 1974).

Surface residues, especially standing stubble of small grain crops, significantly increase overwinter soil water storage in the Pacific Northwest. Water storage was 30 and 50 mm greater with 5 and 11 t ha^{-1} respectively of wheat residue on the surface from September to March than with no residues (Bristow et al., 1986). However, evaporation in summer was greater because of prolonged first-stage drying where residues were present, thus negating about half the gains that had occurred overwinter. To reduce these major losses due to evaporation, Bristow et al. (1986) cultivated the soil as soon as possible after major rains to break capillary continuity to the surface. This reduced water flow to the surface and hastened the onset of second-stage drying.

Overwinter water storage from precipitation was 66 mm greater where standing wheat residues were present than where the land was plowed in the fall. Green pea (Pisum sp.) yields were increased 25% by the additional water (Ramig and Ekin, 1978). In other studies (Massee and McKay, 1979; Ramig and Ekin, 1984), overwinter water storage was significantly greater where stubble was retained on the surface than where it was burned, mainly because of snow trapping by the standing stubble.

Herbicides for no tillage (called direct drilling) were introduced in Australia in 1965. In the winter-rainfall region, most annual pasture species in their early growth stages were controlled by herbicides, but control of larger weeds and clovers (various species) was difficult. In early studies, wheat yields were usually lower with no tillage on a loamy sand, not different on a sandy loam, and higher on a clay loam than with conventional tillage. The introduction and use of improved herbicides in the late 1970s resulted in higher wheat yields with no tillage which, along with higher fuel prices at that time, made no tillage an economically advantageous system. Erosion control was greatly improved by the use of no tillage because the sod remained undisturbed during the pasture phase of the system (Poole, 1987).

Despite the need for erosion control, which is enhanced by surface residues, residues in Australia are often removed, worked into the soil, or burned because of poor drill performance, along with increased weed, insect, and disease problems under high-residue conditions. Although not recommended, burning is the primary disposal method. At various Australian locations, research has shown that soil and water losses were lower where more residues were retained on the surface than where burned. The reduced soil loss was associated with lower peak runoff rates (Tables 15, 16, 17). Water-storage efficiency was higher with no tillage than with cultivation, both where the residues were retained on the surface or where burned (Table 18). The additional water extended the opportunity time for crop planting after a rain, thus further increasing the potential for improving crop yields (Cummins, 1973; Freebairn and Wockner, 1982; Littler and Marley, 1978; Marston, 1980).

Table 15.   Annual soil loss as affected by management practice and soil type in southeast Queensland, Australia.

Soil type and situation	Management		
	Bare fallow	Residue incorporation (average management)	Residue mulch, contour cultivation (good management)
	----------------- t ha^{-1} -------------		
Alluvial, fertile, self-mulching clay; 1-2% slope	60	24	12
Colluvial, red-brown clay; 1-3% slope	76	31	12 (with contour banks)
Colluvial, red-brown clay; 5-8% slope	270	110	10 (with contour banks and crop rotation)

Source: Cummins (1973).

Table 16.   Tillage-system effects on total soil and water loss from simulated rainfall applied at 115 mm h^{-1} in New South Wales, Australia.

Treatment	Soil loss (t ha^{-1})	Water loss (mm)
Residue burned	9.04	22
Residue incorporated	1.36	13
No-tillage fallow	0.32	8

Source: Marston (1980).

Table 17. Soil type and tillage-treatment effects on runoff and soil movement on the Darling Downs, New South Wales, Australia (1978-79 and 1980-81†).

Soil type and tillage treatment	Runoff (mm)	Soil movement (t ha⁻¹)
Greenmount (black earth)		
Residue burned	82	85
Residue incorporated	53	29
Residue mulch	37	3
No-tillage fallow	44	1
Greenwood (grey clay)		
Residue burned	82	48
Residue incorporated	56	12
Residue mulch	43	5
No-tillage fallow	54‡	2

†   Water year = 1 October to 30 September.
‡   Weighed 1-year value.
Source: Freebairn and Wockner (1982).

Table 18. Residue and soil management effects on water-storage efficiency during fallow over 10 years at Warwick, Queensland, Australia.

Residue management	Soil management	Fallow efficiency (%)
Burned	Cultivated fallow	16.8
Burned	No-tillage fallow	20.1
Retained	Cultivated fallow	22.7
Retained	No-tillage fallow	25.8

Source: Littler and Marley (1978).

Because water conservation and erosion control improve with increasing amounts of residue on the soil surface, maximum residue retention generally gives the greatest benefits. However, under rainfed conditions in arid to semiarid regions, even total retention may not provide adequate residues for effective water conservation and erosion control. Under such conditions, Unger and Wiese (1979) showed that managing residues from irrigated winter wheat with no tillage resulted in greater soil water storage during fallow, higher grain yields, and greater water-use efficiency of a following dryland (nonirrigated) grain sorghum crop than where disk or sweep tillage methods were used (Table 19). For an irrigated winter-wheat--fallow--dryland-sorghum-dryland-sunflower (*Helianthus annuus* L.) rotation, water storage during fallow

Table 19. Tillage effects on water storage, sorghum grain yields, and water-use efficiency in an irrigated winter-wheat--fallow--dryland-grain sorghum cropping system, Bushland, Texas, USA (1973 to 1977).

Tillage method	Water storage		Grain yield (t ha^{-1})	Total water use (mm)	WUE‡ (kg m^{-3})
	Amount (mm)	Efficiency (% of precip.)†			
No-tillage	217 a§	35.2 a	3.14 a	350	0.89 a
Sweep	170 b	22.7 b	2.50 b	324	0.77 b
Disk	152 c	15.2 c	1.93 c	320	0.66 c

† Precipitation averaged 347 mm during fallow.
‡ Water use efficiency based on grain yields, growing season precipitation, and soil water changes.
§ Column values followed by different letters are significantly different at P = 0.05 level, based on Duncan's multiple range test.
Source: Unger and Wiese (1979).

after wheat, as well as sorghum grain yield again were highest with no tillage (Table 20) (Unger, 1984a). Wheat and sunflower yields were not significantly affected by the treatments. In both studies, total water use was highest with no tillage, but that was offset by the higher soil water contents at sorghum planting. These results suggest that soil water contents were higher throughout the sorghum growing season with the no tillage than with the other tillage treatments. Analyses of data from these and other residue management studies conducted at Bushland, Texas (USA), showed that sorghum grain yields increased in response to growing-season precipitation as amounts of residues on the surface at sorghum planting increased (Unger et al., 1986).

As in the USA, Canada, and Australia, surface residues effectively conserve soil and water in Asia and Africa (Ali, 1976; Han et al., in press; Heenop, in press; Jalota and Prihar, 1979; Klaij and Serafini, in press; Mane and Shingte, 1982; Verma et al., 1979; Wang et al., in press). However, crop residues are often inadequate for effective soil and water conservation in Asia and Africa because of low production, extensive use for feed, fuel, and other purposes, and (in some cases) widespread destruction by termites and burning.

Table 20. Effect of tillage methods on average soil water storage during fallow† after irrigated winter wheat and on subsequent rainfed grain sorghum yields, total water use, and water-use efficiency at Bushland, Texas, USA (1978-1983).

Tillage treatment	Water storage		Grain yield (t ha⁻¹)	Total water use§ (mm)	WUE# (kg m⁻³)
	(mm)	(%)‡			
Moldboard	89 b††	29 b	2.56 bc	360 bc	0.71
Disk	109 b	34 ab	2.37 cd	363 bc	0.65
Rotary	85 b	27 b	2.19 d	357 c	0.61
Sweep	114 ab	36 ab	2.77 b	386 ab	0.72
No tillage	141 a	45 a	3.34 a	401 a	0.83

† Fallow duration of 10 to 11 months. Fallow precipitation averaged 316 mm.
‡ Based on fallow period precipitation stored as soil water.
§ Includes average growing season rainfall of 301 mm.
# Water use efficiency based on grain yield, growing season precipitation, and soil water changes.
†† Column values followed by the same letter to letters are not significantly different at P = 0.05 level based on Duncan's multiple range test.
Source: Unger (1984a).

Consequently, alternatives to residue management must be relied upon for conserving soil and water, and hence for maintaining crop yields at favorable levels. Examples of practices for conserving soil and water and maintaining crop yields at favorable levels without the use of residues are included in the next section.

Alternatives to residue management for soil and water conservation

In the foregoing section, the emphasis was on using tillage methods for conserving soil and water resources through crop-residue management. However, in many cases, adequate residues to effectively conserve soil and water are not produced, residues are removed for other purposes, or producers choose to use crop production practices that are not based on residue management. Whatever the reason, alternate practices must be used when soil and water conservation is important for successful crop production. Tillage that does not involve the management of crop residues on the surface frequently is referred to as "clean tillage."

Tillage effects

Wind erosion on most clean-tilled soils can be controlled by any tillage that produces an adequately rough, cloddy surface. However, secondary tillage operations and weathering (wetting and drying, freezing and thawing) may decrease surface ridges and cloddiness so that they no longer afford protection against erosion by wind. In such cases, any operation that ridges the soil or increases its cloddiness usually is adequate as an emergency measure to control erosion. Such practices, however, would not be effective on soils that have been pulverized by repeated or intensive tillage or on soils comprised of noncohesive sandy materials. Wind erosion on noncohesive sandy soils is difficult to control by any tillage method, and permanent vegetative cover provides the best protection against wind erosion on such soils.

To control water erosion where clean tillage is used, infiltration must be maintained at satisfactory rates to reduce overland flow (runoff), and any runoff must be conveyed off the land at nonerosive velocities. Retaining water on land to provide more time for infiltration can also reduce runoff. To achieve this, tillage that ridges the soil or roughens its surface is often used in conjunction with graded furrows, contouring, furrow diking (tied ridges), terracing, or land leveling. Increased infiltration is important also for water conservation, provided the soil has the capacity to retain the water for later use by crops.

The influence of tillage *per se* on infiltration, and consequently on water erosion, is related to surface-soil aggregate stability and the roughness, surface detention storage, and pore space that result from the tillage method used. To maintain high infiltration rates, water-stable surface aggregates are essential. These normally result from organic materials maintained on or near the soil surface. In contrast, low-stability soils are easily dispersed by water, which results in surface sealing and increased runoff and erosion.

Temporary water storage on the soil surface can reduce runoff and aid in controlling water erosion when the precipitation rate exceeds the infiltration rate. Ridges formed by tillage (e.g. lister tillage) on the contour have long been recognized for their water-retention benefits (e.g., Dickson et al., 1940; Fisher and Burnett, 1953; Harrold and Edwards, 1972). Runoff may be eliminated also by furrow diking of gently sloping ridge-tilled land. Clark and Jones (1981) showed that all water from a 114-mm rainstorm during a 24-hour period was retained on a Pullman clay loam (Torrertic Paleustoll) by furrow diking at Bushland, Texas. Further studies showed that water retention due to furrow diking increased crop yields (Jones and Clark, 1987).

Not only lister tillage, but also moldboard plows, sweep plows, disks, chisels, rotary tillers, and cultivators affect soil pore space and surface roughness, and therefore water retention (on or near the surface), runoff, and erosion. Burwell *et al.* (1966) illustrated the effects of soil pore space and surface roughness resulting from different tillage methods on the infiltration of simulated rainfall into a Barnes loam (Udic Haploboroll) (Table 7). Cumulative infiltration approached plow-layer total pore space and surface roughness volumes before runoff started, and exceeded those volumes before 25 mm of runoff occurred for the plow treatment. For other tillage treatments, storage volumes were not filled, even though 50 mm of runoff occurred. Smoother surfaces with treatments other than plowing apparently resulted in more rapid soil dispersion and surface sealing, which reduced infiltration.

One system involving clean tillage that has considerable potential for controlling erosion is the plow-plant system. For this system, plowing is delayed until 12 to 24 hours before planting, which leaves the soil covered with residues and growing weeds for a major part of the erosion period. Planting may be in tractor or planter wheel tracks, which results in planting in firm soil. Variations of this system are disking and planting; fall chiseling, then disking; cultivating, or rotary tilling before planting; or rotary tillage (strip or full width) and planting (Griffith *et al.*, 1977). These systems are applicable only in high-precipitation areas where fallowing is not needed for enhancing soil water storage. One major disadvantage of this system, especially for producers with limited resources, is the amount of equipment and labor that must be available during a relatively short period of time to achieve timely planting of the crops.

Water conservation through reduced evaporation without use of surface residues has been achieved by using tillage to create a dust (soil) mulch at the surface (Gill *et al.*, 1977; Karaca *et al.*, in press; Prihar and Jalota, in press; Rao and Kumar, 1975; Smit *et al.*, in press). Dust mulches are effective for conserving water that was stored in soil during a previous rainy period. They are of little value for conserving water during periods of intermittent rainfall because the mulch must be reestablished after each rain, and much of the water may evaporate before tillage can be performed, especially where most rains occur in limited amounts (Unger, 1984b). Consequently dust mulches are effective primarily in regions having distinct wet and dry seasons.

Support practices

When conservation tillage is not used and clean tillage is practiced, various engineering-type or surface management practices (other than tillage) are frequently used to provide soil and water conservation benefits. In this section, these support practices are discussed briefly.

Contouring

Contouring is performing cultural operations in rows perpendicular to the slope of the land. When ridge tillage, as with a lister, is done on the contour, water from low to moderate amounts of precipitation is retained on the land, thus greatly reducing the potential for erosion and conserving water for use by crops. Water from large storms may overtop the ridges, thus resulting in a situation leading to severe erosion. Contouring usually does not control wind erosion unless the ridges increase surface roughness perpendicular to prevailing winds.

Strip-cropping

Strip-cropping is effective for controlling erosion both by water and wind. To control water erosion, the strips should be on the contour so that soil eroded from the cropped area will be trapped in the protective strips. To control wind erosion, protective strips should be perpendicular to prevailing winds.

Terraces

Terraces are earthen ridges constructed with a slight grade across the slope of land and designed to convey runoff from land at nonerosive velocities. If level terraces are used, water is retained on the land until infiltration occurs, thus conserving it for later use by plants. Level terraces also prevent water erosion in most cases. The effectiveness of terraces for conserving both soil and water is enhanced when used in conjunction with contouring, strip-cropping, furrow diking, and conservation tillage.

Maximum water retention for crop use with terraces is achieved with bench or conservation bench terraces. These require land leveling, a costly operation. With bench terraces, the entire interval between the ridges is leveled, and all water, except in extreme cases, is retained on the land. Hence bench terraces are also highly effective for controlling water erosion if they are properly maintained.

For conservation bench terraces (CBTs), only the lower-slope part (usually one-third to one-half) of the interval between ridges is leveled, thus resulting in lower construction costs than those required for bench terraces. Costs are reduced even more when the CBTs are constructed at narrow intervals (two to four equipment widths). For CBTs, runoff from the nonleveled area is retained on the leveled area, thus providing additional water to crops on a portion of the land. As is the case with bench terraces, CBTs effectively reduce erosion. Bench terraces and CBTs have been found effective for increasing crop yields, mainly on slowly permeable soils in semiarid regions (Unger and Stewart, 1988).

Diversion terraces

Diversion terraces have individually designed channels and ridges that are constructed across the slope of the land to divert water out of active gullies, to protect field areas against runoff from nonterraced areas, and to protect farm improvements. Diversion terraces are also used to prevent runoff from adjacent areas from entering terraced fields. Water from these terraces must be channeled through waterways that are adequately protected, normally with vegetation.

Graded furrows

In contrast to contour furrows, which virtually eliminate runoff and erosion, graded furrows are designed to convey runoff water from fields at nonerosive velocities. Each furrow serves as a small graded terrace. Although designed to remove excess water, Richardson (1973) showed that graded furrows also conserve water, as compared with the amount of water conserved on terraced watersheds without furrows. During a 32-month period at Temple, Texas (USA), on a Houston Black clay (Udic Pellustert), runoff was 187 mm from a graded-furrow watershed, and 236 mm from a terraced watershed. The lower runoff with graded furrows was attributed to a more uniform distribution of water over the entire field.

Furrow dikes and soil pitting

Furrow dikes (tied ridges) have been mentioned previously, and consist of small earthen dikes constructed between ridges formed with an implement such as a lister. Pitting is done on tillage-loosened soil, usually on fields that have not been ridged. Furrow diking and soil pitting retain most water on the land, thus conserving soil and water and increasing crop yields (Clark and Jones, 1981; Jones and Clark, 1987; Klaij and Serafini, in press; Morin and Benyamini, in press; Rodriguez, in press; Unger, 1984b; van der Ploeg and Reddy, in press).

Other practices

Other practices that have potential for conserving soil and water, usually on a limited scale or under specialized conditions, include land smoothing, deep tillage, water harvesting (including storage in ponds for later application to crops), water spreading, runoff farming, microwatersheds, mulching, vertical mulching, slot mulching, plastic mulching, and the use of vegetative or artificial barriers. Most of these have been discussed in more detail by Unger (1984b) and Unger and Stewart (1988).

Management techniques to help meet the demand for residues

In many cases, adequate soil and water conservation for sustained crop productivity can be achieved with clean tillage alone or in conjunction with one or more appropriate support practices. However, even more effective soil and water conservation, and therefore the potential for sustaining crop productivity at a higher level, is possible where the management of crop residues is incorporated into the crop production system. Because of continually increasing pressures on land, which can lead to soil degradation, it is imperative that the best management practices be applied to avert land degradation, and thereby sustain the productivity of the land for use by future generations.

Because crop residues can play a major role in soil and water conservation, but are widely used for other purposes, alternate practices that could help meet the demand for residues must be developed and adopted. By adopting one or more of these practices, it should be possible to retain sufficient residues to adequately protect the land against erosion and help conserve water; both are important for sustaining the productivity of the land. The following practices were discussed by Unger (in press).

Limited residue removal

Information regarding the amount of residue required to control erosion under all conditions is not available, but the data shown in Table 1 can serve as a guide. When available residues exceed the indicated amounts, the excess could be removed for other uses without having a major impact on erosion control. However, because protection against erosion increases with the amounts of residue retained on the surface, any residue removal may increase erosion as compared with no removal.

Selective residue removal

Some plant parts are more valuable than others for feed, fuel, building materials, and manufacturing. Examples are chaff, leaves, and seed heads for feed, and woody stems and maize cobs for fuel. Plant parts required for building materials and manufacturing vary widely, depending on the intended use. Because selected parts have the greatest value for a given purpose, removal of only that portion would allow sufficient residues to be retained for conservation purposes.

Other types of selective removal include removing only those residues that pass through the harvester as feed for livestock; allowing animals to forage on fields after crop harvest, but halting grazing while adequate residues for conservation purposes still remain; and removing no or only some residues from highly erodible areas and more from less erodible areas. To help control erosion on highly erodible areas, residues from outside sources, such as cotton-gin trash and sugarcane bagasse could be placed on the land.

Substitution of high-value forages for residues

Residues of many crops have low value as livestock feed, whereas forage crops harvested at the proper growth stage have a high nutritive value. Hence growing crops specifically for feed could allow poor quality residues to be used for conservation. Animal production should not suffer, and may even increase if sufficient forages are produced. Areas devoted to other crops would decrease, and hence total production by these crops may be lowered. However, if residues from these crops are properly managed, improved soil and water conservation and nutrient cycling may result in total yields comparable to those where residues are removed from the entire land area.

Alley cropping

Alley cropping is useful primarily in humid, tropical areas, and consists of growing crops between rows or strips of deep-rooted perennial trees or shrubs. The trees are pruned at the beginning of and/or periodically during the crop growing season to minimize competition with the crop for light and water. The pruned leafy materials and twigs are used as a mulch for the cropped area; some may be used as animal feed. Woody stems and branches may be used as fuel. Usually weeds are not a problem because of shading by the trees or shrubs during the interval between crops, and by the mulch and growing crop during the growing season. The shrub or tree species selected depends on the soil, climate, and crop to be grown. Desirable species grow rapidly, fix nitrogen, have a multipurpose nature (mulch, feed, fuel), and are deep-rooted to minimize competition with crop plants for water and nutrients. The leguminous species, *Leucaena leucocephalia* and *Gliricidia* sp., have performed well in alley-cropping systems.

Utilization of waste areas

In many cases, there is some land on farms or on or near villages that is unsuitable for field crop production. Included is land adjacent to streams or drainage ways, on rocky outcrops, or in low-lying, poorly drained areas. Such areas, if not already used, can be used for growing feed or fuel crops, thus reducing the demand for crop residues for feed or fuel. Closely allied to this would be the establishment of tree-producing areas in a village or community, with all members of the community receiving fuelwood from the wooded area. Such projects could either be cooperative, with all members receiving a share; or an individual could develop the area, then sell wood (fuel) to inhabitants of the village or community. Besides providing feed or fuel, good management of waste or wooded areas would help to control erosion.

Improved balance between feed supplies and animal populations

When animal populations are correctly matched to available feed supplies, it should be possible to maintain adequate crop residues on the land for soil and water conservation. In contrast, excessive animal populations usually cause increased demand for feed (frequently for crop residues), which result in inadequate protection of the land against erosion and poor water conservation. Many cases of land deterioration are attributable to long-term overgrazing and/or residue removal. In such situations, social and economic factors in addition to agricultural factors are involved, and all of these must be considered when developing and implementing strategies for obtaining an improved balance between feed supplies and animal populations. Changes in national policies and priorities may be needed to achieve the improved balance.

Use of alternative fuel sources

Many arid to subhumid regions have an abundance of sunshine, which provides an inexhaustible supply of solar energy during a large part of the year. Although solar units are used for heating and cooking in some countries, their use is limited or nonexistent in others. Solar energy may not be available during cloudy or rainy weather, but even part-time use of solar energy during the year could greatly decrease the demand for crop residues and wood products for fuel purposes. Again, social, economic, and governmental factors must be considered when developing and implementing strategies for using solar energy. Also, technical advances may be required to develop practical, economical, and consumer-acceptable solar energy units, especially for cooking.

References

ALI, M. 1976. Effect of mulches and reflectant on the yield of rainfed wheat. *Indian Journal of Agronomy* 21: 61-63.

ANDERSON, D.T. 1968. Field equipment needs in conservation tillage methods. In: *Conservation tillage in the Great Plains* (Workshop, Nebraska, 1968), 38-91. Great Plains Agricultural Council. Publication no. 32.

ANTONOPOULOS, A.A. 1980. *Illinois biomass resources: Annual crops and residues; canning and food-processing wastes.* Energy fro Agriculture Series ENR, MCA/ BIO 76. Michigan: Michigan State University.

BHAGAT, N., DAVITIAN, H. and POUDER, R. 1979. The potential of crop residues as a fuel for power generation. In: *Sun II: Proceedings of the International Solar Energy Society* (Georgia, 1979), ed. K.W. Bhoer and B.H. Glenn, vol. 3, 2025-2029. New York: Pergamon Press.

BHASKAR, B.V., SUBBA, RAO A. and SAMPATH, S.R. 1980. Urea-molasses enriched crop residue--a source of bulk feed to cattle. In: *Protein and NPN utilization in ruminants: Proceedings of the All India Symposium* (UNDP/ICAR), 130-136. National Dairy Research Institute, Karnal, India: NDRI.

BLACK, A.L. and BAUER, A. 1985. Soil water conservation strategies for Northern Great Plains. In: *Planning and management of water conservation systems in the Great Plains* (Workshop, Nebraska, 1985), 76-86. USDA Soil Conservation Service. Lincoln, Nebraska: USDA/SCS.

BLACK, A.L. and POWER, J.F. 1965. Effect of chemical and mechanical fallow methods on moisture storage, wheat, and soil erodibility. *Soil Science Society of America Proceedings* 29: 465-468.

BLACK, A.L. and SIDDOWAY, F.H. 1977. Winter wheat recropping on dryland as affected by stubble height and nitrogen fertilization. *Soil Science Society of America Journal* 41: 1186-1190.

BOND, J.J. and WILLIS, W.O. 1969. Soil water evaporation: Surface residue rate and placement effects. *Soil Science Society of America Proceedings* 33: 445-448.

BRISTOW, K.L., CAMPBELL, G.S., PAPENDICK, R.I. and ELLIOTT, L.F. 1986. Simulation of heat and moisture transfer through a surface residue-soil system. *Agriculture for Meteorology* 36: 193-214.

BURWELL, R.E., ALLMARAS, R.R. and SLONEKER, L.L. 1966. Structural alteration of soil surfaces by tillage and rainfall. *Journal of Soil Water Conservation* 21: 61-63.

CHEN, C.M. 1984. The use of forest and agricultural residue extracts in the production of exterior phenolic resin adhesives. In: *Adhesives for wood: Research, applications, and needs*, ed. R.H. Gillespie, 52-53. Park Ridge, New Jersey: Noyes Publications.

CLARK, R.N. and JONES, O.R. 1981. Furrow dams for conserving rainwater in a semiarid climate. In: *Proceedings of a Conference on Crop Production with Conservation in the 80/s*, Chicago (Illinois, 1980), 198-206. American Society of Agricultural Engineers. Michigan: ASAE.

COXWORTH, E., KERNAN, J., KNIPFEL, J., THORLACIUS, O. and CROWLE, L. 1981. Review; Crop residues and forages in western Canada: Potential for feed use either with or without chemical or physical processing. *Agriculture and the Environment* 6: 245-256.

CUMMINS, V.C. 1973. A land use study of the Wyreema-Cambooya area of the eastern Darling Downs. Division of Land utilization, Queensland Department of Primary Industries. Technical Bulletin no. 10. Indooroopilly, Queensland: DPI.

DICKSON, R.E., LANGLEY, B.C. and FISHER, C.E. 1940. *Water and soil conservation experiments at Spur, Texas.* Texas Agriculture Experiment Station. Bulletin B-587. Texas: TAES.

FENSTER, C.R. 1973. Stubble mulching. In: *Conservation tillage* (Conference, Iowa, 1973), 202-207. Soil Conservation Society of America. Ankeny, IA: SCSA.

FISHER, C.E. and BURNETT, E. 1953. *Conservation and utilization of soil moisture.* Texas Agriculture Experiment Station. Bulletin B-767. Texas: TAES.

FLAIM, S. and URBAN, D. 1980. *The costs of using crop residues in direct combustion applications.* SERI/TR-353-513. Solar Energy Research Institute. Golden, Colorado: SERI.

FREEBAIRN, D.M. and WOCKNER, G.H. 1982. *The influence of tillage implements on soil erosion* (Conference, Armidale, 1982). Institute of Australian Engineers. Armidale: IAE.

GARRETT, W.N., WALKER, H.G. Jr., KOHLER, G.O., HART, M.R. and GRAHAM, R.P. 1981. Steam treatment of crop residues for increased ruminant digestibility: II. Lamb feeding studies. *Journal of Animal Science* 51: 409-413.

GERTJEJANSEN, R.O. 1976. An evaluation of sunflower stalks for the manufacture of particle board. In: *Proceedings of the Sunflower Forum*, 30-33. Sunflower Association of America. Bismarck, North Dakota: SAA.

GILL, K.S., JALOTA, S.K., PRIHAR, S.S. and CHAUDHARY, T.N. 1977. Water conservation by soil mulch in relation to soil type, time of tillage, tilth and evaporativity. *Journal of the Indian Society of Soil Science* 25: 360-366.

GREB, B.W. 1979. *Reducing drought effects on croplands in the West-Central Great Plains.* U.S. Department of Agriculture. Information Bulletin no. 420. Washington, DC: Government Printing Office.

GREB, B.W., SMIKA, D.E. and BLACK, A.L. 1967. Effect of straw mulch rates on soil water storage during summer fallow in the Great Plains. *Soil Science Society of America Proceedings* 31: 556-559.

GREB, B.W., SMIKA, D.E. and BLACK, A.L. 1970. Water conservation with stubble mulch fallow. *Journal of Soil Water Conservation* 25: 58-62.

GRIFFITH, D.R., MANNERING, J.V. and MOLDENHAUER, W.C. 1977. Conservation tillage in the eastern Corn Belt. *Journal of Soil Water conservation* 32: 20-28.

GUHA, S.R.D. and DHAWAN, R. 1978. Wrapping papers from pigeonpea wood. *Indian Journal of forestry* 1: 287-289.

HAN, S., SI, J. and YANG, C. In press. Research on stubble mulch tillage in rainfed land. In: *Proceedings of the International Conference on Dryland Farming* (Texas, 1988), ed. P.W. Unger, W.R. Jordan, T.V. Sneed and R.W. Jensen. College Station, Texas: Texas A&M University.

HARROLD, L.L. and EDWARDS, W.M. 1972. A severe rainstorm test of no-till corn. *Journal of Soil Water Conservation* 27: 30.

HEENOP, C.H. In press. A successful practice maize production system for arid situations. In: *Proceedings of the International Conference on Dryland Farming* (Texas, 1988), ed. P.W. Unger, W.R. Jordan, T.V. Sneed and R.W. Jensen. college Station, Texas: Texas A&M University.

HEID, W.G. Jr. 1984. Turning Great Plains crop residues and other products into energy. USDA, Economic Res. Serv., Agric. Econ. Rpt. 523. Washington, DC:

HITZHUSEN, F.J. and ABDALLAH, M. 1980. Economics of combustion energy from crop residue. In: *Proceedings of a Conference on Energy from Biomass*, 610-615. London: Allied Science Publishers.

HOFMEYR, H.S., HENNING, P.H. and CRONJE, P.B. 1981. The utilization of crop residues and animal wastes by ruminants. *South African Journal of Animal Science* 11: 111-117.

HORNICK, S.B. and PARR, J.F. 1987. Restoring the productivity of marginal soils with organic amendments. *American Journal of Alternative Agriculture* 2: 64-68.

JALOTA, S.K. and PRIHAR, S.S. 1979. Soil water storage and weed growth as affected by shallow-tillage and straw mulching with and without herbicide in bare-fallow. *Indian Journal of Ecology* 5: 41-48.

JONES, O.R. and CLARK, R.N. 1987. Effects of furrow dikes on water conservation and dryland crop yields. *Soil Science Society of America Journal* 51: 1307-1314.

KACHANOSKI, R.G., DE JONG, E. and RENNIE, D.A. 1985. The effect of fall stubble management on overwinter recharge and grain yield. In: *Proceedings of the Soils and Crops Workshop* (Saskatoon, 1985), 254-261. Saskatoon, Canada: University of Saskatchewan.

KARACA, M., GULER, M., DURUTAN, N., PALA, M. and UNVER, I. In press. Effect of fallow tillage systems on wheat yields in Central Anatolia of Turkey. In: *Proceedings of the International Conference on Dryland Farming* (Texas, 1988), ed. P.W. Unger, W.R. Jordan, T.V. Sneed and R.W. Jensen. College Station, Texas: Texas A&M University.

KEELING, J.W., WENDT, C.W., GANNAWAY, J.R., ONKEN, A.B., LYLE, W.M., LASCANO, R.J. and ABERNATHY, J.R. 1988. Conservation tillage cropping systems for the Texas Southern High Plains. In: *Proceedings of the 1988 Southern Conservation Tillage Conference* (Tupelo, 1988), ed. K.H. Remy, 19-21. Mississippi Agriculture and Forestry Experiment Station. Special Bullentin 88-1. Mississippi: MAFES.

KERNAN, J.A., COXWORTH, E.C., CROWLE, W.L. and SPURR, D.T. 1984. The nutritional value of crop residue components from several wheat cultivars grown at different fertilizer levels. *Animal Feed Science and Technology* 11: 301-311.

KERNAN, J.A., COXWORTH, E.C. and SPURR, D.T. 1981. New crop residues and forages for western Canada: Assessment of feeding value in vitro and response to ammonia treatment. *Animal Feed Science and Technology* 6; 257-271.

KLAIJ, M.C. and SERAFINI, P.G. In press. Management options for intensifying millet-based crop production systems on sandy soils in the Sahel. In: *Proceedings of the International Conference on Dryland Farming* (Texas, 1988), ed. P.W. Unger, W.R. Jordan, T.V. Sneed and R.W. Jensen. College Station, Texas: Texas A&M University.

KLOPFENSTEIN, T. and OWEN, F.G. 1981. Value and potential use of crop residues and by products in dairy rations. *Journal of Dairy Science* 64: 1250-1268.

LEGGETT, G.E., RAMING, R.L., JOHNSON, L.C. and MASSEE, T.W. 1974. Summer fallow in the Northwest. In: *Summer fallow in the western United States*, 110-135. U.S. Department of Agriculture, Agriculture Resources Service. Conservation Resources Report no. 17. Washington, DC: Government Printing Office.

LESSITER, F. 1982. No-tillage defined. No-till Farmer 10(6): 2.

LITTLER, J.W. and MARLEY, J.M.T. 1978. *Fallowing and winter cereal production on the Darling Downs.* Queensland Department of Primary Industries. Progress Report 1968-78. Indooroopilly, Queensland: QPI.

LYLE, W.M. and BORDOVSKY, L.P. 1987. Integrating irrigation and conservation tillage technology. In: *Conservation tillage: toda and tomorrow* (Proceedings, Texas, 1987), ed. T.J. Gerik and B.L. Harris, 67-71. Texas agriculture Experiment Station, Publication no. MP-1636. College Station, Texas: TAES.

MANE, V.S. and SHINGTE, S.H. 1982. Use of mulch for conserving moisture and increasing the yield of sorghum in dryland. *Indian Journal of Agricultural Science* 52: 458-462.

MARSTON, D. 1980 Rainfall simulation for the assessment of the effects of crop management on soil erosion. Unpublished Master of Natural Resources thesis, University of New England, Armidale, New South Wales.

MASSEE, T. and McKAY, H. 1979. Improving dryland wheat production in eastern Idaho with tillage and cropping methods. Idaho Agriculture Experiment Station. Bulletin no. 581. Idaho: IAES.

MILLER, G.E. Jr. 1980. Harvesting crop residues for alcohol production, *Californian Agriculture* 34(6): 7-9.

MORIN, J. and Benyamini, Y. In press. Tillage method selection based on runoff modeling. In: *Proceedings of the International Conference on Dryland Farming* (Texas, 1988), ed. P.W. Unger, W.R. Jordan, T.V. Sneed and R.W. Jensen. College Station, Texas: Texas A&M University.

MORISHITA, D.W., SCHLEGEL, A.J., NORWOOD, C.A. and GWIN, R.E. In press. Comparison of herbicides and tillage in a long-term wheat-fallow rotation. In: *Proceedings of the International Conference on Dryland Farming,* (Texas, 1988), ed. P.W. Unger, W.R. Jordan, T.V. Sneed and R.W. Jensen. College Station. Texas: Texas A&M University.

NORWOOD, C.A., SCHLEGEL, A.J., MORISHITA, D.W. and GWIN, R.E. In press. reduced tillage cropping systems in southwest Kansas. In: *Proceedings of the International Conference on Dryland Farming*, (Texas, 1988), ed. P.W. Unger, W.R. Jordan, T.V. Sneed and R.W. Jensen. College Station, Texas: Texas A&M University.

PAPENDICK, R.I. and MILLER, D.E. 1977. Conservation tillage in the Pacific Northwest. *Journal of Soil Water Conservation* 32: 49-56.

POOLE, M.L. 1987. Tillage practices for crop production in winter rainfall areas. In: *Tillage directions in Australian agriculture*, ed. P.S. Cornish and J.E. Pratley, 24-47. Melbourne and Sydney: Inkata Press.

PRIHAR, S.S. and JALOTA, S.K. In press. Role of shallow tillage in moisture management. In: *Proceedings of the International Conference on Dryland Farming* (Texas, 1988), ed. P.W. Unger, W.R. Jordan, T.V. Sneed and R.W. Jensen. College Station. Texas: Texas A&M University.

RAMIG, R.E., ALLMARAS, R.R. and PAPENDICK, R.I. 1983. Water conservation: Pacific Northwest. In: *Dryland agriculture*, ed. H.E. Dregne and W.O. Willis, 105-124. *Agronomy Monograph no. 23*. Madison, WI: ASA.

RAMIG, R.E. and EKIN, L.G. 1978. Soil water storage as influenced by tillage and crop residue management. In: *Oregon Agriculture Experiment Station Progress Report*, SM 78-4, 65-68. Oregon: OAES.

RAMIG, R.E. and EKIN, L.G. 1984. Effect of stubble management in a wheat-fallow rotation on water conservation and storage in eastern Oregon. In: *Oregon Agriculture Experiment Station Special Report 713*, 30-33. Oregon: AES.

RAO, P. and KUMAR, V. 1975. Effect of mulches and weed control on soil moisture conservation, growth and yield of bajra (*Pennisetum geophytes* S. & H.) under rainfed condition. *Indian Journal of Weed Science* 7: 105-109.

RICHARDSON, C.W. 1973. Runoff, erosion, and tillage efficiency on graded-furrow and terraced watersheds. *Journal of Soil Water Conservation* 28: 162-164.

ROCKWOOD, W.G. and LAL, R. 1974. Mulch tillage: A technique for soil and water conservation in the tropics. *Span* 17: 77-79.

RODRIGUEZ, M.S. In press. Impact of tied ridges, crop residues, and management on maize production in the Sudan savanna zone. In: *Proceedings of the International Conference on Dryland Farming* (Texas, 1988), ed. P.W. Unger, W.R. Jordan, T.V. Sneed and R.W. Jensen. College Station, Texas: Texas A&M University.

RUSSEL, J.C. 1939. The effect of surface cover on soil moisture losses by evaporation. *Soil Science Society of America Proceedings* 4: 65-70.

SANKAT, C.K. and BILANSKI, W.K. 1982. Thermo-ammoniation of maize stover. *Tropical* 59: 62-63.

SMIKA, D.E. 1976a. Seed zone soil water conditions with reduced tillage in the semiarid Central Great Plains. In: *Proceedings of the VIIth Conference of the International Soil Tillage Research Organization* (Uppsala, 1976), 37.1-37.6. Wageningen, The Netherlands: ISTRO.

SMIKA, D.E. 1976b. Mechanical tillage for conservation fallow in the semiarid Central Great Plains. In: *Conservation tillage* (Proceedings, Colorado, 1976), ed. B.W. Greb, 78-91. Great Plains Agricultural Council. Publication no. 77. Colorado: GPAC.

SMIKA, D.E. and UNGER, P.W. 1986. Effect of surface residues on soil water storage. *Advances in Soil Science* 5: 111-138.

SMIKA, D.E. and WICKS, G.A. 1968. Soil water storage during fallow in the Central Great Plains as influenced by tillage and herbicide treatments. *Soil Science Society of America Proceedings* 32: 591-595.

SMIT, H.A., PURCHASE, J.L. and JOUBERT, G.D. In press. Production techniques for dryland wheat production in the summer rainfall region of South Africa. In: *Proceedings of the International Conference on Dryland Farming* (Texas, 1988), ed. P.W. Unger, W.R. Jordan, T.V. Sneed and R.W. Jensen. College Station, Texas: Texas A&M University.

TAYLOR, J.D. 1981. Continuous autohydrolysis, a key step in the economic conversion of forest and crop residues into ethanol. In: *Proceedings of the Conference on Energy from Biomass*, 330-336. London: Allied Science Publications.

THORLACIUS, S.O., COXWORTH, E. and THOMPSON, D. 1979. Intake and digestibility of fababean crop residue by sheep. *Canadian Journal of Animal Science* 59: 459-462.

UNGER, P.W. 1978. Straw mulch rate effects on soil water storage and sorghum yield. *Soil Science Society of America Journal* 42: 486-491.

UNGER, P.W. 1984a. Tillage and residue effects on wheat, sorghum, and sunflower grown in rotation. *Soil Science Society of America Journal* 48: 885-891.

UNGER, P.W. 1984b. *Tillage systems for soil and water conservation.* FAO Soils Bulletin no. 54. Rome: FAO.

UNGER, P.W. In press. Residue management for dryland farming. In: *Proceedings of the International Conference on Dryland Farming* (Texas, 1988), ed. P.W. Unger, W.R. Jordan, T.V. Sneed and R.W. Jensen. College Station, Texas: Texas A&M University.

UNGER, P.W. and PARKER, J.J. Jr. 1968. Residue placement effects on composition, evaporation, and soil moisture distribution. *Agronomy Journal* 60: 469-472.

UNGER, P.W. and PARKER, J.J. 1976. Evaporation reduction from soil with wheat, sorghum, and cotton residues. *Soil Science Society of America Journal* 40: 938-942.

UNGER, P.W., STEINER, J.L. and JONES, O.R. 1986. Response of conservation tillage sorghum to growing season precipitation. *Soil Tillage Research* 7: 291-300.

UNGER, P.W. and STEWART, B.A. 1988. Conservation techniques for Vertisols. In: *Vertisols: their distribution properties, classification and management*, ed. L.P. Wilding and R. Puentes, 165-181. Soil Management Support Services. Technical Monograph no. 18. College Station, Texas: Texas A&M University.

UNGER, P.W. and WIESE, A.F. 1979. Managing irrigated winter wheat residue for water storage and subsequent dryland grain sorghum production. *Soil Science Society of America Journal* 43: 582-588.

VAN DER PLOEG, J. and REDDY, K.C. In press. Water conservation techniques for sorghum growing areas of Niger. In: *Proceedings of the International Conference on Dryland Farming* (Texas, 1988), ed. P.W. Unger, W.R. Jordan, T.V. Sneed and R.W. Jensen. College Station, Texas: Texas A&M University.

VERMA, H.N., SINGH, R., PRIHAR, S.S. and CHAUDHARY, T.N. 1979. Runoff as affected by rainfall characteristics and management practices on gently sloping sandy loam. *Journal Indian Society of Soil Science* 27: 18-22.

WANG, W., LI, D., GAO, X. and WANG, D. In press. Measures for improving water conservation crop yields in rain-fed farming in northern China. In: *Proceedings of the International Conference on Dryland Farming* (Texas, 1988), ed. P.W. Unger, W.R. Jordan, T.V. Sneed and R.W. Jensen. College Station, Texas: Texas A&M University.

WICKS, G.A. and SMIKA, D.E. 1973. Chemical fallow in a winter wheat-fallow rotation. *Journal of the Weed Science Society of America* 21: 97-102.

WILHELM, W.W., DORAN, J.W. and POWER, J.F. 1986. Corn and soybean yield response to crop residue management under no-tillage production systems. *Agronomy Journal* 78: 184-189.

WILLIS, W.O. and FRANK, A.B. 1975. Water conservation by snow management in North Dakota. In: *Proceedings of the Conference on Snow Management on the Great Plains* (North Dakota, 1975), 155-162. Great Plains Agricultural Council. Publication no. 73. Lincoln: University of Nebraska.

WINUGROHO, M., IBRAHIM, M.N.M. and PEARCE, G.R. 1984. A soak-and-press method for the alkali treatment of fibrous crop residues. Calcium hydroxide and sodium hydroxide treatments of rice straw. *Agricultural Wastes* 9: 87-99.

ZENTNER, R.P., CAMPBELL, C.A., SELLES, F., McCONKEY, B., NICHOLAICHUK, W. and BEATON, J.D. 1988. In: *Great Plains Soil Fertility Workshop Proceedings* (Colorado, 1988), ed. J. Havlin, vol. 2, 147-158. Manhattan, Kansas: Kansas State University.

No tillage, minimum tillage, and their influence on soil physical properties

PETER M. AHN and B. HINTZE*

Abstract

The terms no tillage and minimum tillage are often used without adequate definition and need to be defined. In considering their effects on soil physical properties, however, we are concerned more with the absence of the undesirable effects of other forms of tillage, both hand and mechanized, which no tillage and minimum tillage are able to avoid. Soil physical properties affected by tillage include structure and porosity, as repeated tillage and cultivation of tropical soils cause organic-matter contents to decline; but soil structure is influenced by factors other then organic matter, including microaggregation due to iron and aluminium oxides. The latter is very marked in some tropical soils, but absent in others. It is necessary to assess the physical properties of individual soils, and to examine the mechanisms by which they are modified when tilled. Differences in physical properties help to explain the sometimes apparently contradictory results reported. Zero tillage has been successful and can be recommended in some areas, as with the Alfisols at IITA in Nigeria, whereas with soils that have become compacted, and soils such as the compact sandy soils of Senegal, deep ploughing and subsoil loosening facilitate better rooting and thus increase yields.

*   Respectively: IBSRAM Programme Officer for Africa and Coordinator, Vertisol Network; and Coordinator, Land Development Network, Résidence Gyam, 2ème étage, Angle Bd Clozel-Av Marchand, 04 BP 252, Abidjan 04, Côte d'Ivoire.

Introduction

What is meant by "no tillage" and "minimum tillage"? What are the advantages claimed for them? Where are they currently carried out in Africa, and with what results? These are some of the questions evoked, but in a single short presentation all that can be done is to pose some general questions, and to focus attention on some of the mechanisms by which the physical actions of tillage influence and modify soil physical properties.

By tillage we mean any physical loosening of the soil as carried out in a range of cultivation operations, either by hand or mechanized. The soil physical properties with which we are mainly concerned are soil structure, soil porosity, bulk density, and aeration, which in turn affect soil water acceptance and storage, liability to erosion, and other factors.

In recent years a body of opinion has formed which questions the need for ploughing and other soil tillage operations, or at least demands that soil scientists analyze more carefully exactly what is achieved by these operations with a view to assessing whether they are cost-effective and environmentally sound in a context which emphasizes sustainability. The advocacy of no tillage or minimum tillage implies either that tillage operations are unnecessary, in the sense that comparable results can sometimes be obtained without traditional tillage practices, or that they avoid the harmful effects which would occur if these traditional practices were used.

Advocates of no tillage and minimum tillage raise the very pertinent question "why plough?" Some agronomists in Africa, such as Theo Willcocks in relation to Vertisols in the Sudan (Willcocks, 1989), have argued that in some cases cultivation is mainly a weed-control measure. In trials in the central Sudan, shallow wide one-way disc systems were very effective for controlling weeds and preparing a seedbed in one shallow operation which affected the soil only to a depth of 5-8 cm. It should be pointed out, however, that swell-shrink clays have self-loosening properties which noncracking soils do not, so that their physical properties set them apart from other soils. This in turn emphasizes the need to consider individual soils on their merits and not to overgeneralize.

The effects of tillage

When soil is cultivated it is loosened, and any crust or surface seal present is broken: this may be important for soils which crust and seal easily. In a study of crusting in cover sands in northern Nigeria, Sombroek and Zonneveld (1971) found that surface crusting and sealing were widespread, but whereas in most of the soils they examined the breaking of the crust by hand tillage with a hoe was sufficient to allow the soil to be cultivated for a season or more, the crust of some soils reformed after a single rain. These soils could not be used at all for this reason and were traditionally left uncultivated.

The immediate effect of the soil disturbance caused by tillage is to increase the porosity of the soil, which in turn lowers bulk density and increases rainfall acceptance. The effect the farmer probably has most in mind is that he has a looser soil covering in which to plant his seeds, and in which seedlings can emerge without difficulty and extend their root systems without physical obstruction. Peasant farmers in Africa have often observed better germination with a fine seedbed, but in some cases a fine seedbed may not be an advantage. In areas of marginal rainfall in Kenya, it has been observed that rainfall infiltration is usually better and runoff less with a rough, cloddy seedbed, and in years of barely adequate rainfall this factor alone may have significant beneficial effects on crop yields.

The increased aeration of the soil provoked by tillage (helped by higher temperatures if the soil loses its shade or cover) leads to greater microbial activity, and this in turn leads to a faster rate of organic-matter mineralization. This mineralization of organic matter supplies plant nutrients, but leads to the lowering of organic-matter levels and to a weakening of the soil structure. As organic-matter levels fall, the crumbs in the topsoil typically become smaller and of a weaker structural grade. This weakening of the structure leads in turn to a lower porosity, particularly to a lower macroporosity since the macrospaces between the crumbs are reduced in size. Lower porosity in the topsoil in turn leads to lower rainfall acceptance, increased runoff, and surface erosion. As surface structure is lost, the liklihood of crusting and sealing increases.

The nature of structure in tropical soils

The binding forces which contribute to soil structure in the tropics are of two main types - those due to organic matter and those due to the presence of iron and aluminium oxides. Some soils, particularly Vertisols, have little real structural strength, and the structural aggregates (peds) are produced merely as a result of the shrinkage cracks formed when wet soils become drier. This type of apparent structure slakes and is lost as the soil is wetted.

In certain older tropical soils in particular, the attractive forces between clays are reinforced by the presence of iron and aluminium oxides, and this is why certain red soils are traditionally regarded as having good physical properties, though in fact the role of the colourless aluminium oxides may be more important than that of the more obvious iron oxides (Greenland *et al.*, 1968).

The binding effects of oxides often contribute to soil microstructure, which in turn influences the apparent texture of the soil. Clay-sized particles are bound together to form water-stable aggregates, typically of coarse silt to fine sand size (20 to 200 μ in diameter) (Ahn, 1974, 1979). These microaggregates may give even a clay soil a sandy feel. Such well-microaggregated soils have very good physical properties, particularly rainfall acceptance, porosity, available waterholding capacity, and resistance to erosion. In Kenya, the well-known and very productive Kikuyu friable clays form an example of soils with good microaggregation. What is particularly relevant to the effect of tillage on soil physical properties, however, is the fact that this microaggregation is not due to the binding effect of organic colloids and is therefore not lost as organic-matter levels fall.

The effects of heavy machinery on soil compaction

Apart from loss of porosity associated with progressive loss of soil structure due to the lowering of organic-matter contents, porosity can be dramatically lowered as the result of compaction due to the use of heavy machinery. The harmful effects of compaction are one of the major disadvantages of mechanical clearing and land preparation referred to by numerous investigators. Moura Filho and Buol (1972) found that fifteen years of cultivation of a Brazilian Oxisol reduced infiltration rates from 82 to 12 cm h^{-1}. This was associated with a sharp decrease in macropores larger than 0.05 mm in diameter, a decrease considered due mainly to compaction by machinery, though microporosity was unchanged.

A study of the effects of land-clearing methods on soil properties and crop performance of an Ultisol in the Amazon forests of Peru showed that heavy machinery compacted the soil, reduced root penetrability and soil aeration, and decreased infiltration rates and hydraulic conductivity (Seubert *et al.*, 1977). In an Ultisol at Yurimaguas described by Alegre *et al.* (1987), clearing by bulldozing resulted in a compacted zone at 15-45 cm which was still discernible seven years after the initial clearing, whereas land cleared by traditional slash-and-burn methods had lower bulk densities. Of the various management techniques tried out in order to correct the problem of bulldozer-induced subsoil compaction, subsoiling with a chisel plough gave the best yields of test crops of rice, soybean, and maize, and no-till the lowest yields, indicating that no-tillage agriculture is likely to do well only where soil structure and aeration are adequate and have not been adversely affected by inappropriate clearing methods.

Other studies have also indicated that small changes in bulk density may have a marked effect on root development (Sanchez, 1976). In a study of soil compaction caused by tractors in Hawaii, Trouse and Humbert (1961) showed that sugarcane roots became flattened by slight changes in bulk density, while more substantial compaction caused root restriction. The actual bulk density needed to cause restriction varied considerably from soil to soil. In an Andept, roots were restricted at bulk densities of 1.08, but in an Aquept they grew normally at this bulk density, but were restricted when the bulk density had reached 1.76.

Experimental results obtained in Africa with minimum tillage, no tillage, and with ploughing and/or subsoiling

In the African context, the terms no tillage and minimum tillage may not always be employed with the same meaning as in temperate areas of the world, and may also be used differently in the different contexts of shifting cultivation (still the dominant system in most of Africa) and mechanized agriculture. No tillage is often equated with no physical disturbance of the soil other than that provoked by the manual clearing of the original vegetation, usually by slash-and-burn, and by the use of a planting stick to drop in seeds or a hoe or cutlass (machete) to make small planting holes. Similar considerations appear to apply in much of tropical South America (see, for example, Seguy et al., 1989).

In the context of experiments on Alfisols at Ibadan, Nigeria, carried out by IITA, zero tillage means killing a preexisting cover of weeds or cover crop and planting seeds in the dead plant cover by a hand or mechanical dibber, which makes planting holes but which leaves the rest of the ground virtually undisturbed. Similar or comparable techniques are beginning to find favour among highly mechanized farmers in the U.S., as when seeds are drilled into stubble. However, the use of weedicides in the tropics is still expensive, so that some commentators on no-tillage agriculture as recommended by IITA in Nigeria have pointed out that low tillage is not necessarily equated with low inputs. This is in contrast to slash-and-burn methods as practiced in traditional shifting agriculture, where the farmer needs little but his own labour, a cutlass (or machete) and a box of matches. Not only is shifting cultivation genuinely low-input agriculture, but it does less damage to the soil during the initial clearing and land preparation phases than alternative methods of clearing. Lal (1987a) carefully compared the effects of traditional incomplete clearing of forest vegetation at Ibadan with manual clearing, shear blade clearing, and clearing by tree pusher and root rake. His results demonstrated that soil bulk density, penetrometer resistance, runoff, and soil erosion were all most favourable with traditional clearing, and adversely affected by the mechanized clearing methods examined. Lal also emphasized the need to protect surface structures from the full force of tropical rains, and the importance of maintaining macrofaunal activity, especially worms and termites.

The Senegal experience with deep ploughing

In contrast with experiments at IITA, which show good yields with a minimum of tillage and soil disturbance, experience with compact sandy soils elsewhere in Africa has demonstrated the positive effects of deep ploughing on porosity, root development, and hence yields. Nicou *et al.* (1970) showed that deep ploughing produced much higher root densities of rice grown in upland soils (Figure 1), while in 1971 Charreau and Nicou published a series of papers

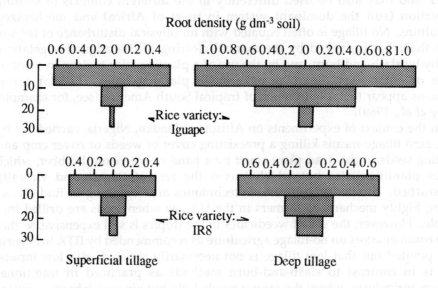

Root density (g dm⁻³ soil)

Figure 1. Influence of plough depth on the root density of two rice varieties grown in upland conditions in South Senegal (data from Nicou *et al.* [1970], as given in Sanchez [1976]).

in *Agonomie Tropicale* which described the beneficial effects of deep tractor ploughing on the soil up to depths of 30 cm. The authors showed an association between lower bulk densities, better root development, and higher yields of peanuts and sorghum (Figure 2).

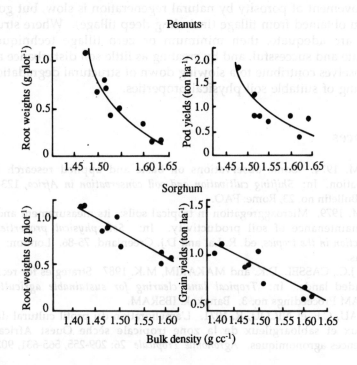

Figure 2. Relationship between bulk density and root weight (left-hand graphs) and between bulk density and yields (right-hand graphs) for peanuts and sorghum grown in sandy Alfisols at Bambey, Senegal (data from Charreau and Nicou [1971], as given in Sanchez [1976]).

Conclusion

The key to the need for tillage in many tropical soils is soil structure, but soil structure as such cannot usefully be measured directly. Instead, as stated by Lal (1987b), from the edaphic point of view soil structure should be defined in terms of porosity (i.e. total volume, pore-size distribution, pore stability, and channel continuity). Soil structure stability, as discussed in this paper, is related to both organic-matter content, which affects mainly topsoil structure and is progressively lowered during cultivation, and the microaggregating effects of iron and aluminium oxides, which contribute to structural stability in some tropical soils but not in others. Where soils have become compacted, then the improvement of porosity by natural regeneration is slow, but good results have been obtained from tillage (including deep tillage). Where structure and porosity are adequate, then minimum or zero tillage techniques can be appropriate and successful, and by creating as little soil disturbance as possible will themselves contribute to a slowing down of structural degradation and the maintaining of suitable soil physical properties.

References

AHN, P.M. 1974. Some observations on basic and applied research in shifting cultivation. In: *Shifting cultivation and soil conservation in Africa*, 123-154. FAO Soils Bulletin no. 23, Rome: FAO.

AHN, P.M. 1979. Microaggregation in tropical soils: its measurement and effects on the maintenance of soil productivity. In: *Soil physical properties and crop production in the tropics*, ed. R. Lal and D.J. Greenland, 75-86. London: John Wiley & Sons.

ALEGRE, J.C., CASSEL, D.K. and MAKARIM, M.K. 1987. Strategies for reclamation of degraded lands. In: *Tropical land clearing for sustainable agriculture*, 77-91. IBSRAM Proceedings no. 3. Bangkok: IBSRAM.

CHARREAU, C.F. et NICOU, R. 1971. L'amélioration du profil cultural dans les sols sableux et sabloargileux de la zone tropicale sèche Ouest Africaine et ses incidences agronomiques. *Agronomie Tropicale* 26: 209-255, 565-631, 903-978, 1183-1237.

GREENLAND, D.J., OADES, J.M. and SHERWIN, T.W. 1968. Electron microscope observations of iron oxides in red soils. *Journal of Soil Science* 19: 123-126.

LAL, R. 1987a. Need for, approaches to, and consequences of land clearing and development in the tropics. In: *Tropical land clearing for sustainable agriculture*, 15-27. IBSRAM Proceedings no. 3. Bangkok: IBSRAM.

LAL, R. 1987b. Surface soil degradation and management strategies for sustained productivity in the tropics. In: *Management of acid tropical soils for sustainable agriculture*, 167-177. IBSRAM Proceedings no. 2. Bangkok: IBSRAM.

MOURA FILHO, W. and BUOL, S.W. 1972. Studies of a Latosol Roxo (Eutrustox) in Brazil. *Experientiae* 13: 201-234.

NICOU, R., SEGUY, L. et HADAD, G. 1970. Comparaison de l'enracinement de quatre variétés de riz pluvial en présence ou absence de travail du sol. *Agronomie Tropicale* 25: 639-659.

SANCHEZ, P.A. 1976. *Properties and management of soils in the tropics*. New York: John Wiley & Sons.

SEUBERT, C.A., SANCHEZ, P.A. and VALVERDE, C.V. 1977. Effects of land clearing methods on soil properties and crop performance in an Ultisol of the Amazon jungle of Peru. *Tropical Agriculture* (Trinidad) 54: 307-321.

SOMBROEK, W.G. and ZONNEVELD, I.S. 1971. *Ancient dune fields and fluvatile deposits in the Rima-Sokoto basin (N.W. Nigeria)*. Netherlands Soil Survey Institute. Soil Survey Paper no. 5. Wageningen: NSSI.

SEGUY, L., BOUZINAC, S., PACHECO, A. et KLUTHCOUSKI, J. 1989. *Des modes de gestion mécanisés des sols et des cultures aux techniques de gestion en semis direct, sans travail du sol, appliquées aux Cerrados du centre-ouest brésilien.* Publication IRAT/EMBRAPA. Montpellier, France: CIRAD.

TROUSE, A.C. Jr. and HUMBERT, R.P. 1961. Some effects of soil compaction on the development of sugarcane roots. *Soil Science* 91: 208-217.

WILLCOCKS, T.J. 1989. Agricultural engineering aspects of Vertisol management. In: *Vertisol management in Africa*, 63-76. IBSRAM Proceedings no. 9. Bangkok: IBSRAM.

The *ASIALAND* management of sloping lands network, and Malaysia's experience in the network

ADISAK SAJJAPONGSE and ZAINOL EUSOF*

Abstract

Five countries, i.e. Indonesia, Malaysia, Nepal, the Philippines, and Thailand are involved in the Management of Sloping Lands in Asia Network. The main objective of the network is to help in conserving soil resources on sloping lands in the countries of the network through research into, and promotion of land development and land management technology which will help to achieve a sustainable form of agriculture. A number of technologies, notably alley cropping, hillside ditch, agroforestry, and cover crops, are currently the subject of IBSRAM investigations in Asia.

Some of the results obtained in Malaysia reveal that more than 30 t ha⁻¹ of soil was washed away as a result of sole cropping of rubber (the farmers' practice), and of rubber intercropped with papaya. This loss was greatly reduced when some form of cover was established. Under peanut for example soil loss was reduced by 65-84% as compared to the loss under papaya, while under legume cover crops the reduction was 60-67%. The results also showed that the rate of soil loss and runoff decreased progressively with a progressive increase in the surface cover.

* Respectively: Coordinator, Sloping Lands Network, International Board for Soil Research and Management, PO Box 9-109, Bangkok 10900, Thailand; and Senior Research Officer, Rubber Research Institute of Malaysia, 260 Jalan Ampang, Kuala Lumpur, Malaysia.

Introduction

Due to population growth, urbanization, and industrialization, the amount of land available for agriculture in Asia has decreased. In order to ensure that an adequate amount of food is produced, land which was once considered as unsuitable for agriculture, because of the risk of erosion or other hazards, is now being cleared and used for cultivation.

The erosion hazard applies particularly to sloping lands, which are widespread and occupy a large area in Asia. Out of the total land area, 35% in Thailand, 63% in the Philippines, and 87% in Nepal is classified as hilly and mountainous land. On this type of land, farmers generally plant their crops up-and-down slope, and cultivate the soil with minimal concern for soil erosion. This causes siltation of valleys and dams, and is increasingly threatening the environment of the area. It is therefore imperative that efforts should be made in order to identify appropriate technology for managing sloping lands so that the threat to the immediate surrounding areas can be avoided.

To reduce soil erosion on sloping lands, many technologies have been suggested and tested. Hoey (1988) stated that cropping between contour grass strips using minimum tillage and surface retention of stubble is a cheap, effective, and easily implemented form of stable agriculture on slopes up to 35%. Among the different alley-forms of cropping which have been tested, Lal (1988) found that hedgerows of leucaena were more effective in reducing runoff and controlling erosion than those of gliricidia. Paningbatan (1987) suggested that the width of alleyways can be varied from 4 m to more than 10 m, depending on the degree of slope - the steeper the slope the narrower the alleyway. Young (1989) suggested the use of agroforestry as a way of maintaining soil fertility and controlling erosion. Bench terraces are known to be one of the most effective mechanical means for controlling erosion on cultivated hillslopes. A mathematical equation for computing the optimum vertical interval and width of the bench has been developed for the construction of the terrace (Jaiswal, 1988).

In an effort to curtail the deterioration and erosion of these lands, the International Board for Soil Research and Management (IBSRAM) has organized a network called the Management of Sloping Lands in Asia Network.

Management of Sloping Lands in Asia Network

The Management of Sloping Lands in Asia Network has been organized to assist the participating national agricultural centres in conducting research on this topic. The main objective of the network is to help in conserving soil

resources on sloping lands in the region through research into, and promotion of, the application of appropriate land development and land management technology, and thereby to achieve a sustainable form of agriculture. The network has the following specific objectives:
- to assist national agricultural institutions in validating or testing improved soil management technologies;
- to facilitate the exchange of research information on soil management among agricultural scientists in the region through meetings, workshops, and publications;
- to strengthen national agricultural institutions by network and information activities; and
- to evaluate and select cost-effective and farmer-acceptable options for agricultural production on sloping lands.

The network started its operations in mid-1988, with financial support from the Asian Development Bank (ADB) and Swiss Development Cooperation (SDC). Five countries - Indonesia, Malaysia, Nepal, the Philippines and Thailand - are currently participating in the network.

Network components

The network is an association of three parties, each of which has different responsibilities:
- **National cooperators.** The cooperators represent the research and development institutions involved in the network. They are responsible for initiating and conducting the national soil management research projects within the framework of the network, which is jointly established by the cooperators and IBSRAM.
- **Funding agencies and donors.** The agencies and donors finance the activities of the network. They also take an active role in planning the research programme.
- **IBSRAM.** Through its coordinator and other staff, IBSRAM helps the national cooperators in the development and execution of the research programme. If the cooperators need additional funds, IBSRAM lends its support by bringing individual needs to the attention of donors.

To ensure that the results from the different network participating countries can be compared, and to facilitate technology transfer from one location to another, IBSRAM's approach is to ensure that a common methodology is followed by all the cooperators taking part in the network's activities. This is done through training workshops, which involve consultations between front-line scientists, network cooperators, and leading specialists.

A Training Workshop on the Management of Sloping Lands was held from 7 to 18 November 1988 in Chiang Mai, Thailand. About 50 participants took an active part in the workshop. They included representatives from China, Indonesia, Malaysia, Nepal, Pakistan, Philippines, Sri Lanka, Thailand, and Vietnam.

Four main activities - classroom lectures, field exercises, group discussions, and a field trip - formed the basis of the workshop. The topics included in the classroom lecture sessions were site selection, site characterization (socio-economic, climatic and soil characteristics), farming systems, experimental design and monitoring, and a session on erosion and runoff.

In the laboratory and field exercises, the participants were divided into groups and were given exercises to familiarize themselves with the methodology required for site selection and characterization. The group discussion sessions were devoted to the presentation of the results of laboratory and field exercises, and to a discussion on the monitoring of chemical, biological, and physical parameters.

An annual meeting is organized for the cooperators to report on the progress of their research activities and findings and to facilitate an exchange of scientific information and ideas. Each participating country hosts the meeting on a rotational basis so that all the experimental fields in the network can be visited by all the cooperators.

Technologies being tested

To achieve the objective of the network, different soil management technologies are being evaluated in the different participating countries. The technologies include alley cropping, which consists of cultivating crops along contours in alleys 4 to 5 m wide and separated by a shrub or legume hedgerow; grass strips, which are 1 m wide and are established every 4 to 6 m along the contours of the slope; hillside ditches, which are built along the contours and across the slope, helping to slow down and contain runoff water and allowing more of the water to perculate into the soil; and agroforestry, which is the association of fruit trees or perennial crops and field crops. The specific technologies to be evaluated in the participating countries are described below.

Indonesia

A cropping system consisting of peanuts from January-April, mucuna or mungbean from May-August, and upland rice from September-December is being used to test the sustainability of crop production on land with a slope of 5-15%. Different soil management techniques are being tried out: no inputs, low inputs, and high inputs - in conjunction with alley cropping and cover cropping.

Malaysia

Five treatments are being used to determine soil fertility changes in relation to the sustainability of intercropping on rubber plantations with slopes ranging from 6-18%. The treatments are: rubber with the establishment of a legume cover, rubber intercropped with papaya, rubber intercropped with papaya as the first-storey crop and corn/peanut as the second-storey crops, and rubber cultivation with minimum inputs (farmers' practice).

Nepal

The cropping system consists of planting corn intercropped with soybean in spring, and mustard or millet in winter. The sustainability of crop production is evaluated by relating yields to different soil management practices, i.e. conventional tillage, minimum tillage, minimum tillage plus a legume cover crop, alley cropping, hillside ditches, strip-cropping, and terracing. The experiment is conducted on land with 50-60% slope.

Philippines

To identify an improved system for sustainable agriculture on sloping land (i.e. land with a slope of 20-40%), four land management techniques are being compared: alley cropping with gliricidia and napier as hedgerows, alley cropping with gliricidia and napier as hedgerows plus high inputs, alley cropping with banana plus sapodilla or cashew nut as hedgerows plus high inputs, and the farmers' practice. The indicator crops to be used are corn and peanuts.

Thailand

Five soil conservation treatments: alley cropping with pigeon pea and leuceana as hedgerows, agroforestry (Chinese plum/coffee), contour grass strips, a combination of grass strips and perennial cash crops, and hillside ditches are being validated at two sites with slopes ranging from 30 to 40%. The effectiveness of these treatments is being assessed in relation to the farmers' practice for sustainable crop production of cereals and grain legumes.

Sustainability evaluation

Two methods employed by the network to identify soil management technologies which can produce sustainable agriculture are (i) to measure yields, and (ii) to evaluate the change of chemical and physical properties of the soil.

In the yield measurement method, the yield components (e.g. the number of seeds/pod, the number of pods/plant, and the number of branches/plant), are also measured in order to identify and understand more about the major factors which cause yield fluctuation. For the soil measurement method, the following soil properties are determined: N, P, Ca, Mg, CEC, pH, organic matter, texture, density, infiltration, compaction, and pF curves. By measuring the change of soil chemical and physical properties and/or the variation of crop yield season after season or year after year, appropriate soil management technologies for specific locations can be identified. Soil loss due to erosion is also measured.

Benefits and expected outputs

The project is designed to produce appropriate technologies for the management of sloping lands, and to encourage their adoption by farmers. The specific outputs of the project will include:

- improved, economical, and farmer-acceptable soil management practices for sloping lands;
- the identification of soil management practices that can lead to sustainable agriculture;
- improved knowledge on sloping-land management;
- institutional strengthening;
- improved competence and confidence of the scientists and other personnel in the project;
- higher income for the farmers; and
- a variety of publications concerning the outcome of the project.

The Malaysian experience

The network study in Malaysia has already provided some interesting results. The experiment in Malaysia was conducted using a randomized complete-block design with four replications. Five treatments involving intercropping under rubber were tested: rubber with the establishment of legume cover crops (a mixture in a 4:1 ratio of *Pueraria phaseoloides* and

Calopogonium caeruleum), rubber intercropped with sequential planting of peanut/corn/peanut (high-input system I), rubber intercropped with papaya (high-input system II), rubber intercropped with a combination of papaya and peanut/corn/peanut planted sequentially (high-input system III), and farmers' practice (rubber with minimum maintenance). Individual plots were 24.6 m long 31.2 m wide. Only two replications were equipped with a soil erosion collection system. The size of the collection system was 10 m wide and 24 m long.

The cropping pattern for the experiment was peanut (March-June) - corn (July-October) - peanut (October-January). The transplanting of young rubber plants, clone RRIM 712, was undertaken in February 1989 with a spacing of 2.4 m within rows and 8.2 m between rows. Papaya seedlings, var. Subang and Esotika, were transplanted on 19-20 May 1989, and spaced 2.4 m x 2.4 m apart. Peanut, var. Matjam, was planted with a spacing of 10 x 30 cm on 22-26 March 1989 and harvested on 28-29 June 1989. The legume cover crop was planted in the first week of March. Corn (supersweet) was planted with a spacing of 25 x 75 cm in July 1989, and harvested in October 1989.

The physical properties of the soil samples taken from the erosion plots prior to planting the first peanut crop are presented in Table 1. The bulk density values range from 1.30 to 1.54 g cm^{-3} with an average value of 1.42 g cm^{-3}. These values appear to be higher than normal topsoil values, which are normally between 1.1 to 1.3 g cm^{-3}. This is because the core samples were taken at the layer below the plough layer. The field had just been ploughed before samples were collected. The total pore spaces were calculated by taking the

Table 1. Some physical properties of the soil from the erosion plots in Malaysia.

Erosion plot	Bulk density (g cm^{-3})	Total pore space (%)	Sat. hydraulic conductivity (cm min^{-1})	Avail. water (10-1500 kPa) (%)
1	1.47	44	1.4	7.6
2	1.50	43	131.3	16.2
3	1.36	49	1.7	6.1
4	1.30	51	0.8	11.8
5	1.32	50	700.7	6.4
6	1.34	49	476.4	11.0
7	1.46	45	6.4	12.2
8	1.47	45	6.3	10.4
9	1.54	42	24.6	10.5
10	1.45	45	1.6	7.7

value of particle density as 2.65 g cm⁻³, and the values ranged from 42 to 51%.
The saturated hydraulic conductivity was highly variable, and plots number 1,
3, 4 and 10 showed extremely low conductivity.

The cumulative soil loss and runoff after 14 rain events are presented in
Table 2. The results showed that farmers' practice and rubber intercropped
with papaya gave the highest amount of soil loss. During the period when
measurements were taken (i.e. two-and-a-half months), a total average of more
than 30 t ha⁻¹ of soil was washed away from both types of cropping systems. It
can be seen from Table 2 that soil loss was greatly reduced when some form of
cover had been established. For example, under peanut soil loss was reduced by
65-84% as compared to the loss under papaya, while under a legume cover crop
the reduction was 60-67%. The amount of runoff followed a similar trend as soil
loss for the various cropping systems, with papaya and farmers' practice giving
the highest amount of runoff.

Table 2. Soil loss and runoff after 14 rain events for the different experimental plots in
Malaysia.

Erosion plot	Treatment	Soil loss (kg ha⁻¹)	Runoff (mm)	% rain
1	Legume	15596	48.2	13.8
7	Legume	13720	95.4	27.3
2	Peanut/corn	16104	83.4	23.8
6	Peanut/corn	6846	51.6	14.7
3	Papaya	36856	170.2	48.6
10	Papaya	41311	129.3	36.9
4	Papaya + peanut/corn	7925	52.8	15.1
9	Papaya + peanut/corn	11452	63.2	18.1
5	Farmers' practice	47233	105.9	30.3
8	Farmers' practice	2402‡	109.4‡	31.3‡

† The total rainfall was 350 mm.
‡ Values do not include the value of the 14th rain event. The data were not available
due to leakage in the sedimentation pit.

Cumulative soil loss and runoff were plotted against cumulative rainfall
(Figures 1 and 2). The shape of the curves showed that the rate of soil loss and
runoff decreased with time, particularly for the peanut and legume treatments.

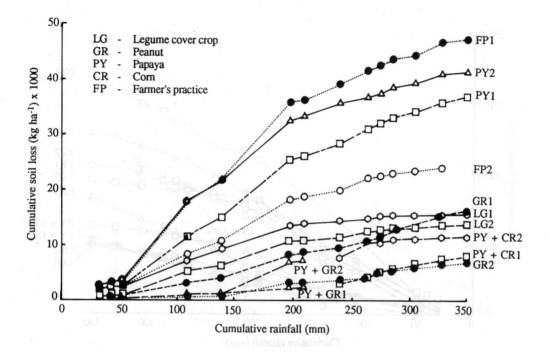

Figure 1. Cumulative soil loss vs. cumulative rainfall for the different treatments.

This suggests that the rate of erosion decreases progressively with a progressive increase in the surface cover. For the farmers' practice and papaya treatments, the increase in surface cover was slow, and thus these treatments not only produced a higher amount of soil loss and runoff, but also resulted in a decrease in the erosion at lower rate when compared to the other treatments.

Another point which is worth noting from the above results is that there is a considerable difference in erosion rates from plots with the same cropping system, especially for the peanut/corn and farmers' practice. This variation is probably a function of the difference in slope steepness and shape, the degree of cover, and the presence of channels. Channelized flow has more detaching and carrying power than even-surface flow.

The results of the Malaysian project show that the soil of the experimental plots is very erodible. In one 56-mm rain event, there was a loss of 14.8 t ha⁻¹ of soil from the farmers' practice, where soil is relatively bare. However, appropriate cropping systems which provide a good soil cover may reduce soil loss considerably. For example, with the same 56-mm storm mentioned above,

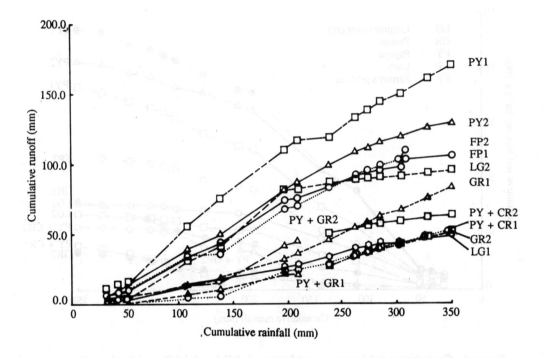

Figure 2. Cumulative runoff vs. cumulative rainfall for the different treatments.

the peanut crop reduced soil loss to only 1 to 2 t ha^{-1}. On the other hand, the introduction of a tree crop like papaya can be as deterimental as the farmers' practice. With a similar storm, soil loss from the papaya plots was between 9 to 14 t ha^{-1}. Thus the selection of a proper crop for intercropping with young rubber not only brings supplementary income to the rubber smallholders but also checks the loss of fertile topsoils.

References

HOEY, M. 1988. Land use, related problems and development strategies for stable agricultural systems for the highland agricultural and social development project. *Land Conservation for Future Generations* 2: 727-734.

JAISWAL, J.P. 1988. Designing bench terraces in north-eastern hill reion of India. *Land Conservation for Future Generations* 1: 237-245.

LAL, R. 1988. Soil erosion control with alley cropping. *Land Conservation for Future Generations* 1: 237-245.

PANINGBATAN, E.P. 1987. Alley cropping in the Philippines. In: *Soil on Management under humid conditions in Asia and the Pacific* (Thailand, 1986), 385-395. IBSRAM Proceedings no. 5. Bangkok: IBSRAM.

YOUNG, A. 1989. Agroforestry for sustainability on steeplands. In: *Soil management and smallholder development in the Pacific Islands* (Solomon Islands,, 1988), 37-49. IBSRAM Proceedings no. 8. Bangkok: IBSRAM.

Soil erosion measurements and the influence of cropping systems on erosion in Nigeria and Cameroon

D. NILL, M. BERNARD, U. SABEL-KOSCHELLA and J. BREUER[*]

Abstract

The joint project of GTZ, IITA, IRA, and the Technical University of Munich under the heading "Soil erosion studies in the humid tropics" has been under way since 1983 in Nigeria and Cameroon. The aim of the project is to study the erosivity of tropical rains (R factor), the erodibility of a wide range of tropical soils (K factor), and the influence of different tillage methods and cropping systems (C factor). For this purpose, data are collected from 500 m² runoff plots, from experiments with two different rainfall simulators, and from various laboratory tests.

K factors proved to be high on an Alfisol and on a soil on volcanic ashes. On the Ultisols, the K factors were rather low, but nevertheless important differences were noticed despite quite similar soil properties. The K factors noted by Wischmeier and Smith (1978) and by Roth et al. (1974) were not the same as the K factors measured in the project. The best results were obtained with a rainfall simulator.

No tillage was an efficient treatment to protect the soil against soil loss, but could not ensure the same yield as a plough with all crops. The maize yields on both an Alfisol and an Ultisol were higher when a plough was used.

---

[*] Technical University of Munich, Lehrstuhl für Bodendunde, D-8050 Freising/Weihenstephan, Munich, Federal Republic of Germany.

Introduction

The joint project of GTZ, IITA, IRA and the Technical University of Munich - "Soil erosion studies in the humid tropics" - started in 1983 in Alore (Nigeria), 25 km outside of the International Institute for Tropical Agriculture, with the establishment of seven 500 m² runoff plots (Sabel-Koschella, 1988), where three different cropping systems where tested and compared to one bare-fallow plot. The aim was to collect data on the erosivity of tropical rainstorms (and notably on the erodibility of an Alfisol), and the performance of different tillage methods and crop rotations on soil erosion and other soil parameters.

In order to collect data on a wider selection of tropical soils, the project included studies in Cameroon, where the existence of very different climatical zones allows the choice of a wide range of representative soils in different ecological environments. Twelve more runoff plots were installed on four different sites. The project started with soil physical tests on aggregate stability (Knobloch, 1987) and the utilization of a small rainfall simulator (Breuer, 1988).

In 1988, phase III of the project started, and to date 21 runoff plots have been established on 7 different sites, and 23 different soils have been selected for trials with a bigger rainfall simulator. Laboratory experiments on soil erodibility have also been conducted (de Neve, 1989), and have yet to be finalized.

The project has three different approaches to acquiring information about the susceptibility of tropical soils to erosion and the influence of different cropping patterns (Figure 1).

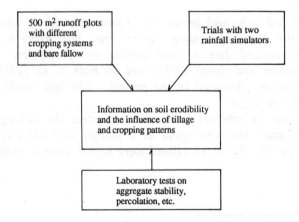

Figure 1. Research approach of the project.

Distribution and choice of soils

The seven different sites (Figure 2) with 500 m² runoff plots were chosen in order to cover the major soil units of the most important parent materials. The aim was to have a good variation of different soil properties, and to represent agriculturally important soils.

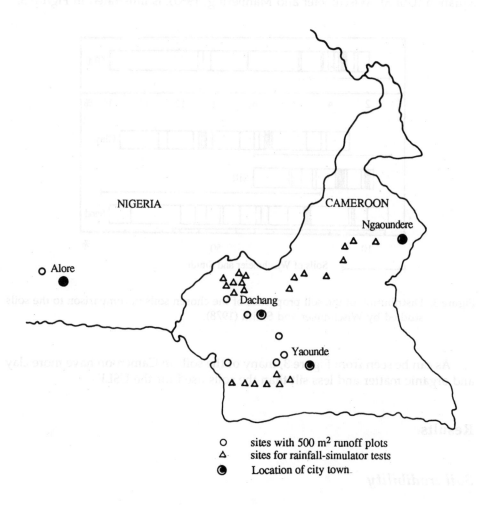

Figure 2. Distribution of experimental sites.

The additional 23 different sites for the rainulator experiments were selected according to the soil units of a soil map prepared by van Ranst and Debaveye (University of Dschang, Cameroon). The main criteria were parent materials, climate and topography. The selected sites represented slopes of between 2-30%, and a wide variability of altitude (100-2000 m), annual rainfall (1200-4000 mm), and parent materials. The range of soil properties, and a comparison with the soils used by Wischmeier for the universal soil loss equation (USLE) (Wischmeier and Mannering, 1969), is illustrated in Figure 3.

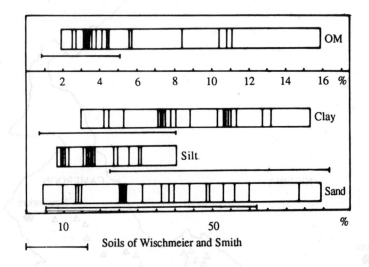

Figure 3. Distribution of the soil properties of the chosen soils in comparison to the soils studied by Wischmeier and Smith (1978).

As can be seen from Figure 3, many of the soils in Cameroon have more clay and organic matter and less silt than the soils used for the USLE.

Results

Soil erodibility

The soil erodibility factor, K, in the USLE is expressed as soil loss in t ha^{-1} per unit of rainfall energy times the maximum 30 minutes of intensity. The cumulative K factors of 5 different soils are shown in Figure 4 (data according to Bernard *et al.*, 1988), and the respective soil properties in Table 1 (data according to Nill, unpublished).

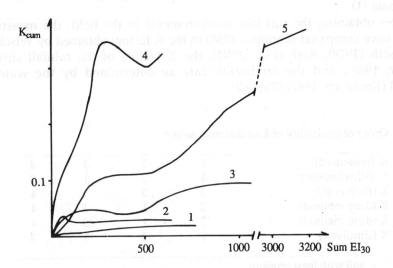

Figure 4. Development of the cumulative K factor (K_{cum}) of three Ultisols (1-3), an Alfisol (5), and a soil on volcanic ashes (4) (Bernard, 1988).

Table 1. Soil properties of the upper 5 cm of five different soils (1989).

	1	2	3	4	5
	Elig Essombala	Nachtigal	Minkoameyos	Bansoa	Alore
	(Ultisol)	(Ultisol)	(Ultisol)	(Volcanic ash)	(Alfisol)
Sand (%)	49	53	45	53	82
Silt (%)	24	9	17	22	9
Clay (%)	27	38	38	25	9
pH H_2O	6.3	5.4	5.4	6.0	5.6
OM (%)	2.2	2.0	2.0	5.7	0.4

The Ultisols in Elig Essombala, Nachtigal, and Minkoameyos have rather low K values. Despite the low K values, there is still a remarkable difference, especially between soil 1 (Elig Essombala) and 3 (Minkoameyos), which are located only about 3 km from one another. They have formed on the same parent material and have rather similar properties. The difference in erodibility seems to be caused by the much better aggregation of the soil in Elig Essombala (1).

After obtaining the soil loss measurements in the field, the measured K factors were compared (Bernard, 1988) to the K factors obtained by Wischmeier and Smith (1978), Roth et al. (1974), the K factors of the rainfall simulator (Breuer, 1988), and the erodibility rate as determined by the waterdrop method (Knobloch, 1987) (Table 2).

Table 2. Order of erodibility of four different soils.†

K (measured)	:	1	2	3	4
K (Wischmeier)	:	4	2	1	3
K (Roth et al.)	:	2	3	1	4
K (drop method)	:	2	1	3	4
K (drop method)	:	2	1	3	4
K (simulator)	:	1	2	3	4

1 = soil with least erosion;
4 = soil with most erosion.

Table 2 shows that there is no relation between the measured K factors and the K factors obtained by Wischmeier and Smith and Roth et al. The waterdrop method indicates the order of erodibility fairly accurately, but the best results were obtained with the rainfall simulator. Even though the order of erodibility was the same with the measured K factor and the rainfall simulator, a statistical correlation was not possible. This could be due to the small plot size (50 x 75 cm), which caused rather big side effects to occur by splash, the drop-size distribution of the rainulator (which forms bigger drops than natural rain), and the absence of the sedimentation processes which occur in bigger plots (Breuer, 1988). All further rainfall-simulator experiments will be undertaken on plots of 6 x 1.8 m.

Influence of treatments on soil loss and runoff

During the experiments on the Alfisol in Alore, Nigeria (Sabel-Koschella, 1988) three treatments were used:

- traditional: a system with staked yam on heaps intercropped with maize, hot pepper, and okra.
- no-till: a rotation of maize and cowpea was chosen; weed control was effected by herbicides.
- plough: same rotation as on no-till; ploughed to 20-cm depth with a disc plough and harrowed.

The influence of the treatments on soil loss and runoff is shown in Table 3.

Table 3. Influence of different treatments on cumulative runoff and soil loss during 1983/84 in Alore, Nigeria.

	Runoff (mm)	Soil loss (t ha^{-1})
Bare-fallow	461	282
Plough	247	13
Traditional	85	6
No-till	77	3

Source: Sabel-Koschella (1988).

The ploughed treatment had the biggest surface flow and soil loss after the bare fallow, followed by the traditional system and the no-till system. The smaller soil loss was accompanied by a higher saturated hydraulic conductivity and an intenser earthworm activity on the no-till and traditional plots.

Influence of treatments on yield

The yield responses on the treatments depended to a considerable extent on the crops. Whereas the maize yield on the Alfisol was significantly lower on the no-till plots than on ploughed plots in 1983/84, the cowpea yield was better on no-till plots (Sabel-Koschella, 1988). The better maize yields on ploughed plots were expected to be reversed to better yields on no-till after a couple of years (Lal, 1979), but yields in 1988 still showed the same behaviour (Figure 5).

The same results were obtained on the Ultisol in Minkoameyos after three years of cropping. The groundnuts yielded better on the no-till plots, whereas maize gave better results on the ploughed treatment.

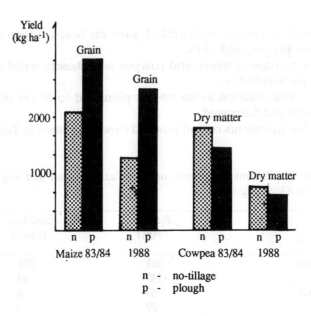

Figure 5. Influence of no-tillage (n) and plough (p) on yields of maize and cowpea 1983/84 and 1988 (kg ha^{-1}) in Alore.

Conclusions

The following conclusions would seem to be in order from the experiments described above:
  (i)   the USLE is not generally applicable on tropical soils;
 (ii)   soil physical tests and simulated rainfall can be of considerable help in distinguishing the erodibility of different soils; and
(iii)   the no-till system is very efficient for diminishing soil loss, but cannot assure the same yields as ploughing with all crops.

References

BERNARD, M., KNOBLOCH, C. and BREUER, J. 1988. *Soil erosion studies in Cameroon and Nigeria.* Lehrstuhl für Bodenkunde. Progress report (February 1988). Munich: Technical University of Munich.

BREUER, J. 1988. Infiltrations - und Bodenabtragsmessungen an tropischen Böden mit einem Regensimulator. M.A. thesis, Lehrstuhl für Bodenkunde, Technical University of Munich. 70p.

DE NEVE, S. 1989. Bijdrage tot de studie van enkele gronden uit Kameroen. M.A. thesis, University of Ghent. 137p.

KNOBLOCH, C. 1987. Erodibilität und K-faktoren ausgewählter Böden Kameruns. M.A. thesis, Lehrstuhl für Bodenkunde, Technical University of Munich.

LAL, R. 1979. *Soil erosion problems on an Alfisol in western Nigeria and their control*, 2d ed. International Institute for Tropical Agriculture. IITA Monograph no. 1. Ibadan, Nigeria: IITA.

ROTH, C.B., DARELL, W.N. and ROMKENS, M.J.M. 1974. *Prediction of subsoil erodibility using chemical, mineralogical and physical parameters*. U.S. Environmental Protection Agency. Report for the Office of Research and Development, Project 15030 HIX. New York: EPA. 111p.

SABEL-KOSCHELLA, U. 1988. Field studies on soil erosion in the southern Guinea Savanna of Western Nigeria. Ph.D. thesis, Lehrstuhl für Bodenkunde, Technical University of Munich. 180p.

WISCHMEIER, W.H. and MANNERING, J.V. 1969. Relation of soil properties to its erodibility. *Soil Science Society of America Proceedings* 33: 131-137.

WISCHMEIER, W.H. and SMITH, D.D. 1978. *Predicting rainfall erosion losses. A guide to conservation planning*. U.S. Department of Agriculture. Agriculture Handbook no. 537. Washington, DC: Government Printing Office.

Section 5: Individual factors and case studies

Drought spells and risk in terms of crop production in Africa

PETER M. AHN[*]

Abstract

The paper discusses the statistical analysis of rainfall and the need to find crops with cropping seasons and water requirements which are adapted to the expected rainfall and its distribution. Even in wetter areas producing perennial crops, short dry periods may seriously reduce yields. In tropical areas of seasonal rainfall further away from the equator, the emphasis is on annual crops adapted to local growing seasons. The rains follow the apparent movement of the sun to give, in areas nearer the equator, two wet seasons separated by two dry seasons, but at about 15° these two wet seasons merge into a single wet season. The latter may be more reliable and productive than the two often poorly defined wet seasons of areas such as the middle belt of West Africa.

The effects of drought depend on the depth and water storage capacity of the soil, the nature of the crop, and the stage it has reached in its growth cycle. The fact that root crops such as yams are less vulnerable than grain crops is reflected in their distribution in the West African yam belt.

The analysis of rainfall patterns in Africa has emphasized the importance of the date of the onset of the rains as affecting both the length of the rainy season and the amount of rain received. This has led to response-farming approaches, which suggest that a farmer can adapt his planting methods and even his crop according to whether the rains start early or late.

[*] IBSRAM Programme Officer for Africa and Coordinator, Vertisol Network, Résidence Gyam, 2ème étage, Angle Bd Clozel-Av Marchand, 04 BP 252, Abidjan 04, Côte d'Ivoire.

Dry spells also influence the soil, particularly the amount of the flush of nutrients produced when a dry soil is wetted, and dry spells, if not damaging to the crop, may therefore have the beneficial effect of increasing nutrients released from humus mineralization.

Introduction

In Africa, as in most of the tropics, the general rule is that the lower the rainfall, the greater its variability. In low-rainfall areas, agronomists and agroclimatologists have to deal with two separate but related problems, low rainfall and low reliability (or high variability) of rainfall.

In East Africa in the 1960s, the East African Agriculture and Forestry Research Organisation (EAAFRO) began a systematic programme of research on the theme 'making the best use of low and unreliable rainfalls', and in the less wet areas of Africa this is still a very major area of concern. Broadly speaking, the approach adopted was (i) to analyse the climate statistically in terms of the probabilities of getting a certain amount of rain in a growing season of a certain length, and (ii) to find or adapt crops whose length of growing season and water requirements fitted in with the statistically analyzed climate. In practice, it proved easier at that time to analyze the climate than to measure with accuracy the exact water requirements of crops. To carry out the needed research on crop water use, lysimeters with accurate weighing devices were constructed to monitor moisture additions and losses, and thus to calculate the water requirements of the crop at different stages of its growth cycle. Research of this type subsequently resulted in a much improved understanding of crop water requirements in the tropics (Doorenbos and Pruitt, 1977; Doorenbos and Kassam, 1979).

What then of those areas of Africa which have moderate or high rainfalls? Here, too, the problem of drought spells and risk is usually present because most tropical rainfalls are seasonal, with one or two wet seasons separated by dry seasons. A period with too much water may be followed by one with too little, and the farmer has to be prepared to deal with both extremes. The fact that even in the drier savanna areas of West Africa crops are often grown on cultivation mounds is partly a way of getting more soil depth and root room, but is also a measure to ensure that the crop is not waterlogged during occasional short periods of heavy rainfall.

Cropping options

Perennial crops

The obvious adaptation of the farmer to the local pattern of rainfall distribution is to plant crops with appropriate lengths of growing season. In very wet forest areas, with rain all or most of the year, the best land use, giving the most protection to the soil, is the planting of perennial tree crops such as oil palm, rubber, coffee, and tea. Even in these wet forest areas, short dry periods may have a marked negative effect on yields: in the wet southwest Côte d'Ivoire, for example, oil palms grow fairly well at Sassandra, but a dry period of one or two months in December-January has a sufficiently marked negative effect on yields for it to be economic to irrigate during this short period. The effects of a short dry spell are not always obvious, since there may be a long time lag between the drought and the resultant lowering of yields. In the case of oil palm, a dry spell will reduce yields 25-30 months later, for example. Tea and rubber are other crops which do best with a high well-distributed rainfall, with no dry months.

Since high rainfalls all the year result in soil leaching and are associated with acid soils, it follows that crops such as tea and rubber which do well in such areas are also adapted to these soil conditions. Tea actually prefers acid soils, being tolerant of high aluminium levels, while rubber is an 'air and water' plant which puts very low demands on soil nutrients and exports very minor amounts in the latex. Cocoa, in contrast, is a relatively demanding crop, doing best on fertile soils with a near-neutral pH, well supplied with divalent bases, and exporting relatively large amounts of nutrients in the beans. In West Africa, the cocoa belt is found in less-wet forest areas of about 1500 mm of rainfall associated with two dry seasons, where the natural vegetation is semideciduous forest rather than rainforest. The main dry season (in November-January) is useful for the drying of the cocoa beans, and one of the problems encountered by would-be cocoa growers in equatorial areas without marked dry seasons (such as the central forest areas of Zaire) has been that of adequate drying of the beans after fermentation.

Annual crops

Most of Africa away from the equatorial belt experiences seasonal rainfall, with rain following the apparent movement of the sun. In lower latitudes the sun passes overhead twice a year, and there are two wet seasons separated by two dry seasons, with the 'summer' dry season generally being shorter than the 'winter' one.

Further away from the equator the short dry season disappears, and the two wet seasons merge into one wet season whose length is closely related to latitude. In practice, the growing season varies from about 5-6 months at about 15°, but becomes progressively shorter going away from the equator until the practical limit of the cultivation of annual crops in well-drained soils without supplemental irrigation is met with a growing season of about two months. In terms of crop possibilities, this means that long-season maize can be grown in areas with longer growing seasons, and shorter-season maize with sorghum and millet can be grown as the growing seasons fall to 3-4 months. Finally, as in southern Niger, a zone is reached where even sorghum cannot be grown, and the only cereal crop is millet with a growing season of about 65-70 days. There is no other grain crop with which the millet can be rotated, though it may be supplemented by some cowpeas and peanuts.

The association of rainfall distribution with latitude is particularly regular and well marked in West Africa, and rainfall belts are in turn closely related to vegetation and soil belts which run east-west, more or less parallel to the coast, so that a traveller moving inland crosses a succession of climate-vegetation-soil zones, each with its adapted agriculture. In East Africa, this pattern related to the apparent movement of the sun is more complex and less regular, being modified by the different direction of the coastline and the presence of high mountain masses. The result is a distribution of rainfalls and growing seasons which is more complicated and less easy to understand, and is much modified by the temperature effects of altitude - with some marked climatic anomalies such as the desert in northern Kenya at latitudes where there is normally a wet season in other parts of the tropics.

Both in West Africa and in East Africa, statistical analysis of rainfall reliability, and the effective length of the growing season that can be expected with any given degree of confidence, have shown the importance of the date of onset of the rains. In the exceptionally dry year of 1973 in the Sahel, Kowal (pers. comm.) concluded that up to a point north of Samaru, northern Nigeria, the date of the start of the wet season and its length were normal, but that north of Samaru the rains started later than normal and were correspondingly short. In other words, the rain-bearing air masses of the Inter-Tropical Convergence Zone (ITCZ) failed to "follow the sun" to the extent that they normally do, arrived late, and failed to go as far north as usual, before retreating on schedule.

Stewart (1988b) in an analysis of rainfall at Niamey, Niger, found evidence for a climatic shift from 1971 onward, with better rainfall in 1954-1970 (average 603 mm) than in 1971-83 (average down to 504 mm), but noted that throughout the period the date of onset of the rains is related both to the amount of rain received and the length of the growing season. The date of onset varies considerably from year to year. In the 30 years to 1983, it occurred between 28 May and 21 July, a span of 55 days, with a median date of onset of 20 June and a median duration of 99 days. The median cropping-season rainfall in those 30 years was 494 mm, but the range was from 275 mm to 771 mm, and the cropping season duration varied from 71 to 154 days. The author's contention, however, is that cropping seasons at Niamey can usefully be divided into two groups, with early onset and late onset. The first group, which occurred in 14 of the 30 years considered, consisted of cropping seasons which began before 19 June. This group had a median date of onset of 10 June, and a median rainfall of 590 mm received in a growing season of 113 days. The second group had a late onset, after 20 June, with a median date of onset of 6 July, nearly a month later than the first group, and a median rainfall of only 351 mm falling in 82 days.

In East Africa, the same author (Stewart, 1988a) analyzed the rainfall at Katumani, in the semiarid areas of Kenya, and also found a close relationship between the date of onset of the main rainy season and the amount of rain and length of the growing season. As shown in Figure 1, good maize crops were associated with rains which began early, before about 5 February, and with rainfalls above about 350 mm, whereas rainfall seasons beginning after 11 March were generally unsuited to maize production.

Stewart and his coworkers (e.g. Stewart and Kanshasha, 1984) contend that an analysis of rainfall patterns can help the farmer to adapt his operations to each particular year. If the rains begin after a date which analysis of past years has shown to form a useful divide between early and late onset, associated with rainfalls which are higher in the first case and lower in the second, then the farmer can adapt his crops and his methods of planting them accordingly. If the rains start early, this might suggest a very good chance of a successful maize crop, but a late onset of rains might suggest that the farmer should plant with wider spacings or change his crop to a less demanding one such as sorghum.

The effects of drought on plant growth

The effects of dry spells occurring within the growing season on plant growth depend on a range of soil and plant factors. These include:
- the length of the dry spell;
- the depth and water storage capacity of the soil, and the amount of water stored at the beginning of the dry spell;
- the plant root system and the volume of soil it exploits; and

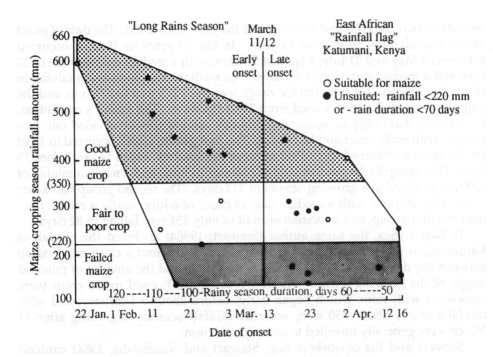

Note: The circles indicate the date of the onset of the long rains (bottom line) and the total amount of rain received in the season (left-hand scale). Open circles and black circles indicate conditions suitable or unsuitable for maize production. Rains which began before 11 March generally gave enough rainfall to grow maize (over 220 mm), but onsets after this date were generally unsuitable because the duration of the season was too short.

Figure 1. Rainfall flag for Katumani National Dryland Farming Research Station, Machakos, Kenya, 1957-83 (adapted from Stewart [1988a]).

-   the nature of the crop and the phenological stage of growth it has reached, and hence its water demand.

Normally the water demand of an annual crop such as sorghum or maize increases to a peak and then falls off as the grain matures, but other crops such as cotton have an indeterminate growing season and continue to take up water, produce flowers, and set seed as long as the water supply allows this. A dry spell at an early stage of growth might do more harm than one at a later phenological stage, partly because the plant is more vulnerable, and partly because its root system is not yet fully in place.

Some plants are physiologically better adapted to drought spells than others. While sorghum can recover from a partial wilt, maize is much less able to do so because of damage to the stomatal mechanism. Drought resistance is helped by deep rooting systems, and various devices which reduce transpiration and water demand, such as small leaves, fewer stomata, waxy leaf surfaces, and even leaves hanging with their edges to the sun.

In general, root crops survive dry spells better than grain crops, a fact which helps to account for the spread of cassava growing in Africa away from its traditional forest-zone areas, and for the distribution of yams (*Dioscorea* spp.) in West Africa, as discussed in the following section on the 'middle belt'.

The West African 'middle belt' of relatively unreliable rainfall

In West African yam-producing countries, principally Côte d'Ivoire, Ghana, and Nigeria, we can distinguish three main climatic belts:

- A wet southern forest belt, with a relatively high rainfall, well enough distributed to support a closed forest vegetation. This can be subdivided into:
  - the wet evergreen forest, generally with rainfalls over 1600-1700 mm, often with a very marked peak in May-June, short dry seasons, and leached and acid soils;
  - the more extensive semideciduous forest areas of moderate rainfall, generally in the range 1200-1600 mm, with more marked dry seasons, particularly in November-January.
- A transitional central area, often termed 'the middle belt' in Ghana and Nigeria, with a moderate rainfall, generally in the range 800-1200 mm, which falls in two wet seasons separated by a poorly defined dry spell in August. The dates of the start of the rains, the occurrence of the dry spell in August, and the date of the end of the rains all vary considerably from year to year. In Côte d'Ivoire, Ghana, and Nigeria, the 'middle belt' is marked by a low population. Although the causes of this low population have been much debated, it seems clear that a major determinant is the unreliability of the moderate rainfall spread out over two ill-defined seasons.
- A savanna area to the north of the middle belt, where a similar total rainfall falls in a shorter but better-defined and more reliable single wet season. Here agriculture is less risky, and population densities higher, or very much higher, than in the relatively empty middle belt.

In central Côte d'Ivoire in the latitudes of Bouaké and Katiola, immediately to the south of the low-populated middle belt, there is a forest-savanna mosaic. Forest is found on the more clay soils of the summits with better water-holding capacities, but savanna (perpetuated by annual burning of the grass vegetation in the dry season) occurs on more sandy soils with lower water-holding capacities developed in local colluvium of the lower slopes. In this area, the rainfall has the typical middle-belt features of unreliability, and the short dry season in August is absent in about two years out of five. In the first wet season, the main crops are maize and peanuts. In the second season the main cash crop is cotton, and some maize may also be attempted. Yields of all of these crops fluctuate markedly from year to year according to rainfall amount and distribution. The production of yams, in contrast, is relatively steady. These are planted at the beginning of the first wet season, and harvested, according to variety, later in the year. The yam tubers are better adapted to survive dry spells than are the aboveground crops such as maize. The result is that although yams are also grown in the forest belt to the south and in the single-season savanna areas to the north, it is this central area of relatively unreliable rainfall which forms the main yam belt.

Traditional yam varieties, such as Krengle, are relatively demanding as regards soil fertility, and are planted immediately after clearing of fallow land. Since a typical farming family of two to four adults normally clears only about half a hectare of new land a year (Ahn and Guillonneau, 1987), the production of traditional yam varieties was restricted accordingly, and farmers were able to market their produce at favourable prices. The recent introduction by the Institut des Savanes (the Ivoirian national agricultural research organization) of less-demanding yam varieties from the West Indies has dramatically changed the traditional pattern of production. The best known of the new varieties is the Florido, which not only grows on poorer soils so that it is not limited to newly cleared land, and can even follow a preceding yam crop, but also produces much higher tonnages per hectare than the traditional varieties. The result is that the new varieties can be produced in larger quantities and sold much more cheaply than the traditional ones, and there is a threat of overproduction. This example from central Côte d'Ivoire illustrates the importance of finding a crop, and even a variety of a crop, adapted to particular soil and climatic conditions.

Drought spells and the soil

Apart from the direct effects of a dry spell on plant growth, dry spells exert indirect effects through their action on the soil. A soil which is subject throughout the year to heavy rainfalls which exceed potential evapo-transpiration is also subject to the effects of downward moving water through the soil profile and the washing out, or leaching, of substances in solution or suspension. Leaching occurs whenever a soil profile already at field capacity receives further rain. Since a soil at field capacity is already, by definition, holding as much water as it can against the forces of gravity, further rain must move down and out of the profile. In the wet evergreen forest of West Africa referred to above, leaching appears to occur mainly during the peak rainfall period in May and June when a metre or more of rain may be received in a six-week period when potential evapotranspiration is hardly likely to exceed 150 mm. The effects of leaching are mainly to remove the divalent bases calcium and magnesium from the soil, and to make the soil more acid. In the semideciduous forest areas of West Africa described above, and in the savanna areas of the middle belt and to the north of the middle belt in West Africa (see above), there are still short periods when rainfall exceeds potential evapotranspiration and leaching from upper horizons to lower horizons may occur, but during dry spells within the wet seasons and during the increasingly long dry seasons the opposite movement occurs as one goes·north, i.e. water moving upwards and evaporating at or near the surface enriches the upper soil horizons with whatever is carried upwards in the soil solution. The result of this movement is that topsoil pH values are commonly only slightly acid to neutral.

A potentially very important effect of dry spells on soil conditions and plant growth which requires further study is the effect of the wetting of a previously dry soil on humus mineralization. It is well known that the first rains of the wet season produce a flush of humus mineralization and the release of a range of nutrients, including nitrogen. The effects of the wetting of a dry soil were studied by Birch in East Africa and published in a series of papers in *Plant and Soil* (Birch, 1959, 1960). Birch concluded that the magnitude of the flush was proportional to the length of time that the soil had been dry. The main flush thus occurred with the first rains after a long dry season, but succeeding wetting and drying produced a series of minor flushes. The implications of this for crops grown in the tropics which rely mainly. on humus decomposition for their nutrient supply have not received the attention they deserve, since it appears that a soil which is more or less continuously wet will not produce the flushes of nutrients that would have resulted from the soil if the topsoil had been through a succession of drying out and wetting cycles. Thus if a crop has a deep root system and can survive topsoil drying out, the drying and then wetting of the topsoil might benefit the plant by increasing humus mineralization.

Although the emphasis has traditionally been on the nitrogen flush, it should be remembered that when humus mineralizes a very wide range of nutrients is produced, and in balanced amounts. Short dry spells not severe enough to limit plant growth may in some cases therefore be beneficial in this respect. Dry spells may also be beneficial if they force plants to extend their rooting systems downwards. The crops most vulnerable to dry spells are those which have enjoyed a good moisture supply in the topsoil and have developed only a shallow rooting system, which they are unable to extend downwards fast enough when conditions change and the soil progressively dries out.

References

AHN, P.M. and GUILLONNEAU, A. 1987. *Le terroir de Katiola : Logbonou.* Centre Vivriers. Rapport Technique. Bouaké, Côte d'Ivoire: IDESSA.

BIRCH, H.F. 1959. Further observations on humus decomposition and nitrification. *Plant and Soil* 11: 262-292.

BIRCH, H.F. 1960. Nitrification in soils after different periods of dryness. *Plant and Soil* 12: 81-96.

DOORENBOS, J. and PRUITT, W.O. 1977. *Crop water requirements.* FAO Irrigation and Drainage Paper no. 24. Rome: FAO.

DOORENBOS, J. and KASSAM, A.H. 1979. *Yield response to water.* FAO Irrigation and Drainage Paper no. 33. Rome: FAO.

STEWART, J.I. 1988a. East Africa: Kenya. In: *Response farming in rainfed agriculture,* ch. 12. Davis, California: The Wharf Foundation Press.

STEWART, J.I. 1988b. East Africa: Niger. In: *Response farming in rainfed agriculture,* ch. 13. Davis, California: The Wharf Foundation Press.

STEWART, J.I. and KASHASHA, D.A.R. 1984. Rainfall criteria to enable response farming through crop-based climate analysis. *East African Agricultural and Forestry Journal* 44: 58-79.

L'examen du profil cultural : un outil pour mieux comprendre le comportement du sol soumis à des travaux aratoires

PHILIPPE DE BLIC*

Résumé

La mise au point des techniques culturales nécessite une évaluation fiable de l'état structural du sol et de son influence sur le comportement des végétaux cultivés. L'anisotropie poussée qui caractérise les horizons supérieurs travaillés conduit souvent à privilégier une approche morphologique prenant en compte les différents niveaux d'organisation macrostructurale du sol en relation avec la colonisation racinaire.

Quelques exemples de description du profil cultural montrent que ce type d'approche peut constituer un outil efficace dans l'étude des modifications de l'état structural du sol sous l'action des travaux aratoires.

Abstract

Study of the cultivation profile as a tool
for a better understanding of soil behaviour in relation to tillage

The development of cultivation techniques requires a reliable evaluation of the soil structure, and of its influence on crop performance. The marked anisotropic character of the upper ploughed horizons often leads to a preference for a morphological approach which takes into account the different levels of macrostructural organization of the soil in relation to the root system.

A few examples of cultivation profiles are given, which show that this approach can be an effective means of studying changes in soil structure after tillage operations.

---

*   Centre ORSTOM de Montpellier.

Introduction

L'un des objectifs majeurs que l'on assigne au travail du sol est la création et le maintien d'un état structural favorisant l'installation et le bon fonctionnement du sytème racinaire des végétaux cultivés.

Cela veut dire que, dans un contexte agricole donné, la mise au point des techniques culturales repose largement sur une évaluation fiable de l'état structural. Or, compte tenu de l'hétérogénéité poussée de la partie supérieure des sols cultivés, on est souvent conduit à multiplier les sites de mesure pour définir un état structural moyen qui, finalement, n'a pas grande signification vis-à-vis du fonctionnement des racines.

C'est ce qui a conduit Hénin et ses collaborateurs à privilégier une approche descriptive et à développer la méthode dite du profil cultural. Cette méthode est fréquemment utilisée dans l'établissement de diagnostics agronomiques reliant des comportements végétaux à des caractères morphologiques du sol; c'est alors un moyen privilégié de dialoggue et de conseil aux agriculteurs.

L'examen du profil cultural est aussi un outil de recherche dans les études ayant trait, d'une part aux modifications de l'état physique du sol sous l'action des techniques culturales (de Blic, 1978, 1987; Manichon, 1982; Monnier, 1986; Papy, 1984), d'autre part aux relations entre état du sol et comportement du peuplement végétal (Tardieu et Manichon, 1986; Tardieu, 1988a, b).

Nous nous attacherons dans cet exposé, tout d'abord à définir le profil cultural en le replaçant dans le cadre plus général de l'organisation du sol, ensuite à présenter les critères de description du profil, à montrer enfin, à l'aide de quelques exemples, comment on peut combiner observations morphologiques et mesures physiques pour aider à la mise au point des technques culturales.

Organisation du sol

Quatre niveaux peuvent être reconnus dans l'organisation macro-structural du sol (figure 1).
- **Elément structural.** Solide géométrique correspondant à l'architecture la plus apparente lors de l'observation du profil. Chaque élément structural est séparé de ses voisins par des surfaces de moindre résistance ou par des vides. Ce terme regroupe à la fois les éléments pédologiques naturels et les mottes et fragments créés par les outils agricoles.
- **Volume structural homogène ou phase structruale.** Volume de sol se distinguant des volumes adjacents par son état structural, c'est-à-dire par la manière dont sont assemblés ses divers constituants et par les conséquences de cet assemblage (porosité, compacité). C'est l'unité de base de la description des sols cultivés.
- **Horizon.** Unité majeure d'organisation formant très généralement une couche parallèle à la surface du sol et pouvant présenter une grande variabilité latérale de l'état structural.

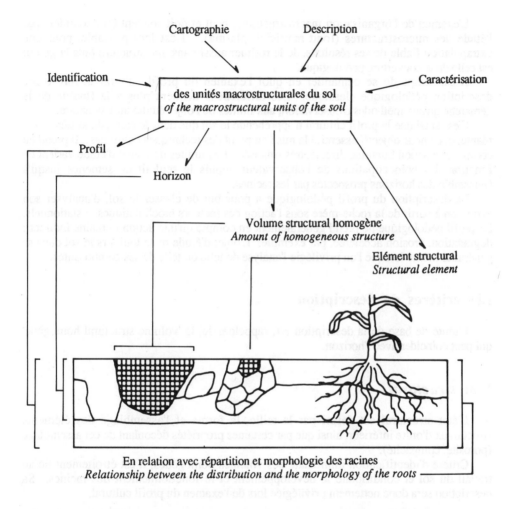

Figure 1. **L'analyse du profil cultural associe étroitement l'approche morphologique et les mesures
in situ.**

*Analysis of the cultivation profile closely associates the methodological approach with
the* in situ *physical measurements.*

- **Profil cultural.** Hénin et ses collaborateurs (Hénin *et al.*, 1969) le définissent
comme "l'ensemble constitué par la succession des couches de terre individualisées
par l'entervention des instruments de culture, les racines des végétaux et les facteurs
naturels réagissant à ces actions".

L'examen de l'organisation macrostructurale peut et doit souvent être complété par l'étude des microstructures (fond matriciel, plasma). Il est indispensable, pour une extrapolation fiable de ses résultats, de le resituer aussi dans une structure plus large qui est celle de la couverture pédologique.

Il est temps de se demander en quoi l'examen du profil cultural diffère d'une description pédologique classique. Les différences tiennent plus à la finalité de la démarche qu'aux méthodes d'observation, aux limites de l'objet étudié qu'à sa nature.

C'est ainsi que le profil cultural n'appréhende le sol que dans le cadre du système sol-plante, avec pour objectif essentiel la mise au point des techniques culturales. Il prend en compte une action humaine directe, très finalisée. Les limites de l'objet d'étude varient en fonction des préoccupations de l'observateur depuis le seul lit de semence jusqu'à l'ensemble des horizons prospectés par les racines.

La description du profil pédologique a pour but de classer le sol, d'analyser son évolution à partir de la roche-mère sous l'action des facteurs bioclimatiques et stationnels. Le profil pédologique ne prend généralement en compte qu'une action humaine indirect : dégradation, érosion accélérée, par exemple. L'objet d'étude reste toujours le sol dans sa généralité même lorsque l'on privilégie l'analyse de telle ou telle de ses composantes.

Les critères de description

L'unité de base de la description est, rappelons-le, le 'volume structural homogène' qui peut coïncider avec l'horizon.

Etat structural

L'état structural est défini par la taille, la forme et l'assemblage des éléments structuraux d'ordre inférieur, ainsi que par certaines propriétés découlant de cet assemblage (porosité, compacité).

Critère d'identification immédiate des volumes homogènes, il est étroitement lié au travail du sol et conditionne le développement et le fonctionnement des racines. Sa description sera donc nettement privilégiée lors de l'examen du profil cultural.

Les éléments structuraux

Lorsque plusieurs éléments structuraux sont assemblés dans un volume homogène il convient de préciser :
- leur taille dominante,
- leur forme
  - soit par une description prenant en compte la planéité des faces, l'émoussé des arêtes, l'allongement préférentiel; ce type d'approche convient bien aux mottes créées par les outils
  - soit par une désignation synthétique empruntée au vocabulaire pédologique habituel (grumeau, polyèdre, prisme, etc.),

- leur état interne, c'est-à-dire leur structure propre qui peut être massive, massive-fissurée, fragmentaire - c'est une variable très importante dans la description des mottes.

Mode d'assemblage

On distingue les types suivants :
- assemblage particulaire,
- assemblage massif avec ou sans amorce de fissures,
- assemblage fragmentaire plus ou moins nettement exprimé en fonction du degré de séparation des éléments qui le constituent.

Porosité

On décrit séparément l'espace poral interne des éléments structuraux et l'importance des vides qui les séparent; des mesures de densité apparente réalisées tant au niveau des éléments structuraux individuels que de leurs assemblages peuvent préciser les descriptions.

Compacité

Elle est estimée au niveau global du volume homogène par la facilité plus ou moins grande que l'on éprouve à enfoncer un objet pointu dans la paroi verticale de la fosse. On note en général une bonne corrélation avec les valeurs de densité apparente.

Les racines

L'examen du système racinaire permet d'apprécier l'influence de l'état structural sur le comportement des plantes cultivées.

On peut effectuer une cartographie détaillée des racines sur des plans verticaux ou horizontaux du sol : leur distribution est ensuite analysée et mise en relation avec les caractéristiques structurales.

On se contente le plus souvent de noter l'abondance des racines, leur taille, leur répartition, leurs relations avec les éléments structuraux : cette description est alors réalisée à l'échelle du volume homogène ou de l'horizon.

Autres éléments de description

Ce sont les variables pédologiques classiques telles que couleur, texture, propriétés mécaniques, matières organiques libres, éléments grossiers, etc.

Transitions entre volumes, limites culturales

On évalue les contraintes qu'elles peuvent représenter vis-à-vis de la circulation de l'air et de l'eau, de la colonisation racinaire.

Réalisation pratique

On effectue les opérations suivantes :
- choix d'une station représentatrice du milieu ou du problème étudié;
- implantation d'une fosse d'observation;
- identification, cartographie et description, sur la face verticale de la fosse, des volumes structuraux homogènes;
- on peut alors faire la somme des surfaces occupées par chaque type de structure ou classe de compacité et effectuer ainsi une quantification des états structuraux présents dans un horizon ou une tranche d'épaisseur donnée.

Inventaire et étude de contraintes - exemple de la Côte d'Ivoire

Lorsque l'on veut recenser et hiérarchiser les contraintes-sol dans un système de culture, il est souvent intéressant de procéder par voie d'enquête. L'analyse du profil cultural est alors bien adaptée à ce type de démarche.

C'est l'approche que nous avons utilisée dans les savanes préforestières du Centre de la Côte d'Ivoire pour aborder l'étude des modifications subies par les sols à la suite du défrichement et de mise en culture mécanisée. Précisons qu'il s'agit de sols ferrallitiques moyennement désaturés appartenant aux groupes typiques et remaniés.

Notre démarche a été la suivante (de Blic, 1978) :

Inventaire des contraintes pédologiques

L'étude morphologique comparée de sols évoluant, les uns en milieu cultivé, les autres sous diverses conditions d'environnement naturel (figure 2) a permis de mettre en évidence un certain nombre de caractères structuraux susceptibles de constituer des contraintes majeures pour les cultures.

Certains de ces caractères, hérités du milieu naturel, tiennent à la forte différenciation verticale des sols.

D'autres sont liés à diverses propriétés des sols mais n'apparaissent que sous l'effet de la mise en culture. Il s'agit en particulier des états structuraux massifs et très compacts que l'on peut observer dans les horizons travaillés. Ces tassements s'accompagnent souvent d'une forte hétérogénéité structurale.

Il est clair, dans le cas présent, que les opérations de travail du sol n'ont pas rempli leur rôle d'optimisation du profil cultural.

Etudes du comportement mécanique

Sur un certain nombre de matériaux pédologiques sélectionnés de façon à couvrir toute la gamme régionale de textures et matières organiques, nous avons étudié le comportement au compactage par la méthode de Proctor.

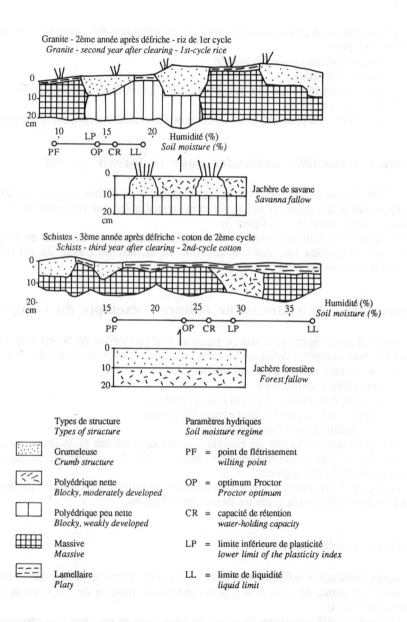

Types de structure
Types of structure

Grumeleuse
Crumb structure

Polyédrique nette
Blocky, moderately developed

Polyédrique peu nette
Blocky, weakly developed

Massive
Massive

Lamellaire
Platy

Paramètres hydriques
Soil moisture regime

PF = point de flétrissement
     wilting point

OP = optimum Proctor
     Proctor optimum

CR = capacité de rétention
     water-holding capacity

LP = limite inférieure de plasticité
     lower limit of the plasticity index

LL = limite de liquidité
     liquid limit

Figure 2. Exemples de transformations morphologiques des sols après mise en culture (Côte d'Ivoire).

Examples of soil morphological changes after cultivation (Côte d'Ivoire).

La caractérisation physico-méchanique des matériaux a été complétée par les déterminations suivantes :
- courbe caractéristique d'humidité,
- limites d'Atterberg,
- capacité de rétention par la méthode de Feodoroff et Bétremieux (Feodoroff et Betremieux, 1964),
- résistance à la pénétration en fonction de l'humidité sur éprouvettes compactées.

Gammes d'humidité optimales pour le travail du sol

A l'aide des paramètres précédents, il a été possible de définir des gammes d'humidité correspondant à des états du sol facilement identifiables sur le terrain (très humide, humide, ressuyé, frais, sec, cf. figure 3).

Ces gammes d'humidité exprimées de façon pondérale ou traduites en hauteurs de pluie, peuvent faciliter les prises de décision en matière de travail du sol (choix des créneaux d'intervention et du type d'outil).

Comparaison de systèmes de culture - exemple du Congo

Un essai a été récemment mis en place au Congo (Vallée du Niari) sous l'égide de l'IBSRAM pour comparer l'effet sur les caractéristiques et la productivité des sols de quatre systèmes de culture, dont trois à base manioc :
- un système traditionnel,
- un système amélioré à bas niveau d'intrants,
- un système amélioré à haut niveau d'intrants,
- un système intensif maïs-soja.

Des profils culturaux ont été examinés en fin de première et de deuxième année de culture. Sur le plan pédologique, nous avons à faire à des sols ferrallitiques jaunes fortement désaturés très argileux.

La figure 4 présente trois profils culturaux examinés sous culture de manioc, en deuxième année, dans trois systèmes de cultures différents.

Système traditionnel

Après rabattage et brûlage de la jachère, le sol est gratté superficiellement à la houe, le manioc est planté en tous sens à la densité approximative de 7.000 pieds ha^{-1}, puis sarclé à la demande.

Le mode de différenciation du profil, les caractères de structure et compacité, ne sont pratiquement pas modifiés par rapport au témoin observé sous jachère naturelle.

Le travail à la houe ne reprend que les 6 à 8 centimètres supérieurs (horizon A 11) où restent localisés les tubercules de manioc.

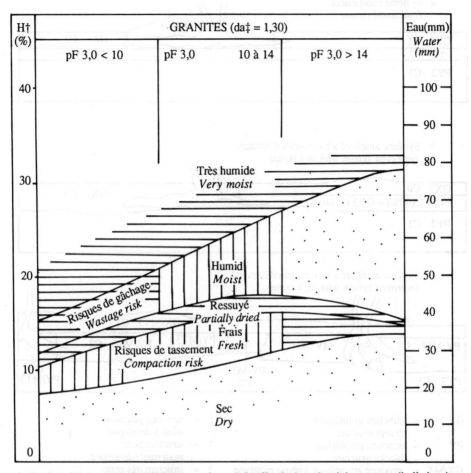

† H = humidité pondérale (%)/*moisture by weight (%)*; ‡ da = densité apparente/*bulk density.*

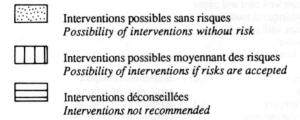

Interventions possibles sans risques
Possibility of interventions without risk

Interventions possibles moyennant des risques
Possibility of interventions if risks are accepted

Interventions déconseillées
Interventions not recommended

Figure 3. Etats d'humidité et travail du sol.

Moisture status and tillage.

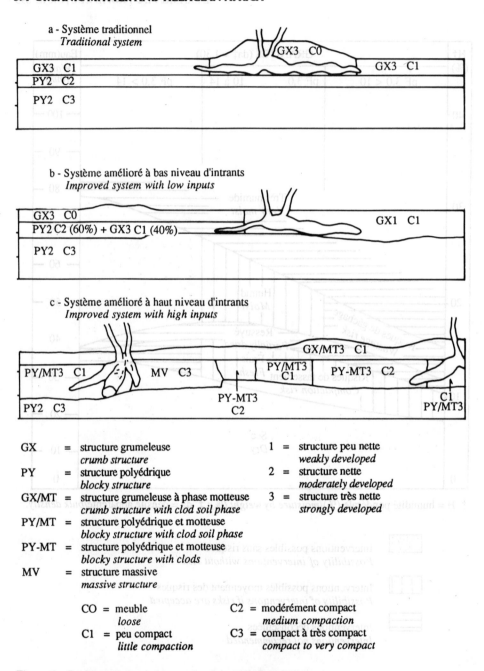

a - Système traditionnel
   Traditional system

b - Système amélioré à bas niveau d'intrants
   Improved system with low inputs

c - Système amélioré à haut niveau d'intrants
   Improved system with high inputs

GX	=	structure grumeleuse	1	=	structure peu nette
		crumb structure			*weakly developed*
PY	=	structure polyédrique	2	=	structure nette
		blocky structure			*moderately developed*
GX/MT	=	structure grumeleuse à phase motteuse	3	=	structure très nette
		crumb structure with clod soil phase			*strongly developed*
PY/MT	=	structure polyédrique et motteuse			
		blocky structure with clod soil phase			
PY-MT	=	structure polyédrique et motteuse			
		blocky structure with clods			
MV	=	structure massive			
		massive structure			

CO = meuble   C2 = modérément compact
     loose   *medium compaction*
C1 = peu compact   C3 = compact à très compact
     little compaction *compact to very compact*

Figure 4. Profils culturaux sous manioc (essai IBSRAM, Congo).

Cultivation profiles under cassava (IBSRAM experiment, Congo).

Système amélioré à bas niveau d'intrants

Un apport de calcaire broyé précède le grattage du sol à la houe. Le manioc est planté en ligne à la densité de 6.250 pieds ha^{-1}, puis sarclé manuellement.

La différenciation verticale du profil est toujours très semblable au témoin. S'y surimpose, dans les 10 à 15 centimètres supérieurs, une différenciation latérale assez variable, souvent suivant des volumes structuraux homogènes étroitement interpénétrés.

Le travail du sol est un peu plus profond qu'en système traditionnel, peut-être en raison de la plantation en ligne.

Système amélioré à haut niveau d'intrants

Un apport de calcaire broyé précède le labour avec une charrue à soc, suivi d'une reprise superficielle aux disques. Le manioc est planté en ligne, à la densité de 6.250 pieds ha^{-1}, puis sarclé manuellement. Les techniques culturales comportent une fumure minérale et des traitements phyto-sanitaires.

Deux horizons travaillés se différencient à la partie supérieure du sol :
- Un horizon de façons superficielles Ap1 bien individualisé, épais de 6 à 8 cm, assez homogène latéralement.
- Un horizon de labour Ap2, caractérisé par la juxtaposition latérale de volumes pédologiques bien contrastés quant à leur état structural et leur compacité.

On observe, par rapport aux deux profils précédents, d'une part une hétérogénéisation structurale avec prédominance nette des structures motteuses et apparition de structures massives, d'autre part une décompaction des 20 centimètres supérieurs avec localisation plus profonde des tubercules de manioc.

La figure 5 présente les histogrammes de répartition des types de structure pour chacun des trois systèmes de culture considérés (regroupement de 4 profils culturaux par système).

Evolution des caractères morphologiques - exemple du Congo

Dans le cadre du même essai IBSRAM-Congo, nous prendrons maintenant l'exemple de la parcelle n°7 du système amélioré à haut niveau d'intrants; son histoire culturelle a été la suivante :

1987-88 - Arachide en première saison après labour à la charrue à soc et reprise aux disques. Maïs engrais vert en deuxième saison, précédé d'un travail aux disques et suivi d'un enfouissement.

1988-89 - Maïs grain en première saison, puis maïs engrais vert en deuxième saison. L'installation des deux cultures a été précédée d'un travail superficiel aux disques.

A la partie supérieure des sols se différencient trois horizons (figure 6) :
- Un horizon Ap1 fréquemment repris par les façons superficielles (5 passages de disques entre décembre 1987 et mai 1989) s'individualise nettement dans les 8 centimètres supérieurs du profil, et ce dès le premier cycle de culture.

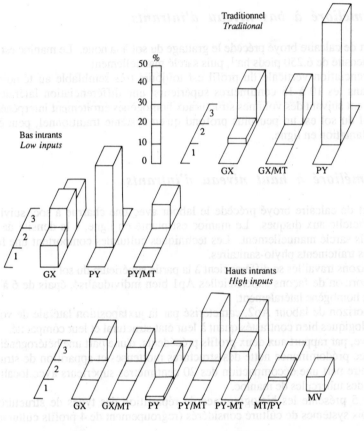

Type de structure
Type of structure

GX = grumeleuse
crumb structure
GX/MT = grumeleuse phase motteuse
crumb structure with clod soil phase
PY = polyédrique
blocky
PY/MT = polyédrique phase motteuse
blocky with clod soil phase
PY-MT = polyédrique et motteuse
blocky with clods
MT/PY = motteuse phase polyédrique
clods with blocky structure
MV = massive
massive

Netteté de la structure
Grade of the structure

1 = peu nette
weakly developed
2 = nette
moderately developed
3 = très nette
strongly developed

Figure 5. Structure des 20 cm supérieurs du sol sous culture de manioc.

Structure of the top 20 cm of the soil under cassava cultivation.

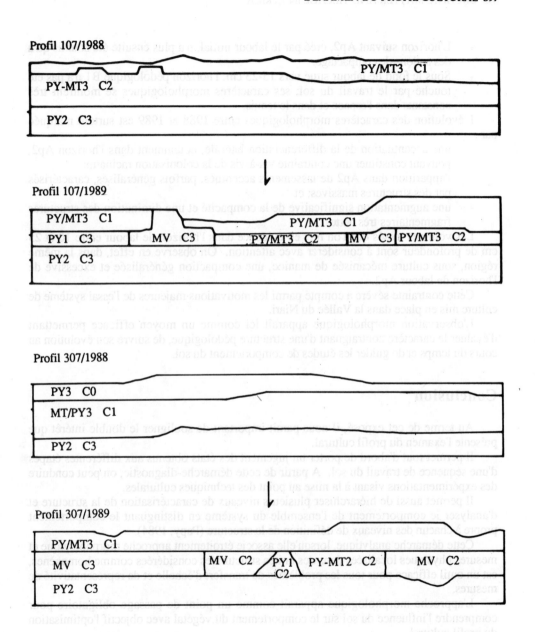

Figure 6. Exemples d'évolution de profils culturaux en culture mécanisée (essai IBSRAM, Congo).

Examples of the development of a cultivation profile using mechanized cultivation (IBSRAM experiment, Congo).

- L'horizon suivant Ap2, créé par le labour initial, n'a plus ensuite été soumis qu'à des actions de compactage.
- Sous le fond de labour situé vers 15-25 cm, l'horizon pédologique B1 n'a pas été touché par le travail du sol; ses caractères morphologiques se montrent très constants dans l'espace et dans le temps.

L'évolution des caractères morphologiques entre 1988 et 1989 est surtout marquée par :

- une accentuation de la différenciation latérale, notamment dans l'horizon Ap2, pouvant constituer une contrainte vis-à-vis de la colonisation racinaire;
- l'apparition dans Ap2 de tassements accentués, parfois généralisés, caractérisés par des structures massives; et
- une augmentation significative de la compacité et une diminution des structures fragmentaires très nettes.

Les volumes tassés que l'on met en évidence dans l'horizon de labour entre 10 et 25 cm de profondeur sont à considérer avec attention. On observe en effet, dans la même région, sous culture mécanisée de manioc, une compaction généralisée et excessive de l'horizon de labour Ap2.

Cette contrainte sévère a compté parmi les motivations majeures de l'essai système de culture mis en place dans la Vallée du Niari.

L'observation morphologique apparaît ici comme un moyen efficace permettant d'évaluer le caractère contraignant d'une structure pédologique, de suivre son évolution au cours du temps et de guider les études de comportement du sol.

Conclusion

Au terme de cet exposé, il nous paraît important de souligner le double intérêt que présente l'examen du profil cultural.

Il permet tout d'abord de porter un jugement des états obtenus aux différentes étapes d'une séquence de travail du sol. A partir de cette démarche-diagnostic, on peut conduire des expérimentations visant à la mise au point des techniques culturales.

Il permet aussi de hiérarchiser plusieurs niveaux de caractérisation de la structure et d'analyser le comportement de l'ensemble du système en distinguant le comportement propre à chacun des niveaux de définition de la structure (Papy, 1984).

Cette démarche analytique, lorsqu'elle associe étroitement approche morphologique et mesures physiques localisées dans des unités structurales considérées comme homogènes, est un outil efficace pour tous les problèmes de transfert d'échelle et de représentativité de mesures.

L'approche morphologique apparaît comme un point de passage obligatoire pour comprendre l'influence du sol sur le comportement du végétal avec objectif l'optimisation du profil cultural.

Bibliographie

DE BLIC P. 1978. *Morphologie et comportement mécanique des sols de la Région Centre en culture semi-mécanisée.* Centre d'Adiopodoumé. Abidjan, Côte d'Ivoire : AVB/ORSTOM. 63p. + annexes.

DE BLIC P. 1987. Analysis of a cultivation profile under sugarcane: methodology and results. In: *Land development and management of acid soils in Africa*, 275-285. IBSRAM Proceedings no. 7. Bangkok: IBSRAM.

FEODOROFF, A. et BETREMIEUX, R. 1964. Une méthode de laboratoire pour la détermination de la capacité au champ. *Science du Sol* 2: 109-118.

HENIN, S. GRAS, R. et MONNIER, G. 1969. *Le profil cultural.* 2è édition. Paris : Mason. 332p.

MANICHON, H. 1982. L'action des outils sur le sol : appréciation de leurs effets par la méthode du profil cultural. *Science du sol* 3: 203-219.

MONNIER, G. 1986. Aspects physiques actuels du travail du sol. *Revista de Agronomia* 20(2-3): 75-84.

PAPY, F. 1984. Comportement du sol sous l'action des façons de reprise d'un labour au printemps. Effet des conditions climatiques et de l'état structural. Thèse Doctorat Ingénieur, Institut National Agronomique, Paris-Grignon. 232p.

TARDIEU, F. et MANICHON, H. 1986. Caractérisation en tant que capteur d'eau de l'enracinement du maïs en parcelle cultivée. II. Une méthode d'étude de la répartition verticale et horizontale des racines. *Agronomie* 6(5): 415-425.

TARDIEU, F. 1988a. Analysis of the spatial variability of maize root density: I. Effect of wheel compaction on the spatial arrangement of roots. *Plant and Soil* 107: 259-266.

TARDIEU, F. 1988b. Analysis of the spatial variability of maize root density: II. Distances between roots. *Plant and Soil* 107: 267-272.

Identification de la microvariabilité après défrichement motorisé d'un sol ferrallitique issu de sables tertiaires

GBALLOU YORO et GNAHOUA H. GODO*

Résumé

    Les caractéristiques morphologique, physique et chimique des microreliefs apparus, après défrichement motorisé, à la surface d'un sol ferrallitique issu de sables tertiaires ont été identifiées.

    L'observation morphologique a permis de reconnaître quatre volumes pédologiques (A, AB, BR, et BD). Les résultats analytiques montrent que ces volumes pédologiques se différencient plus ou moins par leurs caractéristiques physiques (texture, densité apparente, stabilité structurale) et chimiques (MO, S/T, pH).

    Les caractères pénétrométriques révèlent, en revanche, deux types de volumes pédologiques. Ceux à faible résistance et ceux à résistance relativement élevée. Les premiers sont des accumulations alors que les seconds sont des horizons arasés et/ou compactés par le bulldozer.

Abstract

Identification of microvariability after motorized land clearing of a ferrallitic soil derived from Tertiary sand

Morphological, physical, and chemical identification was made on microreliefs which appeared, after motorized land clearing, at the surface of a ferrallitic soil derived from Tertiary sand.

*   Respectivement : Laboratoire de Pédologie, Institut International de Recherche Scientifique pour le Développement à Adiopodoumé (IIRSDA); et Laboratoire d'Agronomie, IIRSDA, BP V 51, Abidjan, Côte d'Ivoire.

After morphological observation, it was found that four pedological volumes (A, AB, BR, and BD) could be distinguished. Results of analyses showed that these pedological volumes were more or less differentiated by such characteristics as soil texture, bulk density, structural stability, OM content, base saturation percentage, and pH.

On the other hand, results of penetrometer resistance distinguished only two pedological volumes: firstly, volumes characterized by low resistance and made of accumulations; and secondly, volumes of high resistance which are horizons levelled down and/or compacted by the bulldozer.

Introduction

Les modifications du sol provoquées par un défrichement motorisé sont connues dans le monde (Kang et Juo, 1982; Lal *et al.*, 1986) et en Côte d'Ivoire (de Blic, 1975, 1976; Moreau, 1983; Roose, 1983; Collinet, 1984). Elles sont étudiées dans leur globalité et généralement comparées aux perturbations ou transformations causées par d'autres modes de défrichement (Yoro, 1979).

En effet, il n'existe presque pas d'étude permettant de différencier les microreliefs créés par l'abattage motorisé et de les caractériser de façon analytique. Et pourant, on relève sur le plan morphologique des hétérogénéités spatiales dues au défrichement mécanisé (Moreau, 1983).

Notre étude, qui est une contribution, se propose d'identifier, après défrichement motorisé d'un sol ferrallitique issu de sables tertiaires, la microvariabilité portant sur les caractères physiques et chimiques.

Méthodologie

Présentation du site

Le site de l'étude s'étend sur un domaine forestier de 30 ha, près de Mafère dans la région d'Aboisso, sud-est de la Côte d'Ivoire. Le climat est de type attiéen caractérisé par quatre saisons (Eldin, 1971) : deux saisons humides qui alternent avec deux saisons sèches.

Les précipitations totales annuelles avoisinent 2000 mm. Elles sont irrégulières (figure 1) et ont tendance à baisser ces dernières années. Mai et juin sont les mois les plus pluvieux. En revanche, janvier est le plus sec (figure 2). La température moyenne oscille autour de 27°C tout au long de l'année.

La végétation appartient au secteur ombrophile du domaine guinéen (Guillaumet, 1971) caractérisé par une forêt dense humide sempervirente. Elle a été fortement entamée par l'agriculture, de sorte qu'elle constitue aujourd'hui un massif hétérogène comprenant des plantations, des jachères, des îlots de forêt relictuelle.

La roche-mère est essentiellement constituée de sédiments sablo-argileux mis en place au tertiaire (Avenard, 1971).

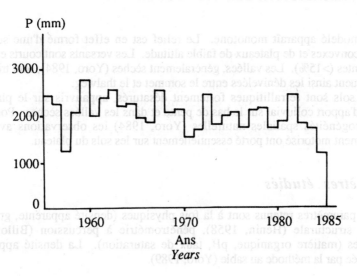

Figure 1. Précipitations annuelles de 1956 à 1985.

Annual rainfall from 1956 to 1985.

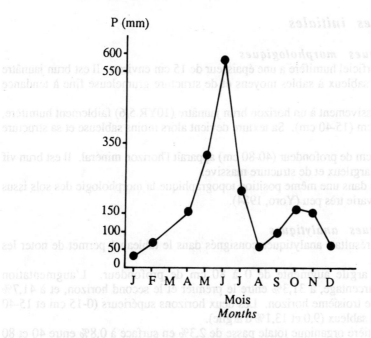

Figure 2. Précipitations moyennes mensuelles de 1956 à 1985.

Average monthly rainfall from 1956 to 1985.

Le modelé apparaît monotone. Le relief est en effet formé d'une succession de collines convexes et de plateaux de faible altitude. Les versants sont courts et accusent de fortes pentes (>15%). Les vallées, généralement sèches (Yoro, 1984), sont très encaissées et accentuent ainsi les dénivelées entre le sommet et le thalweg.

Les sols sont ferrallitiques fortement désaturés, appauvris sur le plateau et peu évolués d'apport colluvial sur le bas de pente et dans les vallées sèches. Pour minimiser les hétérogénéités spatiales naturelles (Yoro, 1984) les observations avant et après défrichement motorisé ont porté essentiellement sur les sols du plateau.

Paramètres étudiés

Les paramètres retenus sont à la fois physiques (densité apparente, granulométrie, stabilité structurale (Henin, 1958), pénétrométrie à percussion (Billot, 1982)) et chimiques (matière organique, pH, taux de saturation). La densité apparente a été déterminée par la méthode au sable (Yoro, 1989).

Résultats et discussion

Caractéristiques initiales

Caractéristiques morphologiques
L'horizon superficiel humifère a une épaisseur de 15 cm environ. Il est brun jaunâtre foncé (10YR 4/4), sableux à sables moyens et de structure grumeleuse fine à tendance particulaire.

Il passe progressivement à un horizon brun jaunâtre (10YR 5/6) faiblement humifère, épais d'environ 25 cm (15-40 cm). Sa texture devient alors moins sableuse et sa structure massive.

Au-delà de 40 cm de profondeur (40-80 cm) apparaît l'horizon minéral. Il est brun vif (7,5YR 5/8), sablo-argileux et de structure massive.

Rappelons que dans une même position topographique la morphologie des sols issus de sables tertiaires varie très peu (Yoro, 1984).

Caractéristiques analytiques
L'examen des résultats analytiques consignés dans le tableau 1 permet de noter les faits suivants.

La teneur en argile augmente de 0 à 80 cm de profondeur. L'augmentation correspond, en pourcentage, à 31,3% entre le premier et le second horizon, et à 41,7% entre le second et le troisième horizon. Les deux horizons supérieurs (0-15 cm et 15-40 cm) sont cependant sableux (9,0 et 13,1% d'argile).

Le taux de matière organique totale passe de 2,3% en surface à 0,8% entre 40 et 80 cm en profondeur.

Tableau 1.   Résultats analytiques du sol à l'état initial.

Paramètres	Horizons (cm)		
	0-15	15-40	40-80
Analyse texturale (%)			
Argile	9	13,1	22,5
Limon fin	2,9	4,1	4,11
Limon grossier	2,6	2,2	2,0
Sable fin	21,4	12,7	10,6
Sable grossier	60,6	65,9	58,8
Instabilité structurale	0,1	0,9	1,8
Paramètres chimiques			
pH	4,4	4,5	4,5
C (%)	13,22	7,46	4,51
N (%)	1,13	0,48	0,38
C/N	11,7	15,54	11,87
MO (%)	2,3	1,3	0,8
Echange cationique (cmol kg^{-1})			
Ca	0,08	0,03	0,01
Mg	0,08	0,02	0,005
K	0,02	0,01	0,005
Na	0,008	0,001	0
S	0,18	0,06	0,02
T	4,6	3,84	2,5
V (%)	3,91	1,56	0,8

MO = matière organique totale; S = somme des bases échangeables; T = capacité d'échange; V = taux de saturation.

L'indice d'instabilité structurale (Is), très faible en surface (0,1), augmente brutalement en profondeur, passant successivement de 0,9 entre 15-40 cm à 1,8 de 40 à 80 cm.

Le taux de saturation, déja faible (3,91%) en surface 0-15 cm) diminue de 2,35% entre 15-40 cm et de 3,11 entre 40-80 cm, soit respectivement un pourcentage de diminution de 60,1% et de 79,5%.

Le pH acide (4,4-4,5) tout le long du profil étudié apparaît être en relation avec le taux de saturation.

Tout ce qui précède autorise à ranger le sol du site comme ferrallitique, fortement désaturé appauvri. Dans la *Soil Taxonomy* on le rangerait dans les kandiudults.

Caractéristiques densitométriques et pénétrométriques

La densité apparente du sol initial augmente avec la profondeur (fiqure 3). Elle est successivement de 1,308 entre 0-15 cm, 1,456 entre 15-40 cm et 1,523 entre 40-80 cm.

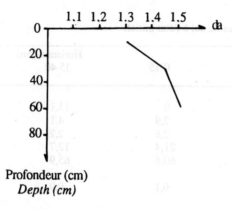

Figure 3. Evolution verticale de la densité apparente.

Vertical evolution of the bulk density.

La résistance à la pénétration (figure 4) croît de 0 à 50 cm de profondeur. L'accroissement devient important (59,6%) à partir de 15 cm.

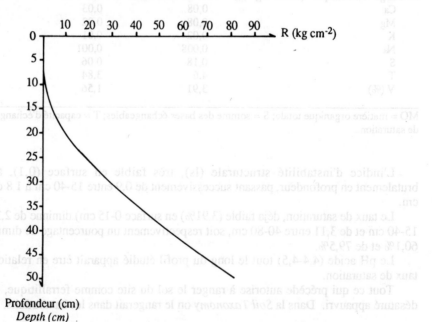

Figure 4. Profil pénétrométrique.

Penetration resistance profile.

Caractéristiques du sol après le défrichement motorisé

Identification morphologique

L'observation de la surface du sol après défrichement motorisé permet, à petite échelle, de distinguer un ensemble de microreliefs constitués de terre et de débris végétaux. Observés de plus près, à grande échelle, on note que ces microreliefs se composent :

- de volumes pédologiques qui apparaissent sous forme de buttes ou de billons discontinus sans orientation préférentielle; ils sont très meubles à boulants; ce sont en fait des accumulations; on en reconnaît trois principaux types selon leur origine :
  - les volumes pédologiques (A) constitués d'horizon A; ils sont humifères, brun jaunâtre foncé (10YR 4/4), riches en débris végétaux (feuilles mortes ou fraîches, branches de bois sec ou frais déchiquetées ou non, racines arrachées...); la texture est sableuse; la structure modifiée est grumeleuse fine à tendance particulaire;
  - les volumes pédologiques (B) formés d'horizon minéral; ils sont brun vif (7,5YR 5/6) à brun jaunâtre (10YR 5/6); sablo-argileux; la structure très modifiée apparaît polyédrique émoussée;
  - les volumes pédologiques (AB) issus du mélange d'horizon A et d'horizon B; ils sont de couleur hétérogène (brun jaunâtre foncé, brun jaunâtre, brun vif) et humifère par endroits; la texture est aussi hétérogène (sableuse à sablo-argileuse). Il faut noter que le mélange n'est pas toujours parfait. Ainsi on arrive à reconnaître quelquefois les matériaux issus de l'horizon A et ceux provenant de l'horizon B. La proportion est variable;
- et des surfaces décapées ou lissées (BD) : elles sont planes, discontinues, sans orientation ni forme préférentielles; elles sont, en effet, soit des ornières des chenilles, soit des zones arasées par la lame du bulldozer. Elles se caractérisent par une couleur homogène, une texture franche sableuse ou sablo-argileuse selon l'épaisseur de la couche de terre enlevée. Elles sont compactes et relativement cohérentes.

Caractéristique analytiques

Vingt prélèvements composites ont été effectués sur chacun des quatre principaux types de microreliefs (A, AB, BR et BD) afin de déterminer les caractéristiques physiques et chimiques. Un mélange homogène a été d'abord réalisé au niveau de chaque microrelief prélevé.

Granulométrie

Le taux d'argile varie d'un type de microrelief à l'autre (figure 5). Il est de 4,8% dans les volumes pédologiques A, 8,0% en AB, 14,3% en BR et de 12,3% dans les surfaces BD.

La différence apparaît ainsi nette entre les volumes pédologiques ou microreliefs d'accumulation (A, AB et BR). Les rapports AB/A; BR/A et BR/AB sont, en effet, respectivement de 1,66; 2,98 et 1,78. Les microreliefs BD, en revanche, se distinguent très peu de BR (BR/BD = 1,16). Ceci tient à l'origine des deux microreliefs formés, en fait, d'horizons de profondeur.

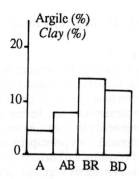

Figure 5. Taux d'argile des microreliefs.

Percentage clay in the different positions in the microrelief.

Le taux d'argile permet donc de distinguer trois types de microrelief : A, AB et BR ou BD.

Si on se réfère aux caractéristiques granulométriques initiales (cf. tableau 1) on remarque que les taux d'argile des divers microreliefs ont diminué par rapport à ceux des horizons dont ils semblent issus. Ces diminutions peuvent être attribuées au lessivage par les eaux de pluies survenues juste après le défrichement. Il faut noter que sur les sols ferrallitiques gravillonnaires issus de schiste (Yoro et Gnamba, 1987) les taux d'argile des microreliefs étaient en concordance avec la texture des horizons initiaux, bien que le défrichement ait été effectué également en début de saison de pluie.

Matière organique totale

Comme le taux d'argile, les teneurs en matière organique totale (figure 6) permettent de distinguer les trois volumes pédologiques A, AB et BR entre eux. Les microreliefs BR et BD ont en effet des teneurs très voisines, voire égales (1,1 et 1,2).

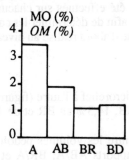

Figure 6. Taux de matière organique des microreliefs.

Percentage organic matter in the different positions in the microrelief.

Les volumes pédologiques A et AB comportant, dans des proportions différentes, des horizons humifères, apparaissent relativement plus riches en matière organique totale que BR et BD provenant des couches sous-jacentes très faiblement humifères. Les teneurs en matière organique totale des types de microrelief reflètent donc la morphologie des horizons initiaux dont ils sont issus.

Par rapport à l'horizon humifère initial (0-15 cm), le taux de matière organique totale des volumes pédologiques A s'est accru de 52% (cf. tableau 1). Cette amélioration, constatée également dans les sols issus de roche granitique ou schisteuse (Gnamba, 1986; Yoro et Gnamba, 1987), pourrait s'expliquer par l'activité biologique qui s'intensifie immédiatement après le défrichement (Akodo, 1977; Moreau, 1983; Gnamba, 1987).

Indice d'instabilité structurale

L'indice d'instabilité structurale (figure 7) permet de différencier les quatre types de microrelief A, AB, BR et BD entre eux. Il est respectivement de 0,1; 0,6; 1,3 et 1,0. Ce sont les microreliefs relativement humifères (cf. figure 6) qui apparaissent les plus stables. On relève ainsi une concordance entre la morphologie et la stabilité structurale des microreliefs.

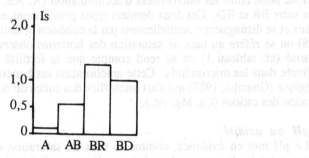

Figure 7.   Indice d'instabilité des microreliefs.

Instability index in the different positions in the microrelief.

La stabilité structurale de A (0,1) n'a pas changé par rapport à celle de l'horizon superficiel humifère initial (0-15 cm). Celle de AB, BR et BD montre que la structure du sol s'est très faiblement améliorée après l'abattage motorisé. Ceci semble (un peu) en contradiction avec la texture sableuse et les teneurs en matière organique (2,3%) des sols étudiés. En effet, sur les sols ferrallitiques gravillonnaires (Yoro et Gnamba, 1987) dont les horizons superficiels (0-10 cm) renferment un peu plus d'argile (12,8%) et de matière organique (7,2%) la structure s'est fortement dégradée après défrichement motorisé.

Taux de saturation

L'examen de la figure 8 montre que le taux de saturation diminue quand on passe successivement des microreliefs A, AB, BR ou BD. Le pourcentage de diminution est de 46,7% entre A et AB, 73,6% entre AB et BR, et 5% entre BR et BD. La différence appa-

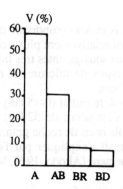

Figure 8. Taux de saturation des microreliefs.

Saturation rate in the different positions in the microrelief.

raît donc nette entre les microreliefs d'accumulation (A, AB, BR) alors qu'elle est très faible entre BR et BD. Ces deux derniers types proviennent, en fait, des même horizons initiaux et se distinguent essentiellement par la cohésion et la structure.

Si on se réfère au taux de saturation des horizons observés avant le défrichement motorisé (cf. tableau 1) on se rend compte que la fertilité chimique s'est fortement améliorée dans les microreliefs. Cette amélioration est à mettre au compte de l'activité biologique (Gnamba, 1987) qui s'est intensifiée des suites du défrichement entraînant une libération des cations (Ca, Mg, etc.).

pH ou acidité

Le pH met en évidence, comme le taux de saturation et les teneurs en matière organique totale, l'existence de trois types de microrelief distincts (figure 9). On a, en

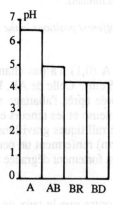

Figure 9. pH des microreliefs.

pH in different positions the microrelief.

effet, les volumes pédologiques faiblement acides (6,5), acides (4,9) et très acides (4,2) qui sont respectivement A, AB et BR ou BD.

Les microreliefs A et AB, relativement riches en matière organique totale et favorables à une intense activité biologique (Gnamba, 1987), ont eu leur pH amélioré de 0,4 ou 2 unités par rapport à celui des horizons initiaux (cf. tableau 1). Ce relèvement du pH s'accorde avec le taux de saturation des microreliefs, même si au niveau des microreliefs BR et BD la relation entre les deux paramètres ne semble pas se vérifier.

Caractéristiques densitométriques et pénétrométriques des microreliefs

Les valeurs de la densité apparente (figure 10) confirment la reconnaissance des quatre types de microrelief définis morphologiquement. Elles sont de 1,04 pour A, 1,16 pour AB, 1,32 pour BR et de 1,50 pour BD.

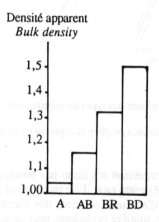

Figure 10.   Densité apparente des microreliefs.

Bulk density in the different positions in the microrelief.

La différence densitométrique observée entre les divers microreliefs est due d'abord à leur forme à ce niveau, on distingue les microreliefs d'accumulation - A, AB et BR - et les microreliefs plans décapés - BD; ensuite aux teneurs en matière organique et en argile. Ce sont celles-ci qui font distinguer les microreliefs d'accumulation entre eux. Il se dégage donc une certaine relation entre la morphologie des microreliefs et leur densité apparente.

Les caractéristiques pénétrométriques (figure 11) révèlent l'existence de deux types de microrelief : les microreliefs d'accumulation (A, AB et BR) et les microreliefs plans décapés (BD). Les premiers, remués et sous forme de billons ou buttes, meubles à boulants, ont de très faibles résistances à la pénétration (<1 kg cm^{-2} à 15 cm de profondeur). Les seconds, en place et arasés ou décapés, cohérents, présentent de fortes résistances à la pénétration dès le cinq premiers centimètres (4,9 kg cm^{-2} entre 0-5 cm).

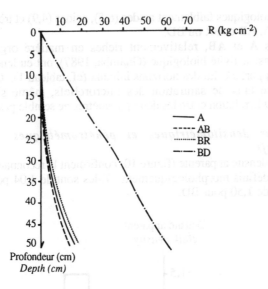

Figure 11. Profils pénétrométriques des types de microreliefs.

Penetrometer resistance profiles in different types of microrelief.

La pénétrométrie à percussion n'a donc pas permis de différencier les microreliefs d'accumulation (A, AB, BR) entre eux. Elle paraît ainsi plus influencée par les caractères extrinsèques (forme, état d'ameublissement) des microreliefs que par leurs propriétés intrinsèques (granulométrie, matière organique, taux de saturation).

Conclusion

Le défrichement motorisé sur le sol issu de sables tertiaires a provoqué des modifications. Ces modifications se manifestent par l'existence de quatre principaux types de microrelief (A, AB, BR et BD) identifiés d'abord morphologiquement (couleur, cohésion, forme, texture), puis caractérisés analytiquement. A ce dernier niveau, les quatre types de microrelief n'ont pas toujours été différenciés. En effet, l'indice d'instabilité structurale (Is), la teneur en argile et la densité apparente cofirment l'identification des quatre types de microrelief alors que la teneur en matière organique totale, le taux de saturation et le pH n'en mettent en évidence que trois A, AB et BR ou BD. La pénétrométrie, quant à elle, révèle deux types de microrelief : les microreliefs d'accumulation (A, AB et BR) et les microreliefs plans décapés ou lissés (BD).

Les principaux microreliefs identifiés et caractérisés confirment les hétérogénéités spatiales ou latérales créées par un défrichement motorisé (Yoro, 1979; Moreau, 1983;

Gnamba, 1986; Boka, 1986) et posent ainsi le problème de prélèvement d'échantillon de sol après un abattage motorisé.

De cette étude on retiendra aussi que :

- le nombre de types de microreliefs dépend de la méthode d'identification ou de caractérisation;
- les caractéristiques des microreliefs concordent avec leur morphologie;
- les microreliefs AB apparaissent comme un sol de type nouveau par rapport aux horizons initiaux.

Bibliographie

AKODO, E.A. 1977. *Etude de l'évolution biologique des sols ferrallitiques forestiers de basse Côte d'Ivoire, sous l'effet du défrichement*. Institut Français de Recherche Scientifique pour le Développement en Coopération. Centre d'Adiopodoumé. Abidjan, Côte d'Ivoire : ORSTOM. 80p.

AVENARD, J.M. 1971. Aspect de la géomorphologie. In : *Le milieu naturel de la Côte d'Ivoire*, 7-72. ORSTOM Mémoire n° 50. Centre d'Adiopodoumé. Abidjan, Côte d'Ivoire : ORSTOM.

BILLOT, J.F. 1982. Les applications agronomiques de la pénétrométrie à l'étude de la structure des sols travaillés. *Bulletin AFES Science du Sol* 3 : 187-201.

BOKA, M.T.A. 1986. *Modifications physiques d'un sol ferrallitique sous l'effet du défrichement lourd motorisé*. Mémoire pour l'obtention du DIPT (Montpellier). Centre d'Adiopodoumé. Abidjan, Côte d'Ivoire : ORSTOM. 51p.

COLLINET, J. 1984. Hydrodynamique superficielle et érosion comparée de quelques sols ferrallitiques sur défriches forestières traditionnelles (Côte d'Ivoire). In: *Challenges in African hydrology and water resources*. International Association of Hydrological Services. IAHS Publication no. 144. Washington, DC: IAHS.

COMBEAU, A. et QUANTIN, P. 1964. Observations sur les relations entre stabilité structurale et matière organique dans quelques sols d'Afrique Centrale. *Cahiers ORSTOM, série Pédologie* 1 :3-12.

DE BLIC, P. 1975. *Comportement des sols après mise en culture mécanisée (région centre Côte d'Ivoire)*. Rapport ORSTOM. Centre d'Adiopodoumé. Abidjan, Côte d'Ivoire : ORSTOM. 27p.

DE BLIC, P. 1976. *Evolution des sols après défrichement et mise en culture semi-mécanisée dans la région centre. Enquête pédologique effectuée en 1975 sur les ensembels de Boyakro et Brikro (secteur de Kounahiri)*. Rapport ORSTOM. Centre d'Adiopodoumé. Abidjan, Côte d'Ivoire : ORSTOM. 20p.

ELDIN, M. 1971. Le climat. In : *Le milieu naturel de la Côte d'Ivoire*, 73-108. Mémoire ORSTOM n° 50. Centre d'Adiopodoumé. Abidjan, Côte d'Ivoire : ORSTOM.

GNAMBA, A. 1986. Quelques effets du défrichement motorisé sur certains types de sols de Côte d'Ivoire. Mémoire pour l'obtention du Diplôme d'Agronomie Approfondie (DAA.), Option Pédologie (ENSA d'Abidjan). Centre d'Adiopodoumé. Abidjan, Côte d'Ivoire : ORSTOM.

GNAMBA, A. 1987. Activité biologique des types de surface observés après différents modes de défrichement des sols du bas Cavaly. Mémoire DEA en Ecologie Tropicale, Université d'Abidjan. Centre d'Adiopodoumé. Abidjan, Côte d'Ivoire : ORSTOM.

GUILLAUMET, J.L. 1971. La végétation. In : *Le milieu naturel de la Côte d'Ivoire*. Mémoire ORSTOM n° 50. Centre d'Adiopodoumé. Abidjan, Côte d'Ivoire : ORSTOM.

HENIN, S., MONIER, G. et COMBEAU, A. 1958. Méthode pour l'étude de la stabilité structurale des sols. *Annales Agronomiques* 1 : 73-92.

KANG, B.T. and JUO, A.S.R. 1982. Effect of forest clearing on soil chemical properties and crop performance. Paper presented at the International Conference on Land Clearing and Development, International Institute of Tropical Agriculture (IITA), Ibadan, Nigeria, 22-26 November 1982.

LAL, R., SANCHEZ, P.A. and CUMMINGS, R.W. 1986. Land clearing and development in the tropics. Rotterdam, Boston: A.A. Balkema.

MOREAU, R. 1983. Evolution des sols sous différents modes de mise en culture en Côte d'Ivoire forestière et préforestière. *Cahiers ORSTOM, série Pédologie* 20(4) : 311-325.

ROOSE, E.J. 1983. Ruissellement et érosion avant et après défrichement en fonction du type de culture en Afrique occidentale. *Cahiers ORSTOM, série Pédologie* 20(4): 327-339.

YORO, G. 1979. *Aperçu sur les modifications du milieu naturel sous l'effet des pratiques culturales.* Rapport ORSTOM. Centre d'Adiopodoumé. Abidjan, Côte d'Ivoire : ORSTOM. 16p.

YORO, G. 1984. *Caractéristiques morpho-pédologiques des types de paysages sur sables tertiaires dans la région du Sud-Est (Bonoua).* Rapport ORSTOM. Centre d'Adiopodoumé. Abidjan, Côte d'Ivoire : ORSTOM. 18p.

YORO, G. 1986. *Appréciation des caractéristiques physico-chimiques des sols selon leur position topographique dans un bassin versant sur schiste, au centre-nord-ouest de la Côte d'Ivoire.* Rapport ORSTOM. Centre d'Adiopodoumé. Abidjan, Côte d'Ivoire : ORSTOM.

YORO, G. et GNAMBA, A. 1987. *Identification et caractérisation des microreliefs créés à la surface du sol par un défrichement motorisé dans la région d'Agboville (sud de la Côte d'Ivoire).* Rapport ORSTOM. Centre d'Adiopodoumé. Abidjan, Côte d'Ivoire : ORSTOM. 24p.

YORO, G. et GODO, G. 1989. *Les méthodes de mesure de la densité apparente : problème de dispersion des résultats dans un horizon donné.* Rapport IIRSDA. 10p. Also in: *Tillage and organic-matter management in humid and subhumid Africa.* These proceedings.

Les méthodes de mesure de la densité apparente : problème de dispersion des résultats dans un horizon donné

GBALLOU YORO et GNAHOUA H. GODO[*]

Résumé

Il existe plusiers méthodes de mesure de la densité apparente : celles de laboratoire et celles du terrain. L'utilisation des dernières sur quatre types de sol, selon un dispositif en carré latin, a permis de montrer qu'il existe une interaction entre le sol et la méthode, et que la valeur de la densité apparente dépend non seulement du sol mais également de la méthode de mesure.

Abstract

Methods for measuring bulk density: a problem of density of results in a given horizon

There are many methods to measure bulk density, both laboratory and field medthods. Applying field methods to four types of soil in a latin square scheme showed that there is interaction between soil and method, and the value of bulk density depends both on the soil and the method of measurement.

[*] Respectivement : Laboratoire de Pédologie, Institut International de Recherche Scientifique pour le Développement à Aciopodoumé (IIRSDA); et Laboratoire d'Agronomie, IIRSDA, BP V 51, Abidjan, Côte d'Ivoire.

Introduction

La densité apparente est l'un des paramètres les plus importants dans les études portant sur la structure du sol. Elle est, en effet, liée à la nature et à l'organisation des constituants du sol (Chauvel, 1977). Elle permet, en outre, de calculer la porosité et d'apprécier ainsi indirectement la perméabilité - la résistance à la pénétration des racinces (Maertens, 1964), la cohésion des horizons (Yoro, 1983; Yoro et Assa, 1986) et la réserve en eau du sol (Henin *et al.*, 1969).

Son étroite relation avec le type de structure fait qu'elle est faible dans "les sols à structure grumeleuse stable" comme le chernozem, dont les agrégats sont en grumeaux (Duchaufour, 1970). Dans les sols tropicaux, les horizons humifères, relativement structurés (de Boissezon, 1965), se caractérisent par des densités apparentes plus faibles que celles des horizons minéraux sous-jacents à structure massive (de Blic, 1975; Kouakou, 1981; Yoro, 1983).

Sa connaissance peut permettre de déterminer ou d'orienter les travaux de préparation du sol tels que le labour, le sous-solage, le pulvérisage, le hersage.

Elle peut être évaluée au laboratoire sur des échantillons prélevés ou sur le terrain dans des horizons en place. Au laboratoire on utilise principalement la méthode à la paraffine et à l'eau, et la méthode au pétrole. Ces deux méthodes conviennent surtout pour la mesure de la densité apparente des mottes (Vizier, 1971), des petits agglomérats (Monnier *et al.*, 1973) et des agrégats (Yoro, 1983). Pour les mesures *in situ*, il existe quatre méthodes : la méthode au cylindre, la méthode au sable, la méthode au densitomètre à membrane et la méthode par gammamétrie.

Les descriptions, les avantages et les limites d'utilisation de ces méthodes sont repris dans le Bulletin de Groupe de Travail édité par l'ORSTOM (Audry *et al.*, 1973). Ce bulletin n'aborde malheureusement pas le problème de l'uniformité ou de la dispersion des densités apparentes déterminées dans un même sol ou horizon à partir de plusieurs méthodes. En outre, divers travaux (Berger, 1964; Collinet, 1988) mentionnant la porosité calculée à partir de la densité apparente, ne précisent pas laquelle des quatre méthodes de terrain a été utilisée. Certains travaux comme ceux de de Blic (1987) s'appuient sur les densités apparentes obtenues en utilisant alternativement deux méthodes. Or au cours de quelques unes de nos études (Yoro, 1983; Godo *et al.*, 1989) nous avons remarqué que pour un même horizon nous obtenons des densités apparentes différentes lorsque nous utilisons simultanément la méthode au sable et la méthode au densitomètre à membrane. Nos constatations n'étant pas fondées sur une analyse statistique ne permettent pas de dégager l'influence de la méthode sur la valeur de la densité apparente.

Notre étude, qui est une contribution, veut se prononcer sur la variation de la densité apparente d'un sol ou d'un horizon donné en fonction des méthodes de mesure.

Remerciement : Laboratoire de Pédologie, Institut International de Recherche Scientifique pour le Développement à ... agronomie (IIRSDA), et Laboratoire d'Agronomie, IIRSDA, BP V 51, Abidjan, Côte d'Ivoire.

Matériel et méthodes

Rappel des principes des méthodes de terrain

Les principes de la méthode au cylindre (C), de la méthode au sable (S) et de la méthode au densitomètre à membrane (M) sont basés sur la détermination du poids spécifique apparent d'un volume de sol prélevé. Le volume est estimé immédiatement sur le terrain alors que le poids est évalué au laboratoire après séchage et pesée. La connaissance de ces deux variables permet de calculer la densité apparente selon la relation :

$$da = P/V$$

où P = le poids sec de l'échantillon

V = le volume de l'échantillon prélevé et séché

Des deux variables, le volume apparaît le plus important car sa détermination nécessite beaucoup d'attention et de doigté (Audry et al., 1973). Un geste en plus ou en moins et le volume est surestimé ou sousestimé. Le souci de l'avoir plus précis est d'ailleurs l'une des raisons des multiples recherches qui ont conduit à la mise au point des méthodes que nous connaissons. La précision du volume dépend cependant de la spécificité de chaque méthode.

En ce qui concerne la méthode par gammamétrie (G), le principe se fonde sur la mesure de l'intensité atténuée N, qui s'exprime selon la formule :

$$N = N_o e^{-uPl}$$

où N = intensité atténuée (nombre de photons gamma émis dans le sol)

N_o = intensité initiale (nombre de photons initiaux)

u = coefficient massique d'atténuation du matériau (coefficient d'absorption massique du sol)

P = masse volumique du matériau

l = longueur du parcours (épaisseur de sol traversée par les photons)

e = base des logarithmes népériens

La connaissance de u, l et N_o permet de calculer P, la densité apparente.

Application des quatre méthodes sur différents types de sol

Au cours de certaines de nos études (Yoro, 1983; Godo et al., 1989) nous avons constaté que pour un même sol ou un même horizon, la densité apparente variait plus ou moins selon la méthode utilisée. Ces observations portant sur les résultats de deux méthodes (méthode au sable et méthode au densitomètre à membrane) s'avèrent insuffisantes pour conduire à une conclusion crédible. Nous avons donc entrepris d'appliquer, sur les mêmes sols, les quatre méthodes permettant de mesurer, in situ, la densité apparente. Cette démarche devra nous amener à déterminer l'influence de la méthode de mesure sur la valeur de la densité apparente.

Pour atteindre cet objectif, les conditions suivantes ont été nécessaires :
- choix de quatre types de sol dont les caractères morphologiques essentiels sont consignés dans le tableau 1;
- mesures effectuées dans les 15 premiers centimètres;
- utilisation simultanée des quatre méthodes sur un type de sol;
- un même manipulateur pour les quatre répétitions de chaque méthode;
- sur un type de sol, les mesures s'effectuent sur une surface de 16 m², soit 1 m² pour une répétition et 4 m² pour une méthode; le dispositif de collecte des données est un carré latin (4 x 4), qui a été adopté pour essayer de réduire au niveau d'une même méthode "la variabilité des mesures liée à l'hétérogénéité dans l'espace" (Audry et al., 1973); et
- traitement à l'ordinateur des données brutes selon deux logiciels, STAT VIEW 512⁺ et NDMS, pour tester les facteurs sol et les facteurs méthode d'une part, et les interactions sol-méthode d'autre part.

Tableau 1. Caractères morphologiques essentiels des sols (horizon 0-15 cm).

Type de sol	Couleur	Texture	Eléments grossiers (%)	Structure	Occupation
L_{11}	Brun grisâtre très foncé (10YR 3/2)	Limono-sableuse	0	Fragmentaire grumeleuse	Jachère 15 ans
L_2	Brun foncé (10YR 3/3)	Limono-argileuse	65	Fragmentaire grumeleuse	Jachère 15 ans
F	Brun foncé (10YR 3/3)	Sableuse	0	Grumeleuse fine à tendance particulaire	Forêt secondaire
J	Brun jaunâtre foncé (10YR 4/4)	Sableuse	0	Grumeleuse fine à tendance particulaire	Jardin arboré

Résultats et discussion

Valeurs de la densité apparente en fonction des méthodes de mesure

L'analyse de la variance effectuée sur l'ensemble des données obtenues sur les quatre types de sol par les quatre méthodes de mesure (tableau 2) montre une différence hautement significative (α = 1% entre les sols étudiés. Elle confirme ainsi leur distinction fondée essentiellement sur la texture, la structure et l'itinéraire cultural (cf. tableau 1).

Tableau 2. Analyse de variance.

Source de variation	Degré de liberté	Somme des carrés des écarts	Carré moyen	F (calculé)	S α = 1%†
Facteur sol	3	0,88709	0,29570	1037,9	HS‡
Facteur méthode	3	0,53044	0,17681	620,6	HS
Interaction sol-méthode	9	0,21834	0,02426	85,1	HS
Variance résiduelle	48	0,01368	0,00028		

† S = signification
‡ HS = hautement significatif

Les résultats de cette analyse de variance renforcent le rôle de la densité apparente comme un paramètre d'identification ou de caractérisation des sols. La densité apparente permet, en effet, non seulement de distinguer les types de sol entre eux, mais de suivre l'évolution physique d'un sol soumis à diverses techniques culturales (de Blic, 1987).

L'examen du tableau 2 révèle également que les moyennes des densités apparentes (tableau 3) qui sont respectivement de 1,303 pour la méthode S, 1,214 pour M, 1,444 pour C et 1,417 pour G, diffèrent entre elles de manière significative pour α = 1%. La valeur de la densité apparente est donc liée, non seulement au type de sol, mais aussi à la méthode de mesure. Il existe d'ailleurs une forte interaction entre les deux paramètres (tableau 2).

Tableau 3. Densité apparente de 4 types de sol en fonction des méthodes de mesure.

Méthodes de mesure	Types de sol				
	L_2	L_{11}	F	J	Moyenne
S	1,248	1,291	1,159	1,515	1,303
M	1,063	1,143	1,150	1,502	1,214
C	1,394	1,517	1,343	1,524	1,444
G	1,280	1,517	1,295	1,579	1,417

S = méthode au sable
M = méthode au densitomètre à membrane
C = méthode au cylindre
G = méthode par gammamétrie

La méthode M donne les plus faibles valeurs. En revanche, la méthode C donne des densités apparentes élevées. Celles déterminées par les méthodes S et G sont inter-médiaires. Il se dégage cependant des affinités entre M et S de même qu'entre C et G.

L'influence de la méthode de mesure sur la valeur de la densité apparente a été mieux appréciée au niveau de chaque type de sol (tableau 4). Dans le sol L_2, les moyennes 1,248, 1,063, 1,394 et 1,280 obtenues respectivement par les 4 méthodes S, M, C et G (cf. tableau 3) diffèrent entre elles de manière significative ($\alpha = 5\%$). Dans les sols L_{11} et F seules les différences $L_{11}C$-$L_{11}G$ et F_S-F_M ne sont pas significatives. Dans le sol J sur les six différences calculées, trois sont significatives (tableau 4).

Tableau 4. Comparaison des moyennes au niveau de chaque type de sol (logiciel STAT VIEW 512[+])

Comparaison	Différence x	Fisher PLSD
L_2S-L_2M	.185	.0270*
L_2S-L_2C	-.145	.027*
L_2S-L_2G	-.032	.027*
L_2M-L_2C	-.331	.027*
L_2M-L_2G	-.217	.027*
L_2C-L_2G	.113	.027
$L_{11}S$-$L_{11}M$	.147	.03*
$L_{11}S$-$L_{11}C$	-.226	.03*
$L_{11}S$-$L_{11}G$	-.226	.03*
$L_{11}M$-$L_{11}C$	-.374	.03*
$L_{11}M$-$L_{11}G$	-.374	.03*
$L_{11}C$-$L_{11}G$	2.500E-4	.03
F_S-F_M	.009	.009
F_S-F_C	-.184	.009*
F_S-F_G	-.136	.009*
F_M-F_C	-.193	.009*
F_M-F_G	-.145	.009*
F_C-F_G	.048	.009*
J_S-J_M	-.009	.036
J_S-J_C	-.009	.036
J_S-J_G	-.064	.036*
J_M-J_C	-.022	.036
J_M-J_G	-.077	.036*
J_C-J_G	-.055	.036*

* Significatif pour $\alpha = 5\%$.

Il se révèle que d'un sol à un autre, les différences observées entre les densités apparentes déterminées par les quatre méthodes ne sont pas toutes significatives. Et ceci semble être en relation avec l'interaction très marquée qui existe entre les méthodes et les sols. D'une manière générale, il ressort que la valeur de la densité apparente est liée à la fois au type de sol et à la méthode de mesure utilisée.

Il importe donc de préciser désormais, dans toute étude de densité apparente, la méthode ayant permis de collecter les données. En outre, dans une étude d'évolution des caractères densitométriques il faut utiliser la même méthode au cours des différentes périodes de mesure. A ce propos il convient de souligner que la méthode au sable bien que "longue et minutieuse" (Audry *et al.*, 1973) nous apparaît mieux adaptée que les autres pour le suivi de la densité apparente des sols non immergés sous culture. Elle est en effet applicable sur les sols labourés, les sols gravillonnaires et sur des buttes (Boka, 1986; Gnamba, 1986; Yoro et Crodo, 1989).

Précision des méthodes

Les coefficients de variation (tableau 5) nous ont permis de cerner la précision d'une part et la sélectivité d'autre part de chacune des méthodes, d'abord au niveau de chaque sol, ensuite sur l'ensemble des quatre sols.

Tableau 5. Coefficient de variation (%).

Méthode de mesure	Types de sol				Les sols groupés
	L_2	L_{11}	F	J	
S	0,786	0,829	0,358	1,526	10,105
M	0,943	0,684	0,167	0,777	13,870
C	1,026	1,666	0,584	1,166	5,542
G	1,535	0,765	0,352	0,876	9,364

Dans chacun des sols étudiés les coefficients de variation diffèrent d'une méthode à l'autre. La méthode S dans le sol L_2 donne la plus faible valeur (0,782). Dans le sol L_{11} c'est plutôt la méthode M qui permet d'obtenir le plus faible coefficient de variation (0,684). Un classement par ordre croissant des valeurs de ce paramètre donne S M C G dans L_2; M G S C dans L_{11} et F et M G C S dans J. Ainsi la méthode M, avec des coefficients de variation très souvent faibles, se révèle plus précise que les autres. La précision de S est variable selon les sols. D'une façon générale, la précision ou la dispersion des résultats est liée à la fois à la méthode et au sol comme la valeur de la densité apparente.

Les coefficients de variation calculés sur l'ensemble des données (cf. tableau 5) sont supérieurs à 5. Ils induisent ainsi une dispersion des moyennes par rapport à ce que nous avons observé au niveau d'un type de sol où ils oscillent entre 0,2 et 1,7. Cette dispersion traduit, quant à elle, les différences entre les sols étudiés, et concorde avec les valeurs de la densité apparente liées non seulement à la méthode mais aussi au sol. La méthode M dont les résultats conduisent à un coefficient de variation relativedment élevé (13,870) permet de mieux caractériser un sol et de l'identifier par rapport aux autres milleux édaphiques. Elle est suivie, par ordre décroissant, des méthodes S ($v = 10,105$), G ($v = 9,364$) et C ($v = 5,542$). Ainsi la méthode C qui donne des moyennes permettant de calculer un faible coefficient de variation, rend moins compte de la différence entre les propriétés densitométriques des sols.

Conclusion

L'étude sur les méthodes de mesure de la densité apparente a permis de mettre en évidence les points suivants.
-   Au niveau d'un sol donné, les quatre méthodes de mesure de terrain donnent des résultats statistiquement différents. La valeur de la densité apparente est donc liée non seulement au type de sol mais aussi à la méthode de détermination.
-   Les méthodes M et S permettent d'obtenir des densités apparentes relativement faibles alors que G et C en donnent des valeurs élevées.
-   La méthode M se révèle plus précise que les autres et permet, en outre, de mieux traduire les différences entre les propriétés densitométriques des sols. Elle présente cependant des limites et ne peut pas être conseillée dans le cadre d'un suivi de l'évolution de la densité apparente d'un sol sous culture motorisée. Elle est, en effet, difficilement utilisable dans les horizons trop meubles ou boulants et sur des billons enclins à se déformer sous l'effet de la pression exercée (Audry *et al.*, 1973).

A partir de ce qui précède il ressort qu'il faut :
-   utiliser la même méthode au cours des diverses périodes de mesure;
-   préciser la méthode de mesure et ne plus se contenter seulement de donner la valeur de la densité apparente;
-   comparer uniquement les densités apparentes obtenues par la même méthode; et
-   pour le suivi de l'évolution de la densité apparente d'un sol sous culture motorisée, préférer la méthode au sable : elle est proche de M, et en outre elle a l'avantage d'être la moins coûteuse et donc à portée de tout le monde.

Remerciements

Je remercie mes collègues de l'Institut, G. Gnounhouri et B. Assienan, pour leur collaboration, surtout pour le traitement des données.

Bibliographie

AUDRY, P., COMBEAU, A., HUMBEL, F.X., ROOSE, E. et VIZIER, J.F. 1973. *Essai sur les études de dynamique actuelle des sols*. Bulletin de Groupe de Travail. Bondy : ORSTOM.

BERGER, J.M. 1964. Profils culturaux dans le Centre de Côte d'Ivoire. *Cahiers ORSTOM, série Pédologie* 2(1) : 41-68.

BOKA, T. 1986. Modifications physiques d'un sol ferrallitique sous l'effet du défrichement lourd motorisé. Mémoire pour l'obtention du DIAT (ESAT-Montpellier). Centre d'Adiopodoumé. Abidjan, Côte d'Ivoire : ORSTOM. 50p.

CHAUVEL, A. 1977. Recherches sur la transformation des sols ferrallitiques dans la zone tropicale à saisons contrastées. Evolution et réorganisation des sols rouges de moyenne Casamance. Thèse, Université de Strasbourg. Collection Travaux et Documents no. 62. Paris : ORSTOM. 532p.

COLLINET, J. 1988. Comportements hydrodynamiques et érosifs de sols de l'Afrique de l'Ouest. Evolution des matériaux et des organisations sous simulation de pluies. Thèse, Université de Strasbourg. 513p.

DE BLIC, P. 1975. *Comportement des sols après mise en culture mécanisée (Région Centre Côte d'Ivoire)*. Centre d'Adiopodoumé. Abidjan, Côte d'Ivoire : ORSTOM. 27p.

DE BLIC, P. 1987. Analysis of a cultivation profile under sugarcane : methodology and results. In: *Land development and management of acid soils in Africa II*. IBSRAM Proceedings no. 7. Bangkok: IBSRAM.

DE BOISSEZON, P. 1965. Les sols de savane des plateaux Batéké. *Cahiers ORSTOM, série Pédologie* 3(4) : 291-304.

DUCHAUFOUR, P. 1970. *Précis de pédologie*. 3è édition. Paris : Masson.

GNAMBA, A.S. 1986. Quelques effets du défrichement motorisé sur certains types de sols de Côte d'Iviore. Mémoire pour l'obtention du Diplôme d'Agronomie Approfondie (D.A.A.), Option Pédologie (ENSA d'Abidjan). Centre d'Adiopodoumé. Abidjan, Côte d'Ivoire : ORSTOM.

GODO, G., YORO, G., GOUE, B. et AFFOU, Y. 1989. Caractérisation physique et socio-économique du site expérimental IBSRAM de Bécédi, Sous-Préfecture de Sikensi. Abidjan, Côte d'Ivoire : ORSTOM-IBSRAM. Mimeo. 33p.

HENIN, S., MONNIER, G. et GRAS, R. 1969. *Le profil cultural*, 2è éd. Paris : Masson.

KOUAKOU, K. 1981. Etude de la dynamique actuelle d'un sol ferrallitique sous jachère de 3 ans, sous jachère de 20 ans et sous forêt en basse Côte d'Ivoire. Rapport de stage (ENSA d'Abidjan). Centre d'Adiopodoumé. Abidjan, Côte d'Ivoire : ORSTOM. 16p.

MAERTENS, C. 1964. La résistance mécanique des sols à la pénétration : ses facteurs et son influence sur l'enracinement. *Annales Agronomiques* 24(5) : 533-545.

MONNIER, G., STENGEL, P. et FIES, J.C. 1973. Une méthode de mesure de la densité apparente de petits agglomérats terreux. Application à l'analyse des systèmes de porosité du sol. *Annales Agronomiques* 24(5): 533-545.

VIZIER, J.F. 1971. Etude des variations du volume spécifique apparent dans les sols hydromorphes au Tchad. Allure des phénomènes. *Cahiers ORSTOM, série Pédologie* 9(2) : 133-145.

YORO, G. 1983. Contribution à l'étude de caractérisation de la structure. Identification et évolution des paramètres structuraux de deux types de sols du nord-ouest de la Côte d'Ivoire. Incidences agronomiques. Thèse de Docteur-Ingénieur, Université d'Abidjan. 279p.

YORO, G. et ASSA, A. 1986. Modifications structurales de deux sols ferrallitiques du nord-ouest de la Côte d'Ivoire sous l'effet du piétinement par l'homme. *Cahiers ORSTOM, série Pédologie* 22(1) : 31-41.

YORO, G. et GODO, G.H. 1989. Identification de la microvariabilité après défrichement motorisé d'un sol ferrallitique issu de sables tertiaires. Centre d'Adiopodoumé. Abidjan, Côte d'Ivoire : IIRSDA. Also in: *Organic-matter management and tillage in humid and subhumid Africa* (these proceedings). Bangkok: IBSRAM.

Un site d'expérimentations du réseau IBSRAM à Minkoameyos, Yaoundé

R. AMBASSA-KIKI

Résumé

Les caractéristiques du milieu physique sont décrites, notamment du point de vue de l'hydrographie, du climat, des sols et de la végétation du site, compte tenu de sa vocation de support des essais agronomiques. Par ailleurs, l'environnement socio-économique de la région est présenté à travers une description succincte de sa population, de son système foncier et de son système de production. Cette orientation vise à faciliter l'utilisation optimum dudit site aux fins agricoles.

Abstract

An IBSRAM experimental site at Minkoameyos, Yaoundé

Biophysical features relating to site hydrography, climate, soils, and vegetation are described, insofar as they are relevant to agronomic trials. The socioeconomic characteristics of the region are given through a concise description of its population, its land-tenure conventions, and its farming system. This background information should facilitate optimum agricultural utilization of the site.

Introduction

Les deux principaux critères auxquels répond le choix d'un site à vocation agricole sont d'ordre socio-économique et physique (milieu physique) (Latham, 1987). Ces critères prennent encore plus d'importance quand il s'agit d'un site d'expérimentation, les résultats obtenus ici avant d'être appliqués ailleurs dans des endroits similaires. Il est par conséquent nécessaire d'avoir une idée claire des habitudes culturales, ainsi que des caractéristiques fondamentales des terres concernées, avec en vue une mise en valeur des sols qui en maintiendrait le potentiel productif. Si cela est vrai pour tous les types de sol, cela l'est certainement encore plus pour les sols tropicaux dont la fragilité des propriétés intrinsèrques (principalement la capacité d'échanges et la teneur en matières organiques et en bases échangeables) sont connues (Duchaufour, 1970; Bernard et Ambassa-Kiki, 1989).

Cette communication décrit de façon succincte les systèmes de production en vigueur dans la zone et les caractéristiques pédologiques, physiques et chimiques des principales unités de sol répertoriées dans le site retenu pour les essais agronomiques des réseau IBSRAM. Une interprétation de ces caractéristiques est donnée, compte tenu de la vocation de la parcelle. Les données sur le milieu physique (hydrographie, climat, végétation) sont destinées à aider à cette interprétation.

Milieu physique

Situation géographique et topographie

Le site IBSRAM occupe à Minkoameyos (localisé à environ 10 km à l'ouest de Yaoundé) l'extrême-sud d'un domaine grossièrement triangulaire de 300 ha (figures 1 et 2) appartenant à l'Institut de la Recherche Agronomique (IRA). Situé entre les latitudes 3°51' et 3°53'N et les longitudes 11°25' et 11°27'E (Mimbe, 1985) ce domaine est un vaste interfluve en forme de plateau aux versants échancrés par les axes de drainage et par les affluents des rivières qui le bordent au nord et au sud. Son point culminant (813 m) présente une faible variation topographique (2-10%) sur plus de 10 ha. Cependant, à l'approche des talwegs, les pentes, convexes, deviennent fortes et atteignent parfois 70% (Bidzanga Nomo, 1986). Ce type de pen'es est caractéristique du modelé des sols ferrallitiques (Humbel, 1972).

Hydrographie

Minkoameyos appartient au bassin du fleuve Nyong. Au niveau du domaine IRA, les principaux affluents de ce fleuve sont la Mefou au et le Nga au sud. Leurs affluents respectifs découpent des vallons plus ou moins profonds tout autour du plateau central de l'interfluve. Au sud, où le Nga borne le domaine et le site, le réseau hydrographique développe un paysage particulier de zones mal drainées.

Climat

La région de Yaoundé, et en particulier le site de Nkolbisson/Minkoameyos, sont situées dans la zone guinéenne qui subit l'influence des masses d'air équatoriales venant du sud et maritime en provenance du sud-ouest. La résultante est un climat subéquatorial du type guinéen forestier (Vallérie, 1973). La distribution annuelle des pluies est bimodale

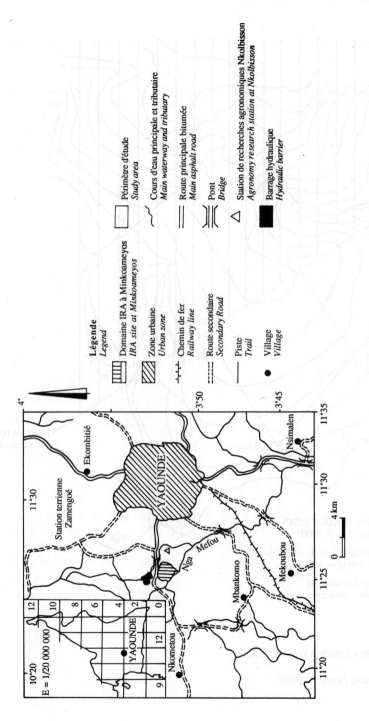

Figure 1.   Carte de situation: extrapolation à partir de la carte topographique Yaoundé NA 32 XXIV 1952, revisée en 1971.

Map of the site: adapted from the topographical map Yaoundé NA 32 XXIV, 1952, revised in 1971.

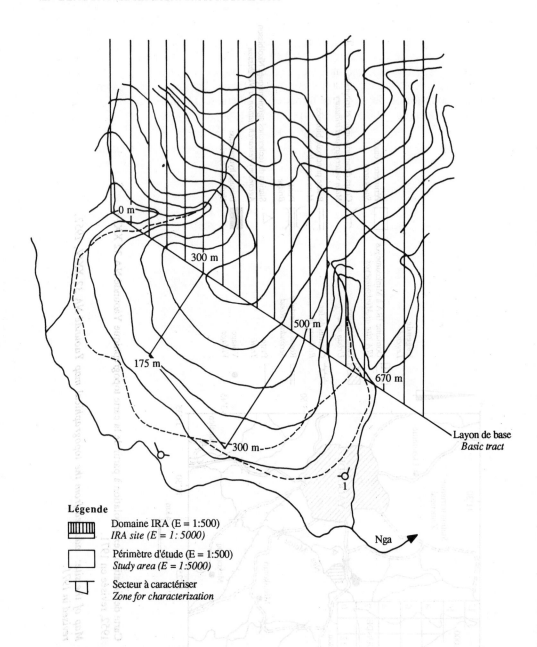

Figure 2. Perimètre d'étude à l'échelle 1:5000.

Study area (scale: 1:5000).

avec deux pics des pluies) et octobre (petite saison des pluies) et une période de faibles précipitations en juillet (petite saison sèche) (figure 3). Les mois de décembre et de janvier, et dans une certaine mesure celui de février, constituent la période majeure de grande séchèresse (grande saison sèche). Une bonne proportion des pluies qui tombent viennent sous forme de violentes tornades ou averses. La pluviométrie annuelle a peu varié sur 11 ans (de 1400 à 1600 mm) (Omoko, 1984).

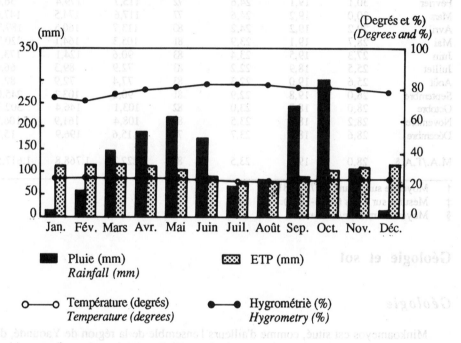

Figure 3. Données climatiques à Nkolbisson (1970-1981).

Climatic data for Nkolbisson (1970-1981).

Les températures annuelles moyennes varient de 19,0°C (minimum) à 28,0°C (maximum). On a cependant observé pendant la période considérée des extrêmes journaliers minimum et maximum respectivement de 17,0°C et 32,1°C.

L'hygrométrie moyenne mensuelle varie relativement peu. Elle atteint son minimum en février (grande saison sèche) et son maximum se situe en petite saison sèche (juillet-août), période de faible insolation. Celle-ci correspond également au moment où l'ETP est à son plus bas niveau, tandis qu'en février-mars elle atteint sa plus forte valeur. Ces relevés climatiques donnés dans le tableau 1 ont été rassemblés et calculés pour Nkolbisson (Omoko, 1984) et sont supposés applicables à Minkoameyos qui n'en est éloigné que d'environ 1,5 km à vol d'oiseau.

Tableau 1. Données climatiques à Nkolboisson (Omoko, 1984).

	Temp. max.† (°C)	Temp. min.† (°C)	Temp. moy. (°C)	Hygr. moy. (%)	ETP moy.‡ (mm)	Insn. moy.‡ (h)	Pluv. moy. (mm)
Janvier	29,0	19,0	24,0	75	112,4	195,0	13,7
Février	30,1	19,1	24,6	72	115,7	179,4	56,6
Mars	30,0	19,2	24,6	77	117,6	174,5	147,6
Avril	29,2	19,2	24,2	80	113,7	160,5	187,7
Mai	28,7	19,1	23,9	81	105,3	164,1	220,7
Juin	27,3	19,5	23,4	83	90,6	124,1	173,1
Juillet	25,5	18,9	22,2	83	72,9	89,3	66,0
Août	25,6	19,0	22,3	83	77,4	72,9	82,3
Septembre	26,0	19,8	22,9	81	90,0	103,3	245,0
Octobre	28,0	18,0	23,0	82	103,1	146,4	302,3
Novembre	28,2	18,8	23,5	80	108,4	161,9	106,7
Décembre	28,6	18,8	23,7	78	115,6	196,9	15,8
M.A./T.A.§	28,0	19,0	23,5	80	1.222,7	1.768,8	1.617,5

†   Moyenne sur 11 ans (1970-1981).
‡   Mesures sur 3 ans (1978-1981).
§   Moyenne annuelle/total annuel.

Géologie et sol

Géologie

Minkoameyos est situé, comme d'ailleurs l'ensemble de la région de Yaoundé, dans une zone appartenant au socle ancien constitué essentiellement de roches métamorphiques. La roche-mère repérée dans le site est une embréchite à grenats. La rareté des affleurements rocheux (moins de 10% de l'ensemble du domaine de l'IRA, Mimbe, 1985) fait penser à une altération rapide de cette roche comme cela a été observé par ailleurs dans certains sols du Nigéria (Moormann et al., 1975). La profondeur la plus fréquente de l'horizon C (6 à 10 m) (Onguéné Mala, données non publiées) tend à confirmer cette hypothèse en même temps qu'elle montre le peu d'influence de cet horizon sur les caractéristiques des sols.

Sols

Minkoameyos comporte trois grandes classes de sol d'importance inégale : les sols minéraux bruts (environ 10% du total) que l'on rencontre dans les endroits accidentés (sols sur embréchite) ou plus rarement sur certains versants (sols sur cuirasse); les sols hydromorphes (10%) concentrés dans les bas-fonds; les sols ferrallitiques de divers groupes, occupant l'essentiel (80%) du domaine. Ces derniers, riches en minéraux 1:1, comportent différentes variantes suivant leur teneur en oxydes ou en hydroxydes de fer. On a ainsi des sols ferrallitiques à faciès rouges qui contiennent essentiellement des oxydes de fer, ceux à faciès jaunes qui renferment surtout des hydroxydes de fer, et ceux à faciès ocres, contenant autant des deux (Onguéné Mala, données non publiées).

Au niveau du site, seuls les sols ferrallitiques typiques modaux à faciès rouges sont représentés. Les caractéristiques du profil représentatif sont données dans les tableaux 2 (Mimbe, 1985) et 3, mais il existe des variantes locales au niveau de l'épaisseur du A_{11} (6

Tableau 2. Caractéristiques du profil représentatif (Mimbe, 1985).

Caractéristiques	Paramètres	Horizons (cm)			
		A_{11} 00-14	A_{12} 14-29	B_{21} 29-82	B_{22} 82-200
Roche-mère		Embréchite à grenats			
	Gravier	00	00	00	00
	Sables	45	39	33	26
Granulométrie	Limons	24	18	14	15
(%)	Argile	31	43	52	59
	Classe texturale	LAS	A	A	A
Acidité	pH H_2O	4,8	5,0	5,3	5,2
	pH KCl	4,3	4,4	4,4	4,4
	C	2,0	1,2	0,6	0,5
Matières	M.O.	3,4	2,1	1,0	0,9
organiques (%)	N total	0,17	0,12	0,09	0,08
	C/N	12,4	10,2	7,5	6,2
P_2O_5 (%)	P_2O_5 total	0,11	0,10	0,08	0,08
	Ca^{++}	0,84	0,90	0,76	0,60
	Mg^{++}	0,50	0,46	0,64	0,52
Cations	K^+	0,08	0,03	0,03	0,03
échangeables	Na^+	0,01	0,01	0,01	0,01
(cmol kg^{-1}	S	1,43	1,40	1,44	1,16
	CEC (T)	9,76	9,44	8,32	7,84
	V %	15	15	17	15
Densité apparente (Mg m^{-3})		1,09	1,37	1,44	1,48
	CPCS (1967)	Sol ferrallitique typique modal à faciès rouges			
Classification	USDA	Rhodic Kandiudult			
	FAO	Ferric acrisol			

à 10 cm) en fonction de l'utilisation des sols. C'est un sol fortement désaturé (CPCS, 1967), assez sableux en surface et de plus en plus argileux en profondeur. L'horizon de surface (A_{11}), peu épais, est moyennement humifère malgré la présence fréquente d'un couvert forestier. Ce sol est d'autre part assez sensible au tassement et la circulation inconsidérée d'engins agricoles lourds qui peuvent avoir sur lui des conséquences agronomiques graves (tableau 3). Cette homogénéité pédologique ne semble pas perturbée par le gradient de pente de direction dominante NE-SO enregistré dans le site et variant beaucoup entre 5 et 23%. Dans ces conditions, la différenciation des unités cartographiques s'est faite sur la base de l'utilisation des sols associée aux classes de pente.

Tableau 3. Autres caractéristiques des horizons A du profil représentatif.

Paramètres		Valeurs	Auteurs
Humidités caractéristiques	Capacité au champ (Hc)	24%	Ambassa-Kiki, 1988
	Point de flétrissement (Hf)	16%	---- " ----
	Point d'adhésivité (Ha)	32%	--- " ---
	Optimum pour le travail du sol (Ht)	21-24%	--- " ---
Limites d'Atterberg	Limite de liquidité (W_1)	53%	Ambassa-Kiki, 1988
	Limite de plasticité (W_p)	28%	--- " ---
	Limite de retrait (W_r)	17%	--- " ---
Facteur d'érodibilité (K)	Forêt secondaire défrichée manuellement	0,01	Bernard, 1988
	Forêt secondaire défrichée au bull	0,08	--- " ---
Conductivité hydraulique (K(θ))	Sous forêt secondaire	35 m/jr	Eveng, 1988
	Sous culture†	5 m/jr	--- " ---

† Verger à haut niveau d'intrants.

Végétation

Le site IBSRAM se trouve dans la zone à climat sub-équatorial décrite par Martin (1959) et primitivement occupée par une forêt humide. A l'heure actuelle, des lambeaux de forêt presque primaire (jachère de plus de 30 ans) caractérisée par un sous-bois clairsemé et facile à défricher occupent certaines plages de ce site. La végétation climacique y est représentée par *Entandrophagma utile, Entandrophagma camdollei, Terminalia superba, Mansonia altissima, Terminalia altissima, Triplochyton scleroxylon*, etc. Le reste du site est occupé par :
- Des plages de forêt secondaire (jachère de plus de sept ans) à *Elaeis guineensis, Musa acuminata, Ceiba pentandra, Musanga cecropioides, Albizzia zygia* plus quelques représentants de la végétation climacique. La densité des arbres y est plus lâche que précédemment et le sous-bois plus touffu et dominé par *Aframomum* sp., *Costus*

affer, Megaphrinium macrostachyum, Haumania danckelmaniana, etc. Les plantations de cacaoyers, non entretenues par ailleurs, sont comptées dans ce groupe.

- Des jachères à *Chromolaena odorata* (de 3 à 7 ans) où on ne rencontre plus que quelques bananiers plantains (*Musa acuminata*) et de manioc (*Manihot esculenta*).
- Des parcelles sous culture, de quelques mois à deux à trois ans (parfois plus), plus ou moins envahies d'herbes notamment *C. odorata*, d'où émergent encore beaucoup de pieds de bananiers plantains et de manioc.

Caractéristiques socio-économiques

De nombreuses études ont été consacrées, tout ou partie, à l'environnement socio-économique du Centre-Sud camerounais (Weber, 1974; Mutsaers *et al.*, 1981; Leplaideur, 1985; Ay *et al.*, 1986; Tonyé *et al.*, 1986; Depommier, 1988; Djimde et Raintree, 1988). Les thèmes abordés ont fréquemment trait aux caractéristiques de la population et au système foncier dont l'appréhension est essentielle dans l'évaluation de l'efficacité des systèmes de production, de leurs potentialités et de leurs tendances futures.

Population

La population dans la zone est principalement composée du groupe ethnique Beti et comporte 7% de ruraux (dont environ 67% ont plus de 45 ans) pour 93% de citadins (recensement national, 1976). Cette inégalité dans la répartition et cette structure s'expliquent par la proximité de la ville de Yaoundé qui accueille l'essentiel du flot migratoire composé surtout de jeunes admis à différents établissements scolaires, ou à la recherche du travail. Cette situation influence directement la production agricole basée sur le travail manuel fourni par une main-d'oeuvre essentiellement familiale. Déjà on remarque une assez forte pression démographique sur les terres le long des routes où sont concentrés la plupart des paysans. Cette pression va décroissant à mesure que l'on avance vers l'intérieur du domaine en direction du site où on ne trouve plus guère de maisons d'habitation.

Système foncier

Le système traditionnel foncier qui prévalait autrefois et reposait sur un contrôle communautaire de la terre dont seuls les fruits étaient objets d'appropriation est en train de disparaître. Avec l'introduction du cacaoyer durant la période coloniale et, par la suite, l'augmentation de la population, la terre est devenue objet d'appropriation privée et a pris une valeur marchande. Cette valeur s'est étendue non seulement à la cacaoyère, mais aussi à l'ensemble des sols défrichés pour les autres cultures (Weber, 1974), notamment les cultures annuelles et pluriannuelles (C.A.P.).

En matière d'héritage, l'ensemble des terres du défunt est divisé en nombre de parts égal au nombre des épouses ayant eu des enfants mâles, lesquels se distribuent chaque part. Les épouses du défunt continuent cependant de joir de l'usufruit des champs vivriers qu'elles cultivaient jusque là (Depommier, 1988).

Officiellement, bien qu'elles continuent à faire l'objet d'un contrôle communautaire, l'ensemble des terres non dotées d'un titre foncier appartiennent au domaine de l'Etat qui peut les réquisitionner à tout moment contre indemnisation des membres impliqués de la communauté, en réparation du préjudice subi. Le site IBSRAM entre dans ce cas et ne peut plus faire l'objet d'aucune revendication bien qu'il soit encore exploité par les paysans.

Le système de production

Le système de production est tributaire de la structure de la population en ceci qu'elle détermine les types de cultures pratiquées, le système dominant d'utilisation des terres et les dimensions des parcelles cultivées.

Cultures pratiquées

Cultures arbustives. Les arbres sont un élément important dans le système de production du Centre-Cameroun. Ay *et al.* (1986) estiment même à plus de 60% la valeur de leur part dans la production agricole. Malgré la baisse des prix des matières premières, le cacaoyer (*Theobroma cacao*) reste encore pour le moment le plus important pourvoyeur de revenu monétaire. Quant aux autres arbres fruitiers (palmier à huile [*Elaeis guineensis*], safoutier [*Dacryodes edulis*], avocatier [*Persea americana*], agrumes [*Citrus* sp.], manguier [*Mangifera indica*], goyavier [*Psidium guajava*], etc.) et au papayer (*Carica papaya*) éparpillés dans la forêt, cultivés autour des cases ou en une lâche association avec le cacaoyer auquel ils peuvent éventuellement servir comme plante d'ombrage, leur valeur marvhande ne cesse d'augmenter.

Cultures annuelles et pluriannuelles (C.A.P.). Il s'agit des plantes désignées selon la terminologie coloniale par l'expression "cultures vivrières". La plus importante est le manioc. En termes de fourniture de calories, Ay *et al.* (1986) classent en premier lieu le manioc, suivi du plantain, de la banane douce (*Musa sinensis*), de l'arachide (*Arachis hypogea*), du mais (*Zea mays*) et du macabo (*Xanthosoma sagittifolium*). L'arachide a également l'avantage d'être la plus importante source (25% du total) des protéines (Ay *et al.*, 1986).

Il existe bien entendu une longue liste d'autres cultures annuelles pratiquées dans la zone, notamment la patate douce (*Ipomea batatas*) et une gamme variée de légumes et de plantes rampantes comme les courges (*Cucumeropsis manii, Citrullus lanatus*, etc.), mais elles sont considérées comme d'importance moindre, sauf peut-être *Cucumeropsis manii* (voir ci-dessous).

Séquence traditionnelle des C.A.P. chez les Beti. La séquence commence par le défrichement de la forêt suivi d'un brûlis. La courge (*Cucumeropsis manii*[1]) est alors semée en semis direct entre les arbres abattus. Ceci peut se passer indifféremment en première ou deuxième campagne. Plus tard, elle en recouvre (envahit) les troncs avant d'être récoltée au bout d'une année environ. Le mais, le manioc et le plantain sont ajoutés à faibles densités. En seconde année, l'arachide est mise en place dans le champ en association avec du manioc, du macabo, du plantain, plus éventuellement une vingtaine d'autres cultures (légumes, tomates, etc.). Après la récolte de l'arachide, le manioc, le macabo et le plantain restent dans le champ. Ils seront récoltés progressivement pendant une période pouvant excéder 5 ans si la fertilité du sol ainsi que le niveau d'infestation des cultures le permettent. Graduellement la parcelle revient à la jachère. Dans le système traditionnel, une nouvelle parcelle est défrichée et la séquence recommence avec la courge (Ay *et al.*, 1987; Tonyé *et al.*, 1986).

Ce schéma traditionnel n'est pas appliqué tel quel dans le site bien que la terre ne manque pas. Malgré également l'existence de tronçonneuses, les nouvelles défriches forestières sont rares et les parcelles de courge aussi. On a plutôt affaire la plupart du temps à des jachères à *Chromolaena* suivies de cultures (C.A.P.) à dominance d'arachide ou de maïs.

Les systèmes d'utilisation des terres

Le système d'utilisation des terres est dominé par la cacaoculture et les cultures annuelles (C.A.P.), spéculations quasi indissociables au niveau de l'exploitation paysanne (Depommier, 1988). Ce système est caractérisé par une division sexuelle du travail, une cacaoculture essentiellement paysanne, la culture itinérante des C.A.P.

Division sexuelle du travail. En général, les opérations les plus pénibles (essartage des forêts, dessouchage des sissonghos (*Pennisetum purpureum*)) sont exécutées par les hommes[2] tandis que les labours et les semis sont laissés aux femmes et aux enfants (quand ils sont disponibles). De plus en plus cependant, on voit les hommes labourer et les femmes exécuter certains défrichements, notamment ceux des jachères à *Chromolaena*. Dans la plupart des ménages, le mari et la femme sont les seuls à travailler régulièrement dans l'exploitation. Les enfants étant la plupart du temps à l'école, n'interviennent que ponctuellement avec les personnes âgées pour la récolte des arachides et du maïs et l'entretien des cultures (désherbage surtout).

La main-d'oeuvre payée par les services de la recherche est surtout constituée d'hommes, les femmes étant trop occupées à entretenir leurs champs de C.A.P. pour se rendre disponibles à plein temps.

Cacaoculture. Essentiellement affaire des hommes, la cacaoculture, à l'origine d'une semi-sédentarisation de l'agriculture itinérante, est peu représentée dans le site. Elle

---

[1] Courge à grains à coque blanc-laiteux, appelée "ngon" dans la langue locale. Ces grains sont accommodés dans les sauces après séchage et décorticage.

[2] Les fermes qui ne peuvent pas disposer d'une main-d'oeuvre familiale mâle sont obligées d'en louer à l'extérieur ou de se contenter d'une petite parcelle qu'elles peuvent défricher elles-mêmes.

se fait généralement en monoculture excepté à la création de nouvelles plantations où quelques C.A.P., notamment le mais, l'arachide, le plantain ou le macabo, sont associées (Tonyé *et al.*, 1986). Elle se fait également sous ombrage d'arbres dont certains produisent fruits, bois d'oeuvre et divers autres produits. Les engrais ne sont pas utilisés pour le cacao. Seules la lutte phytosanitaire et la commercialisation bénéficient de l'attention des pouvoirs publics.

Culture itinérante. Généralement réservée aux femmes, la culture itinérante qui est pratiquée en deux campagnes par an correspondant aux deux saisons des pluies, concerne principalement les cultures annuelles et pluriannuelles (C.A.P.). Au début de la saison des pluies, presque toutes les C.A.P. se retrouvent dans un même champ, mais en proportions variables en fonction de la «tête de liste» qui elle, est en position dominante (tableau 4). Les «têtes de liste» les plus fréquentes dont la parcelle porte d'ailleurs automatiquement le nom sont la courge (voir ci-dessous), l'arachide et le mais.

C'est donc un système de cultures associées dans lequel si l'on excepte la cendre issue des brûlis[3] , la seule technique de restauration de la fertilité des sols est la jachère qui peut être courte (ne dépassant pas 4 ans), longue (ne dépassant pas 9 ans) ou très longue (plus de 10 ans). Aucun intrant moderne (engrais, herbicide, pesticides) n'est utilisé. La préparation du terrain ainsi que l'entretien des cultures se font généralement avec des outils manuels (hache ou tronçonneuse, machette, houe) (Tonyé *et al.*, 1986).

- **Le champ de courge** : Le champ n'est pas labouré et la courge ainsi que le maïs sont semés directement. Le manioc et le macabo qui suivront plus tard à faibles densités peuvent être buttés localement. La courge est rarement plantée deux fois de suite dans le même champ. La raison donnée par les paysans est la forte chute de ses rendements.
- **Le champ d'arachide** : Un labour à plat léger (à la houe) précède le semis à raison de 100 000 pieds environ à l'hectare (Ay et al., 1986). Le maïs, le manioc et le plantain lui sont associés à des densités différentes (tableau 4). Ces chiffres semblent néanmoins différents de ceux observés dans le site (tableau 5). Les deux dernières cultures peuvent survivre à l'arachide et au maïs pendant 3 à 5 ans, les boutures de manioc étant souvent rajoutées au fur et à mesure de la récolte de cette plante. D'autres cultures, notamment les légumes, peuvent être ajoutées dans des sous-paracelles où il y a eu accumulation de cendres par exemple.
- **Le champ de maïs** : Le maïs comme plante dominante et "tête de liste" se retrouve dans des parcelles plus petites que celles des cultures précédentes. La densité de semis, sur terrain labouré ou non, est de l'ordre de 40 000 à 60 000 pieds à l'hectare. Il faut préciser que ces chiffres sont atteints en tenant compte du fait que le nombre de poquets à l'hectare est d'environ 10 000 et qu'il y a de 3 à 6 graines par poquets, le maïs n'étant pas démarié. Il vient généralement après une jachère à *Chromolaena* ou une jachère plus courte. Il se cultive fréquemment dans les vallées,

---

[3]   La cendre en elle-même est riche comme le montrent les caractéristiques chimiques de celle récoltée en 1989 dans une défriche brûlée à Minkoameyos : C organique : 5,44%; N total : 0,30%; P assimilable : 14 ppm; bases échageables (cmol kg^{-1}) : Na$^+$ = 0,63; K$^+$ = 238,63; Mg^{2+} = 144,36; Ca^{2+} = 66,13. Mais les quantités à l'hectare sont trop faibles pour avoir un effet notable sur les rendements. De plus, dès les premières pluies elle est vite emportée par les eaux de ruissellement.

en contre-saison. Il lui est alors associé des légumes qui constituent de la nourriture fraîche en saison sèche (Ay *et al.*, 1986).

Tableau 4. Principaux types de champ et leurs caractéristiques (Ay *et al.*, 1986).

Type de champ	Superficie (ha)	Densité à l'ha	Type de jachère	Age jachère (ans)	Plantes associées
Nouveau champ de *Cucumeropsis*	0,50-1,5	Variable	Forêt	>20	Un an de *Cucumeropsis* ≠ maïs (3000 pieds ha⁻¹) et manioc (300 à 600 pieds ha⁻¹). Ensuite champ d'arachide.
Champ traditionnel de *Cucumeropsis*	0,1-0,2	Variable	Forêt	>20	Idem. Puis séquence d'arachide + maïs + manioc, *Chromolaena*, brousse, forêt.
Arachide	0,03-0,3	10 000	Brousse / *Chromolaena* / Herbe	8-15 / 3-5 / 2-5	Maïs (5000 p ha⁻¹) Manioc (3000 p ha⁻¹) Plantain (500 p ha⁻¹)
Maïs	0,01-0,05	40 000 à 60 000	*Chromolaena*	2-7	Macabo et legumes à faible densité.

Tableau 5. Champ d'arachide dans le site (arachide + plantes associées).

Français	Anglais	Scientifique	Densité de plantation (pieds ha⁻¹)
Arachide	Groundnut	*Arachis hypogea*	200 000
Maïs	Maize	*Zea mays*	10 000
Manioc	Cassava	*Manihot esculenta*	10 000
Macabo	Cocoyam	*Xanthosoma sagittifolium*	2 500
Plantain	Plantain	*Musa acuminata*	50

Le jardin de case. Etant donné l'absence d'habitations dans le site, le jardin de case n'y est pas représenté, mais il fait partie intégrante du système de production dans la région.

Le jardin ou champ de case est un système caractérisé par des parcelles de petites dimensions autour des cases (Tonyé *et al.*, 1986). On y retrouve la plupart des C.A.P. (maïs, arachide, tubercules, etc.) ainsi que des légumes (aubergines [*Solanum* sp.], tomate

[*Lysopersicon esculentus*], gombo [*Hibiscus esculentus*], amarante [*Amaranthus* sp.], etc.), diverses plantes aromatiques, du tabac (*Nicotiana tabacum*) et de la canne à sucre (*Saccharum officinarum*). Ses bouquets de bananiers plantains et d'arbres fruitiers dispersés (manguier, safoutier, agrumes) ainsi que le papayer (*Carica papaya*) peuvent constituer un étage supérieur (Depommier, 1988).

Ces parcelles sont plus ou moins fumées par les ordures ménagères et divers résidus de récolte de l'exploitation. Le petit bétail y est rarement intégré (étables ou enclos dont le lisier sera épandu dans les jardins de case). Les divagations du petit bétail causent cependant plus de dégâts que celui-ci n'apporte d'avantages. Dans certains cas et pour cette raison les champs ou jardins de case sont enclos (haies mortes ou parfois vives) (Depommier, 1988).

Conclusion

La connaissance des sols et de la végétation du site ainsi que celle de la topographie du terrain nous permettent déjà de suggérer une affectation des deux types d'essais prévus dans les différentes zones prospectées. Ainsi, les essais de mise en valeur des sols acides peuvent être menés dans la section du site où la pente est au plus égale à 15% et où les champs cultivés et les jachères à *Chromolaena* sont plus fréquents, tandis que les essais de défrichement se verraient bien placés dans la section la plus abrupte (pente = 16 à 23%), mais la moins exploitée. La connaissance de la structure de la population et du système de production de la région quant à elle nous permettra de choisir la meilleure combinaison possible dans la définition des traitements dits traditionnels. Elle permet également d'espérer une bonne disponibilité en main-d'oeuvre. Enfin, en attendant des données plus fines, les caractéristiques mécaniques et chimiques du sol décrites ici nous donnent une bonne indication quant aux mesures à prendre et aux erreurs à éviter lors des essais agronomiques.

Le site ainsi caractérisé à travers la description de son milieu physique et de son environnement socio-économique apparaît somme toute représentatif de la région. Il l'est en ceci que les sols et les gradients de pente rencontrés ici sont courants dans la zone comme y est courante à l'échelle paysanne la mise en culture sur des pentes allant jusqu'à 18%. Toutes choses égales par ailleurs, l'on peut objectivement penser que les essais menés ici sont appelés à trouver une application heureuse en milieu paysan.

Remerciements

Les figures 1 et 2 sont adaptées de Tchienkoua que nous remercions vivement en même temps que tous ceux qui ont contribué à la mise en oeuvre de ce travail.

Bibliographie

AMBASSA-KIKI, R. 1988. Travail du sol en relation avec le type de sol et la production agricole. Cours de perfectionnement organisé par TLU/IRA à Yaoundé (Centre Jean XXIII), 22 février - 4 mars 1988. 9p.

AY, P., FOAGUEGUE, A., NOUNAMO, L., BERNARD, M., MANKOLO, R. and THO, C. 1986. *Report on exploratory survey and establishment of IRA-IITA-IDRC "on farm" research work in villages of Yaoundé and Bikok areas*, pt II, 10-57. Yaoundé: IRA.

BERNARD, M. 1988. Soil erosion studies in Cameroon and Nigeria. GTZ/IRA/IITA/TU-Munich progress report. Mimeo. 23p.

BERNARD, M. et AMBASSA-KIKI, R. 1989. Première approche de l'étude de l'érosion hydrique en zone humide camerounaise. Première Conférence Annuelle du CCB, Ngaoundéré (Cameroun), 4-9 décembre 1989.

BIDZANGA NOMO, L. 1986. Etude pédologique et évaluation des terres pour les cultures du manioc (*Manihot esculenta*) et des agrumes (*Citrus* spp.). Mémoire de fin d'études, ENSA, Dschang, Cameroun. 122p.

CPCS (Commission de Pédologie et de Cartographie des Sols). 1967. *Classification des sols*. Grignon, France : Laboratoire de Géologie et de Pédologie, Ecole Supérieure d'Agronomie. 87p.

DEPOMMIER, D., ed. 1988. *Utilisation du sol et potentiel agroforestier dans le système à cacaoyer et cultures vivrières du plateau sud-camerounais : cas de la Lékié et du Dja-et-Lobo*. Rapport Afrena no. 17. 53p.

DJIMDE, M. and RAINTREE, R., eds. 1988. *Agroforestry potential in the humid lowlands of Cameroon*. Afrena Report no. 12. 96p.

DUCHAUFOUR, P. 1970. *Précis de pédologie*. Paris : Masson et Cie. 482p.

EVENG, J.D. 1988. Mesures d'infiltration dans le domaine IRA à Minkoameyos. Rapport de stage pré-professionnel. Cameroun : Centre Universitaire de Dschang. 32p.

HUMBEL, F.X. 1972. *Initiation à la pédologie et aux sols camerounais*. Yaoundé : ORSTOM. 153p.

LATHAM, M. 1987. Clearing and management of lateritic soils - first steps in the implementation of IBSRAM projects. In: *Land development - management of acid soils*, 7-14. IBSRAM Proceedings no. 4. Bangkok: IBSRAM.

LEPLAIDEUR, A. 1985. Indicateurs de la mobilisation potentielle des paysans dans un projet de développement. Cas du Centre et du Sud du Cameroun. *Agronomie Tropicale* 40(40):357-371.

MARTIN, D. 1959. *Etude pédologique du Centre de Recherche Agronomique de Nkolbisson*. Rapport no. P104. Yaoundé : ORSTOM-IRCAM. 12p.

MIMBE, C. 1985. Le domaine IRA à Minkoameyos : Etude pédologique et évaluation des terres pour les cultures de l'arachide, du cacaoyer et du mais. Mémoire de fin d'études, ENSA, Dschang, Cameroun. 122p.

MOORMANN, F.R., LAL, R. and JUO, A.S.R. 1985. *The soils of IITA*. IITA Technical Bulletin no. 3. 48p.

MUTSAERS, H.J.W., MBOUEMBOUE, P. and MOUZONG BOYOMO. 1981. Traditional food crop growing in the Yaoundé area (Cameroon). *Agro-Ecosystems* 6: 273-278.

OMOKO, M. 1984. Dynamique de l'eau dans un sol ferrallitique et étude comparée entre l'évapotranspiration mesurée et calculée en climat équatorial. Thèse de Doctorat 3ème Cycle, Université de Bordeaux. 130p.

TONYE, J., AMBASSA-KIKI, R. et NSANGOU, M. 1986. *Identification des systèmes d'utilisation des terres dans la zone forestière du Cameroun : possibilités d'amélioration*. Yaoundé : IRA. 25p.

VALLERIE, M. 1973. *Contribution à l'étude des sols du Centre-Sud Cameroun.* Rapport no. P29. Yaoundé : ORSTOM. 15p.

WEBER, J. 1974. *Structures agraires et évolution des milieux ruraux. Le cas de la région cacaoyère du Centre-Sud Cameroun.* Paris : ORSTOM. 50p.

Appendixes

Appendixes

Appendix I

DISCUSSION GROUP REPORT: TILLAGE AND CROP-RESIDUE MANAGEMENT

Chairman: P. UNGER,
Cochairman: C. OFORI
Secretary: J. INGRAM

Summary

The potential for improving soil management through tillage and crop-residue management is as applicable to semiarid regions as it is to the humid tropics. The many off-field uses for crop residues were noted, and substitutes would be needed if such residues are to be retained on fields. It was also noted that residue management should entail residues from any sources, not just those of crops. Tillage development must not overlook different situations in different countries, and must be aimed at sustainable production. Definitions for various tillage practices have already been given, but an alternative approach may be to base the definitions on the results obtained from a given tillage operation rather than on the actions taken. FAO is to call for country papers regarding tillage practices (possibly tillage/residue practices) in the various countries. FAO will then assemble the reports and distribute them to research and ministry of agriculture personnel in the countries. It was stressed that residues should be managed to minimize tillage requirements.

Body of the report

To facilitate the discussion, tillage and residue management were discussed individually at first, then jointly toward the end of the discussion period.

Tillage

The overall goal of tillage is to increase crop production while conserving resources (soil and water) and protecting the environment.

The benefits of tillage were listed, not necessarily in order of importance, as (i) seedbed preparation, (ii) weed control, (iii) evaporation suppression, (iv) water infiltration enhancement, and (v) erosion control.

Some constraints to tillage were also listed. They are: (i) labour shortages and inefficient use of labour, (ii) poor extension services, (iii) lack of appropriate tools for different soil conditions, (iv) limitations in draft power (animals are weak after the dry season, and must regain strength after the rains start before they can be effectively used for tillage), and (v) problems in performing deep tillage, where it is required.

The constraints to tillage were classified as: (i) farmer-related, (ii) technology-related, and (iii) environment-related.

The question was raised "Is enough known about different soils to maximize current tillage knowledge?" It was agreed that there is a need to collate present soil information and tillage practices as they are related to each other. FAO is willing to call for country papers collating such information and to disseminate the information to the scientific communities as well as to personnel at the ministry level in the different countries. Such a document would supplement FAO Soils Bulletin no. 54 (Tillage systems for soil and water conservation). Definitions always cause problems and a suggestion was made to define tillage operations from the result point of view rather than from the action taken point of view.

Residue management

Points stressed during the residue management discussion were:
- Residues from all sources should be considered, not just crop residues.
- Crop residues serve many off-field purposes and, if they are to be retained on the field, then provisions must be made for these off-field functions that the residues normally would serve.
- Semiarid and arid regions often have a very short supply of residues.
- Sometimes crop residues are a nuisance, and are considered to be a problem.
- Residue management can include management of mulching crops as well as off-field materials.
- A small amount of residue left in a field can have a marked effect on soil properties.
- In some cases, termites destroy virtually all available crop residues.

From a synthesis of these points (and variations thereof), it was concluded that crop residues should be left in the field as much as possible, and must always be thought of as a resource, never as a waste. Supplements to crop residues can be very beneficial, especially in arid regions, and an integrated approach to their combined management is required. Such supplements could be produced specifically for management on fields. Whether or not residues should be incorporated into the soil needs further analysis, but it is important to capitalize on the role played by the farmer. In many cases, residues management in relation to animal husbandry practices is important. Also, the animals are often not owned by the farmers, but by nomadic people who have grazing rights after the crop is harvested.

Tillage and residue management

Discussion of this topics jointly led to the following observations:
- Crop residues may impede tillage operations.
- Present practices are not necessarily ideal or satisfactory. If not, the need for further development should be assessed.
- The development of practices should not be restricted to current equipment.
- Tillage should be restricted to when and where it is required.

The key statement appears to be "Residues should be managed so as to minimize tillage requirements". Possible tillage options range from maximum to zero. Building on the promise that tillage should be minimized to limit oxidation of soil organic matter, residue management should be optimized. It was recognized, however, that how far the farmer can go along the continuum from maximum to zero tillage depends on many factors. One thesis put forward was that it should go as far as possible. Considering the earlier statement that a relatively small quantity of residues can be beneficial, preservation and/or incorporation of residues should reduce tillage energy requirements. Cost/benefit analyses for alternatives should be conducted. The group agreed that IBSRAM and TSBF collaboration should be encouraged to further understand tillage and residue management implications.

Appendix II

PROGRAMME OF THE WORKSHOP

Wednesday 10 January

Discours d'inauguration - *Opening addresses*
- FOFIFA Director General
- FAO Representative
- IBSRAM Director   Marc Latham
- Minister of Scientific Research and Development
    Techniques

Session I: FAO and IBSRAM programmes
The FAO programme on tillage and organic matter   C.S. Ofori
    management
The IBSRAM *AFRICALAND* networks
    - Management of acid soils   O. Spaargaren
    - Land development for sustainable agriculture   B. Hintze
    Discussion

Session II: State of the art reports on tillage and residue management - ongoing and future programmes I

Session III: State of the art reports on tillage and residue management - ongoing and future programmes II

Session IV: The Madagascar experience on soil organic-matter management and tillage for acid soils
    Organic matter and soil amelioration (FOFIFA)   J. Rakotoarisoa
                                                            J.L. Rakotomanana
    Variety selection for acid soils (FIFAMANOR/NORAD)   Rakotondramanana
    Discussion

Thursday 11 January

Session V: Progress reports on national projects for the AFRICALAND acid soils network: Burundi, Cameroon, Côte d'Ivoire, Ghana

Session VI: Progress reports on national projects for the AFRICALAND acid soils network: Madagascar, Nigeria, Zambia

Session VII: Progress reports on national projects for the AFRICALAND land development network: Cameroon, Côte d'Ivoire, Ghana, Nigeria

Session VIII: Presentation of new projects and discussion

Friday 12 January

Session IX: The role of organic matter in tropical soils
The role of organic matter for nutrient cycling   E. Fernandez
The TSBF approach to determine soil organic matter   J. Ingram
   management options
Le role de la matière organique dans les sols tropicaux   G.H. Godo
Discussion

Session X: Organic-matter management I
Maintenance and management of organic matter   S.R. Juo
   in tropical soils with emphasis on the effect of low
   activity clays
Manure and compost as organic inputs   C.S. Ofori and
                                        R. Sant'Anna
Organic-matter and soil fertility management in   A.A. Agboola
   humid tropics
Discussion

Session XI: Organic-matter management II
The use of cover crops and mulch to improve   J. Parr
   soil moisture and weed control
Biological activities and soil physical properties   H.W. Scharpenseel
Agroforestry for the management of soil organic-matter A. Young
Discussion

Session XII: Why do we till?
Tillage and residue management under rainfed   P. Unger
   conditions - options for the present and future
No tillage, minimum tillage, and their influence   P. Ahn and B. Hintze
   on soil physical properties
Adapted tillage equipment   R. Ambassa-Kiki

Saturday 13 January

Session XIII: Soil conservation
Soil conservation and sustainable agriculture: Adisak Sajjapongse
   management of sloping land in Asia (*ASIALAND*)

Reclamation of degraded land   B. Hintze
Soil erosion measurements and the influence of   D. Nill
   cropping systems on erosion in Nigeria and Cameroon

Session XIV: Special factors affecting crop production
Drought spells and risk in terms of crop production   P. Ahn
The study of the cultivation profile as a tool to evaluate   P. de Blic
   tillage and organic-matter management
Les méthodes de mesure de la densité apparente :   G. Yoro
   problème de dispersion des résultats

Session XV: Network Coordinating Committee

Sunday 14 January

Field trip to Beferona
Visit to the FOFIFA station and the site of the IBSRAM project on the management
   of acid soils.

Plenary Session
Reports from discussion groups and NCC

Closing Session

Appendix III

LIST OF PARTICIPANTS

(in order of country work base)*

Botswana

MONAGENG, Kgopisano
Senior Technical Officer
Department of Agricultural Research
Postbag 4 0033
Gaborone

Burundi

KIBIRITI, Christine
Chercheur
Institut des Sciences Agronomiques
 du Burundi (ISABU)
B.P. 795
Bujumbura

Cameroon

AMBASSA-KIKI, Raphaël
Research Officer
IRA/CNS
B.P. 5578
Yaoundé

BINDZI-TSALA, Joseph
Chef du CNS
B.P. 5578
Yaoundé

NILL, Dieter
Agronomiste
IITA/IRA/GTE
B.P. 7814
Yaoundé

NILL, Elke
Biologiste
IITA/IRA/GTE
B.P. 7814
Yaoundé

Côte d'Ivoire

GODO, Gnahoua H.
Laboratoire d'Agronomie
IFRSDA
B.P. V 51
Abidjan

YORO, Gballou René
Chercheur
IIRSDA
B.P. V 51
Abidjan

Ethiopia

REGASSA, Hailu
Research Officer
Institute of Agricultural Research
P.O. Box 2003
Addis Ababa

---

* Representatives from international organizations are listed separately at the end.

Federal Republic of Germany

SCHARPENSEEL, H.W.
Professor of Soil Science
Institut für Bodenkunde der
  Universität Hamburg
Allendeplak 2
D-2000 Hamburg 13

Ghana

LAMPTEY, Daniel
Teaching/Research Assistant
School of Agriculture
University of Cape Coast
Cape Coast

Kenya

ODUOR, Paul
Soil Chemist
National Agricultural Research
Laboratories
P.O. Box 14733
Nairobi

Madagascar

ANDRIAMAMPIANINA, Nicolas
Collaborateur Technique
DRFP
B.P. 904
Tel: 40321

BENOIT, Clerget
Selectionneur
DRAAE
B.P. 1444
Tananarive 101

DE GUIDICI, Pascal
Chercheur en agronomie
LRI/Université de Tana
B.P. 3383
Antananarivo 101

DUPUY, Jacques
Chef de Service
LRI/Université de Tana
B.P. 3383 Route d'Andraisoro
Antananarivo 101

NAESS, Ola
Conseiller Technique Agroforesterie
FIFAMANOR
B.P. 198
Antsirabe 110

RABEMANANJARA, Dorothée
Collaborateur Technique
FOFIFA
B.P. 1690
Antananarivo 101

RABEMANANTSOA, J. Odon
Chef de Programme (arbustes fourragers)
DRZV
B.P. 4 Ampandrianomby
Antananarivo

RABESON, Raymond
Pédologue
FOFIFA/IRRI
B.P. 1444
Antanarivo

RAFANOMEZANTSOA, Samuel J.
Documentaliste
CIDST
B.P. 6224
Tel: 33288

RAJAONARIVO, Andriantahina
Chercheur
FOFIFA
B.P. 1444
Antananarivo 101

RAKOTOARISOA, Benjamin E.
Chef de Section Recherche
  (pommes de terre)
FIFAMANOR
B.P. 198
Antsirabe 110

RAKOTOARISOA, Jacqueline
CDRR/FOFIFA
B.P. 1690
Antananarivo 101

RAKOTOMANANA, Henri
Directeur National
PEM Nanisana
B.P. 1028
Antananarivo

RAKOTOMANANA, Jean Louis
Chef de Programme de Recherches
Projet Terre Tany
B.P. 904
Antananarivo

RAKOTONDRAMANANA
Chef du Département Recherche
FIFAMANOR
B.P. 198
Antsirabe 110

RAKOTONIAINA, Pierrot
Chercheur
DRFP
B.P. 904
Ambatobe

RALIMANGA, Viviane
Consultante de la Banque Mondiale
Immeuble IEC 2ᵉ Etage
Anosy
Antananarivo

RAMALANJAONA, Gabriel
Chef DRT
B.P. 254 Ambatobe
Tananarivo 101

RAMAMONJY, Adriana
Assistante
LRI Ampandrianomby
B.P. 3383

RAMARILALAO, Sahondra
Ingénieur Agronome
Lot II I 72 Ankadivato
Antananarivo

RAMIARAMANANA, Daniele
Collaborateur Technique
FOFIFA/DRD
B.P. 1444
Antananarivo

RAMILISON, Rodolphe
Agronome
DRAAE
B.P. 1444
Antananarivo 101

RAMPANANA, Léa
Chercheur
DRA/FOFIFA
B.P. 1444
Antananarivo

RANDRIAMAMPIANINA, J.A.
Collaborateur Technique
PRD Lac Alaotra
B.P. 80
Ambatendrasaka 503
Tel: 81372

RANDRIANAMPY, Joseph
Pédologue
FOFIFA
DRR Mahitsy FOFIFA
B.P. 1444

RAHERIMANDIMBY, Joseph L.
Ingénieur
PEM
B.P. 1028
Antananarivo

RASAMBAINARIVO, John Henri
Chef de Division (alimentation animale)
DRZV-FOFIFA
B.P. 4 Ampandrianomby

RAVELOMANANA
Chercheur
DRT-FOFIFA
B.P. 254
Ambatobe

RAVOHITRARIVO, Clet Pascal
Directeur Scientifique du FOFIFA
B.P. 1690
Antananarivo

RAZAFIMAHALEO, Martial
Technicien
FOFIFA
B.P. 1690
Antananarivo

RAZAFINDRABE, Andrianirina Joseph
Chef du Laboratoire de Pédologie
B.P. 4096
Tsimbazaza

RAZAFINDRAZAKA, Jean Alfred
Chercheur
DRZV-FOFIFA
B.P. 254
Ambatobe

RAZAFINDRATSITA, Roger
Responsable de l'Agrostologie
DRZV-FOFIFA
B.P. 4
Ampandrianomby

RAZAFINDRAZAY, Marie Angèle
Ingénieur d'Agriculture
FIFAMANOR
B.P. 198
Antsirabe 110

ROLLIN, Dominique
Responsable Equipe Recherche
    Développement
ODR/PPI
B.P. 82
Antsirabe 110

TORSKENAES, Erick
Ingénieur d'Agriculture
FIFAMANOR
B.P. 198
Antsirabe 110

Malawi

KUMWENDA, Wells Frank
Farm Machinery Unit
Ministry of Agriculture
Chitedze Research Station
PO Box 30134
Lilongwe

Nigeria

AGBOOLA, Akinola A.
Professor of Soil Fertility and
    Farming Systems
Department of Agronomy
University of Ibadan
Ibadan

OBI, A. Olu
Lecturer and Researcher
Department of Soil Science
Faculty of Agriculture
Obafemi Awolowo University
Ile-Ife

OGUNKUNLE, A.O.
Lecturer
Department of Agronomy
University of Ibadan
Ibadan

OPARA-NADI, Oliver A.
Soil Physicist
Imo State University
P.M.B. 2000
Okigwe

République Centrafricaine

NGOUANZE, Fidele
Directeur
Bureau National de Pédologie et
  de Conservation des Sols
B.P. 1374
Bangui
Tel: 61-62-75

Senegal

KHOUMA, Mamadou
Chef de Division de l'Agriculture
  OMGV
B.P. 2353
Dakar

Tanzania

ANTAPA, Paul Léonard
Senior Agricultural Research Officer
Selian Agricultural Research Institute
PO Box 6024
Arusha

MSAKY, John Joseph
Senior Lecturer
Sokoine University of Agriculture
P.O. Box 3008
Morogoro

O'KTING'ATI, Aku
Senior Lecturer
Sokoine University of Agriculture
PO Box 3000
Morogoro

Togo

WOROU, Kodjo Soklou
Chef Division Etudes et Cartographie
  des Sols
Institut National des Sols/ORSTOM
B.P. 1026
Lome
Tel: 213096

Uganda

Y KITUNGULU ZAKE, Julius
Makarere University
PO Box 7062
Kampala

USA

FERNANDES, Erick C.M.
North Carolina State University
P.O. Box 7619
Raleigh, NC 27695

JUO, Anthony S.R.
Professor of Agronomy and Tropical Soils
Department of Soil and Crop Sciences
Texas A&M University
College Station
Texas 77843

Zambia

CHISHALA, Benson Hosten
Soil Chemist
MT Makulu Central Research Station
P/B 7
Chilansa

International Organizations and Institutes

CIRAD

RAUNET, Michel
Pédologue
CIRAD
BP 5035
34032 Montpellier
France

REBOUL, Jean Louis
Délégué CIRAD à Madagascar
4 Rue Rapiera
B.P. 853
Antananarivo
Madagascar

CSD

STOCKLI, Anton
Technical Adviser
Coopération Suisse
Lova Soa
B.P. 1278
Fianarantsoa

CSIRO

DENNIS, Edward Albert
Senior Research Officer
CSIRO
Academy Post Office
Kwadaso
Kumasi
Ghana

FAO

NABHAN, H.
Coordonnateur de Projet
Programme Engrais Malagasy
c/o FAO Antananarivo
B.P. 3971
Antananarivo
Madagascar

OFORI, Charles Seth
Senior Officer (Soil Management)
FAO
Via Delle Terme di Caraccala
Rome
Italy

SANT'ANNA, Racim
Fonctionnaire Régional des Ressources
  en Sols
Bureau Régional FAO
B.P. 1628, Accra
Ghana

IBSRAM

ADISAK Sajjapongse
ASIALAND Coordinator
P.O.Box 9-109
Bangkhen
Bangkok 10900
Thailand
Tel: 579-7590, 5794012, 579-7753, 580-5958
Telex: 21505 IBSRAM TH
Fax: 66-2-5611230
E-mail: CGI134

AHN, Peter M.
Coordinator
IBSRAM-*AFRICALAND*
04 BP 252
Abidjan 04
Republic of Côte d'Ivoire
Tel: (225) 22-19-11, 22-19-12
Telex: 214235 FATTN MIX : IBSRAM
Fax: (225) 22-17-60

CHALINEE Niamskul
Administrative Officer
P.O.Box 9-109
Bangkhen
Bangkok 10900
Thailand
Tel: 579-7590, 5794012, 579-7753, 580-5958
Telex: 21505 IBSRAM TH
Fax: 66-2-5611230
E-mail: CGI134

HINTZE, Bernhard
Coordinator
IBSRAM-*AFRICALAND*
04 BP 252
Abidjan 04
Republic of Côte d'Ivoire
Tel: (225) 22-19-11, 22-19-12
Telex: 214235 FATTN MIX : IBSRAM
Fax: (225) 22-17-60

LATHAM, Marc
Director
P.O.Box 9-109
Bangkhen
Bangkok 10900
Thailand
Tel: 579-7590, 5794012, 579-7753, 580-5958
Telex: 21505 IBSRAM TH
Fax: 66-2-5611230
E-mail: CGI134

SPAARGAREN Otto
Network Coordinator
IBSRAM-*AFRICALAND*
04 BP 252
Abidjan 04
Republic of Côte d'Ivoire
Tel: (225) 22-19-11, 22-19-12
Telex: 214235 FATTN MIX : IBSRAM
Fax: (225) 22-17-60

ICRAF

YOUNG, Anthony
Principal Scientist
Box. 30677
Nairobi
Kenya

IRRI

HOOPPER, Jim
Chef de mission
B.P. 4151
Antananarivo
Madagascar

ORSTOM

DE BLIC, Philippe
Pédologue
Centre ORSTOM
B.P. 5045
Montpellier
France

TSBF

INGRAM, John
ISBF Program Coordinator
c/o Unesco-ROSTA
P.O. Box 30592
Nairobi
Kenya

USAID

PARR, James F.
Coordinator USDA/USAID
 Dryland Agriculture Project
Agriculture Research Service USDA/
 USAID
Beltsville, MD
USA

POTTER, Christopher S.
AAAS Fellow
Agricultural Research Service USDA/
 USAID
Washington, DC 20523
USA

UNGER, Paul W.
Agricultural Research Service USDA/
 USAID
PO Box Drawer 10
Bushland
TX 79012
USA